Manuel D. Ortigueira and Duarte Valério
Fractional Signals and Systems
De Gruyter Studium

Fractional Calculus in Applied Sciences and Engineering

Volume 7

Manuel D. Ortigueira and Duarte Valério

Fractional Signals and Systems

—

DE GRUYTER

Mathematics Subject Classification 2010
34A30, 34M03,34M03,35F35,35F40,35G40,26A33,34A08,34N05,93C15,93C55,93C57

Authors
Prof. Manuel Duarte Ortigueira
Centre of Technology and Systems — UNINOVA
Department of Electrical Engineering
NOVA School of Science and Technology of NOVA University of Lisbon
Quinta da Torre
2829-516 Caparica
Portugal

Prof. Duarte Valério
Universidade de Lisboa
Instituto Superior Técnico
Dep. de Engenharia Mecânica
Av. Rovisco Pais 1
1049-001 Lisboa
Portugal

ISBN 978-3-11-077716-1
e-ISBN (PDF) 978-3-11-062458-8
e-ISBN (EPUB) 978-3-11-062132-7
ISSN 2509-7210

Library of Congress Control Number:2019951786

Bibliographic information published by the Deutsche Nationalbibliothek
The Deutsche Nationalbibliothek lists this publication in the Deutsche Nationalbibliografie;
detailed bibliographic data are available on the Internet at http://dnb.dnb.de.

To Joana, Eduardo José, and Laurinda

To Teresa

Acknowledgement

From Manuel D. Ortigueira to

- Juan Trujillo, with whom I had great discussions. With him, I could understand what "great mathematician" means. He is one among a few that I know.
- J. Tenreiro Machado for his critical esprit, cleverness, and imagination.
- My colleagues from the Centre of Technology and Systems and from the Dept. of Electrical Engineering of Nova Faculty of Sciences and Technology of Nova University of Lisbon for their friendship and good environment.
- Many colleagues that collaborated or had great discussions with me (to those I may forget I beg their pardon; the order is arbitrary): V. Martynyuk, G. Bengochea, A. M. Lopes, M. Duarte, R. Magin, V. Tarasov, R. Herrmann, A. Oustaloup, P. Melchior, F. Mainardi, V. Kiryakova, C. Ionescu, Y. Luchko, J. Sabatier, Y. Podlubny, Y.Q. Chen, D. Baleanu, D. Torres, R. Garrappa, I. Petras, C.P. Li, J. Sá da Costa,...

From Duarte Valério to

- Professor José Sá da Costa, my MSc and PhD supervisor, with whom twenty years ago I first learned about the existence of fractional derivatives.
- My daughter Teresa, far more challenging than any mathematical problem.

This work was funded by Portuguese National Funds through the FCT — Foundation for Science and Technology, under the project PEst- UID/EEA/00066/2019, and through IDMEC, under LAETA, project UID/EMS/50022/2019.

DOI 10.1515/9783110624588-202

Preface

This book has a twofold objective: to bring the methodologies of *Signal Processing* (SP), namely the theory of linear systems, to the *Fractional World*, and to introduce fractional derivatives and anti-derivatives in the theory and practice of *Signals and Systems*.

In a simplistic vision, Signal Processing is Mathematical Engineering. It is a like a "marriage" of many mathematical areas with engineering concepts and tools. This brought generality to SP, and allowed it to become the base of many application areas. It currently uses: transforms (Laplace, Fourier, Hankel,...), vectorial, matricial, and tensorial calculus, complex variable functions, set theory, Galois fields, differential and difference equations, graph theory, stochastic processes theory, Statistics, etc.. But, in parallel, it became also a source of new mathematical theories, like Information theory, Wavelet transform and multiresolution. Many numerical tools currently used were developed in SP.

We can trace back the origins of SP to the 17th century, but the most interesting tools were developed by Gauss (he was the first to consider a fast way of computing the Fourier transform) and Fourier (the Fourier series is one of its most important tools), as well as Prony (who proposed an algorithm for decomposing a function as sum of exponentials, in 1795), Galois (his group theory is the base of the coding in Telecommunications), Euler (he is everywhere; the most interesting transforms are associated to his name), Laplace (probability, transform), Schuster (periodogram, 1898), Heaviside, Carson, Wiener, Kolmogorov, Levinson, van der Pol, and many others.

Meanwhile, and since the forties in the last century, SP evolved and established the bases for the development of many application areas, namely: Telecommunications, Acoustics, Speech, Image, Mechanics, Biomedical Engineering, Economics, Finance, Seismology, Genomics, etc.. Using Mathematics with all the generality and implementing many mathematical tools, it opened the doors of a new world that came in the last years.

Concerning Fractional Calculus (FC), it dates back to the Leibniz letter exchanges with Bernoulli and others, but its true beginning was with Liouville in 1832 and its developments in the 19th and 20th centuries. During the last 30 years it evolved considerably and became popular in many scientific and technical areas. The progresses in applications vis-à-vis theoretical developments motivated the re-evaluation of past formulations. The concepts of fractional derivative (FD) and fractional integral (FI) assumed various forms not always equivalent and also not compatible with each other. Twenty years ago, the Riemann-Liouville (RL) derivative was currently used; since the last years of past century, the Caputo (C) derivative became preferred, supposedly due to the initial conditions; the Grünwald-Letnikov (GL) derivative was considered a mere approximation to those.

Anyway, there was an intrinsic incompatibility between FC, based on RL and C, and the assumptions currently made in SP. RL and C are unable to support the

DOI 10.1515/9783110624588-203

generalisations of the formalism used in SP. Here, we will try to face the questions involved, and see how to solve the problems.

We separate the text into three parts. The first is dedicated to the study of continuous-time signals and systems, while the second replicates the procedure for the corresponding discrete-time. The third part is dedicated to the presentation of advanced topics.

For both continuous and discrete-time cases, we introduce the concepts of *Signals* and *Systems* and present their characterisations. For the case of *linear systems*, we describe their properties and introduce their input-output relations, stated in terms of the *impulse response*, *transfer function*, and *frequency response*. This is done progressively from the integer order to the fractional order systems. The suitable definitions of fractional derivatives emerge naturally from the backward integer/fractional coherence demanded for the systems. Well-known important tools used in daily applications are introduced: feedback, Bode diagrams, stability tests, state-space representations, Laplace transforms, Fourier series and transforms, and Z transform.

In the third part we make a brief study of the system output characteristics when the input is a stochastic process. We include the two-sided fractional derivatives and the fractional Brownian motion. A generalisation of the integer order difference equations is presented to consider fractional delays. Finally, variable order systems are introduced, and a very useful tool for them is presented: the variable order Mittag-Leffler function. The appendices include material on several short topics for reference purposes.

Lisbon, July 2019   *Manuel Ortigueira and Duarte Valério*

Contents

Part I: Continuous-time

Part II: **Discrete-time**

Part III: Advanced topics

Appendixes

Fractional Calculus in Applied Sciences and Engineering

Volume 6
Yingjie Liang, Wen Chen, Wei Cai
Hausdorff Calculus. Applications to Fractal Systems, 2018
ISBN 978-3-11-060692-8, e-ISBN (PDF) 978-3-11-060852-6,
e-ISBN (EPUB) 978-3-11-060705-5

Volume 5
JinRong Wang, Michal Fečkan
Fractional Hermite-Hadamard Inequalities, 2018
ISBN 978-3-11-052220-4, e-ISBN (PDF) 978-3-11-052362-1,
e-ISBN (EPUB) 978-3-11-052244-0

Volume 4
Kecai Cao, YangQuan Chen
Fractional Order Crowd Dynamics. Cyber-Human System Modeling and
Control, 2018
ISBN 978-3-11-047281-3, e-ISBN (PDF) 978-3-11-047398-8,
e-ISBN (EPUB) 978-3-11-047283-7

Volume 3
Michal Fečkan, JinRong Wang, Michal Pospíšil
Fractional-Order Equations and Inclusions, 2017
ISBN 978-3-11-052138-2, e-ISBN (PDF) 978-3-11-052207-5,
e-ISBN (EPUB) 978-3-11-052155-9

Volume 2
Bruce J. West
Nature's Patterns and the Fractional Calculus, 2017
ISBN 978-3-11-053411-5, e-ISBN (PDF) 978-3-11-053513-6,
e-ISBN (EPUB) 978-3-11-053427-6

Volume 1
Dingyü Xue
Fractional-Order Control Systems. Fundamentals and Numerical
Implementations, 2017
ISBN 978-3-11-049999-5, e-ISBN (PDF) 978-3-11-049797-7,
e-ISBN (EPUB) 978-3-11-049719-9

Part I: **Continuous-time**

1 Introduction to signals and systems

1.1 Signals

1.1.1 Signals and their characteristics

Signals are the media we use to interact with systems or observe them. From both the practical and physical points of view, they assume a great multitude of forms, that we translate mathematically [25, 66, 97, 102].

Definition 1.1.1. *We define a* signal *as any function — real or complex — of time, space, speed, ..., associated to a phenomenon — physical, chemical, biological, economic, social...— and that has, or carries, some kind of information about it.*

This means that a signal is a quantified description of a given phenomenon, or gives support for its transmission. There are some useful functions that are frequently employed as signals, and are sometimes by extension called signals themselves. It is the case, for instance, of the Heaviside function. A signal may depend on only one independent variable, or on several. Signals that depend on spatial coordinates are an example of the latter case. In many applications, there is only one independent variable, that is frequently the time. To exemplify, let us refer several important signals:

- Speech — this is a function of time.
- Wind speed — this can be a function of time or direction.
- Price of a good — this can be a function of time, or of location, or both.
- Air temperature — this can be a function of time, location, and/or height.
- Electrical currents in a neural system — these are a function of time and node in the network.
- Trade index — this is a function of time.
- ECG, EEG — these are functions of time.

In the following, we will consider only the single variable case, assumed to be "time". In the *continuous-time* case the domain is $\mathbb{R}$ and the signal will be called "continuous", using an abusive, but simple language. If the signal is defined on a discrete sub–set of $\mathbb{R}$, the signal is *discrete-time*, or simply "discrete". In most applications, such domain is defined by the set $t_n = nh$, $n \in \mathbb{Z}, h \in \mathbb{R}^+$. The h parameter is the intersample interval and its inverse is the *sampling rate*. When the independent variable is time, the intersample interval is also called sampling period, since it is the period of the delta comb that is considered as the ideal sampler. This nomenclature was borrowed from sampling theory and digital signal processing. Frequently, we write $x_n = x(t_n) = x(nh)$. Discrete signals are also called *sequences*.

There are several ways to classify signals, according to their characteristics.
1. Amplitude

DOI 10.1515/9783110624588-001

Analog signal   Continuous-time signal with an amplitude that can assume values in a continuous range.

Quantised signal   Continuous-time signal with an amplitude that can assume values only in a discrete range.

Sampled signal   Discrete-time signal with an amplitude that can assume values in a continuous range.

Digital signal   Discrete-time signal with an amplitude that can assume values only in a discrete range.

Example 1.1.1. *Some examples are easily found considering a power plant with many turbines. The rotation speed of a turbine is an analog signal. The number of turbines in a power plant that are working at each instant is a quantised signal. The power produced in each day is a sampled signal. The number of maintenance actions carried out in each week is a digital signal.*

2. Duration

Finite duration signal   Signal that is non null only during a bounded interval of time. The same designation can be attributed to signals that assume small values, outside a given interval. For any signal, it is possible to define a duration, but the theme exceeds our objectives [97, 100].

Window   The same as a finite duration signal; also used for those signals that depend, not on time, but on other independent variables, and are non-null only on a bounded support.

Pulse   Signal with short duration.

Right signal   Signal null for $t < \tau \in \mathbb{R}$.

Causal signal   Right signal null on $\mathbb{R}^-$. The designation that comes from the notion of causal system (as we will see below in section 1.2).

Left signal   Signal null for $t > \tau \in \mathbb{R}$.

Anti-causal signal   Left signal null on $\mathbb{R}^+$.

3. Periodicity

Continuous periodic signal   Continuous signal $x_p(t)$ for which there is a $T \in \mathbb{R}^+$ such that

$$x_p(t) = x_p(t \pm T) \ \forall t \in \mathbb{R} \tag{1.1}$$

Period   Any value of T in (1.1). The period is not unique: any integer multiple thereof is also a period.

Fundamental period   Least positive value that T can assume, in order to verify (1.1). In what follows, when T or the period are mentioned without further qualification, the fundamental period must be understood.

Linear combination of continuous periodic signals   Notice that, if x_1 and x_2 are periodic, $x(t) = a_1 x_1(t) + a_2 x_2(t)$ may not correspond to a periodic signal. A necessary and sufficient condition for $x(t)$ to be a periodic signal with period T is that periods T_1 and T_2 be commensurable: $n_1 \cdot T_1 = n_2 \cdot T_2 = T$, $n_1, n_2 \in \mathbb{Z}^+$.

Sinusoidal signals   The simplest and more important examples of periodic continuous signals are given by the functions $\cos(\omega_o t + \phi)$, $\sin(\omega_o t + \phi)$, and $e^{i(\omega_o t + \phi)}$ $t, \omega_0, \phi \in \mathbb{R}$.

Fundamental frequency The real number $f_0 = \frac{1}{T}$.

Angular frequency The real number $\omega_0 = 2\pi f_0 = \frac{2\pi}{T}$.

Harmonics of a sinusoid with fundamental frequency f_0 Other sinusoids with frequencies nf_0, $n \in \mathbb{Z}$.

Discrete periodic signal Discrete signal x_n for which there is a positive integer $N \in \mathbb{Z}^+$ verifying

$$x_n = x_{n\pm N} \ \forall n \in \mathbb{Z} \tag{1.2}$$

or, which is the same,

$$x(nh) = x((n \pm N)h) \ \forall n \in \mathbb{Z} \tag{1.3}$$

The above considerations about the period remain valid in the discrete case, with two exceptions:

Linear combination of periodic discrete signals This is always periodic.

Discrete sinusoids Given by functions $\cos(\omega_0 n + \phi)$, $\sin(\omega_0 n + \phi)$, and $e^{i(\omega_0 n + \phi)}$, $n \in \mathbb{Z}$, $\omega_0 \in [0, 2\pi)$, $\phi \in \mathbb{R}$. Their behaviour is somehow different from that of continuous sinusoids. If we set $\omega_0 = 2\pi f_0$, we can prove that *the above sinusoids are periodic if and only if f_0 is a rational number*. In this case, it can be shown that:

- $f_0 = \frac{M}{N}$, $M, N \in \mathbb{Z}^+$,
- $M < N$
- If the least common multiple of M and N is 1, then N is the period.
- There are only N linearly independent sinusoids with frequencies $f_k = \frac{k}{N}$, $k = 0, 1, \cdots, N - 1$. This is very important in the definition of the Discrete Fourier Transform (DFT) [3, 31, 61, 100].

To finish this short study, it is important to refer that any periodic signal can be considered as a sum of delayed versions of a basic function defined on a period:

$$x_p(t) = \sum_{n=-\infty}^{\infty} x_0(t - nT) \tag{1.4}$$

where $x_0(t)$ is a pulse with duration T.

1.1.2 Important signals in continuous-time

The following functions are very useful, and often found as signals:

1. *Heaviside unit step*

$$\varepsilon(t) = \begin{cases} 1 & t > 0 \\ 0 & t < 0 \end{cases} \tag{1.5}$$

Normally, it is not important to fix $\varepsilon(0)$. For coherence with the inverse Laplace transform and with other definitions like the signum function, it is convenient to make $\varepsilon(0) = 1/2$. (Another option, often found, is $\varepsilon(0) = 1$, but this is not so convenient mathematically.)

2. *Unit ramp, $r(t) = t\varepsilon(t)$*

$$r(t) = \begin{cases} t & t \geq 0 \\ 0 & t < 0 \end{cases} \tag{1.6}$$

3. *Unit impulse, Dirac impulse, Dirac delta, or delta*
   We can introduce $\delta(t)$, which is not strictly speaking a function, without caring about mathematical correctness, as (see appendix A)

$$\delta(t) = \begin{cases} \infty & t = 0 \\ 0 & t \neq 0 \end{cases} \tag{1.7}$$

and so that

$$\int_{-\infty}^{+\infty} \delta(t)dt = \int_{0^-}^{0^+} \delta(t)\,dt = 1. \tag{1.8}$$

The unit impulse is related with the unit step $\varepsilon(t)$ by

$$\varepsilon(t) = \int_{-\infty}^{t} \delta(\tau)\,d\tau \Rightarrow \delta(t) = \frac{d\varepsilon(t)}{dt} \tag{1.9}$$

This explains why we represent the delta by a unit vertical arrow at $t = 0$. We can show that $k\delta(t)$ corresponds to an impulse with area k (measured by its integral); say

$$\int_{-\infty}^{t} k\delta(\tau)\,d\tau = k\varepsilon(t) \tag{1.10}$$

Generalising this result we verify that, given a signal $x(t)$, we have

$$x(t)\delta(t - \tau) = x(\tau)\delta(t - \tau). \tag{1.11}$$

4. *Periodic comb*
   The periodic repetition of impulses is called a *comb*. It is defined by

$$p(t) = \sum_{n=-\infty}^{\infty} \delta(t - nh), \ \ h \in \mathbb{R}^+, \tag{1.12}$$

and is called ideal sampler, since, from (1.11),

$$x_p(t) = p(t)x(t) = \sum_{n=-\infty}^{\infty} x(nh)\delta(t - nh), \tag{1.13}$$

reason why h is called sampling interval or sampling period (we can consider almost periodic combs too [85]).

5. *Fractional powers*
   Let $0 < \alpha \leq 1$. We define a sequence of fractional functions by

$$\phi_n(t) = \frac{t^{n\alpha}}{\Gamma(n\alpha + 1)}\varepsilon(t), \ \ n \in \mathbb{Z} \tag{1.14}$$

Tab. 1.1: Important continuous signals

Signal	Definition
Unit impulse or Dirac delta	$\delta(t) = \begin{cases} \infty & t = 0 \\ 0 & t \neq 0 \end{cases}$
Heaviside unit step	$\varepsilon(t) = \begin{cases} 1 & t \geq 0 \\ 0 & t < 0 \end{cases}$
Signum or sign function	$\mathrm{sgn}(t) = 2\varepsilon(t) - 1 = \begin{cases} +1 & t > 0 \\ 0 & t = 0 \\ -1 & t < 0 \end{cases}$
Periodic comb	$p(t) = \sum_{-\infty}^{\infty} \delta(t - nh)$
Ramp	$r(t) = t.\varepsilon(t)$
Absolute value	$|t| = t.\,\mathrm{sgn}(t)$
Rectangular pulse	$\mathrm{Rect}_T(t) = \varepsilon(t + T) - \varepsilon(t - T)$
Triangular pulse	$\mathrm{tri}_T(t) = \left(1 - \frac{|t|}{T}\right).\mathrm{Rect}_T(t)$
Fractional power	$\phi(t) = \frac{t^a}{\Gamma(a+1)}\varepsilon(t)$
Exponential	$x(t) = e^{at} \quad a \in \mathbb{C}$
Causal exponential	$x(t) = e^{at}.\varepsilon(t) \quad a \in \mathbb{C}$
Cisoid or complex exponential	$x(t) = e^{i\omega t} = \cos(\omega t) + i\,\sin(\omega t)$
Laplace function	$x(t) = e^{-|t|}$

with the convention that [20]

$$\frac{t^{-n}}{\Gamma(-n+1)}\varepsilon(t) = \delta^{(n-1)}(t), \quad n \in \mathbb{Z}^+ \tag{1.15}$$

Function $\Gamma(\cdot)$ is the Euler gamma function, that will be studied below in appendix B.

Table 1.1 presents the definitions of these and other functions often found as continuous signals.

1.1.3 Important signals in discrete-time

Most of the above described signals have direct correspondent in discrete-time. However, in some situations we introduce modifications that are related to results we want to obtain in the transform domain.

1. *Discrete-time impulse or Kronecker delta*

$$\delta(n) = \begin{cases} 1 & \text{if } n = 0 \\ 0 & \text{if } n \neq 0 \end{cases} \tag{1.16}$$

A discrete comb can also be defined, but we will not do it, since it will not be important in the following developments.

Tab. 1.2: Important discrete signals

Signal	Definition
Unit impulse or Kronecker delta	$\delta(n) = \begin{cases} 1 & n = 0 \\ 0 & n \neq 0 \end{cases}$
Unit step	$\varepsilon(n) = \begin{cases} 1 & n \geq 0 \\ 0 & n < 0 \end{cases}$
Signum	$\text{sgn}(n) = \begin{cases} +1 & n \geq 0 \\ -1 & n < 0 \end{cases}$
Ramp	$r(n) = n.\varepsilon(n)$
Absolute value	$\|n\| = n.\,\text{sgn}(n)$
Rectangle pulse	$\text{Rect}_N(n) = \begin{cases} 1 & \|n\| < N \\ 0 & \|n\| \geq N \end{cases}$
Triangle pulse	$\text{tri}_N(n) = \left(1 - \frac{\|n\|}{N}\right).\,\text{Rect}_N(n)$
Exponential	$x(n) = z^n \quad z \in \mathbb{C}$
Causal exponential	$x(n) = z^n.\varepsilon(n) \quad z \in \mathbb{C}$
Cisoid or complex exponential	$x(n) = e^{i\omega n} = \cos(\omega n) + i\sin(\omega n)$
Laplace function	$x(n) = e^{-\|n\|}$

2. *Discrete unit step*

$$\varepsilon(n) = \begin{cases} 1 & \text{if } n \geq 0 \\ 0 & \text{if } n < 0 \end{cases} \tag{1.17}$$

Notice the difference for $n = 0$. This definition keeps valid a relation similar to (1.9):

$$\varepsilon(n) = \sum_{k=-\infty}^{n} \delta(k) \tag{1.18}$$

3. *Discrete signum or sign*
   The continuous signum from Table 1.1 verifies an interesting relation: $\text{sgn}(t) = 2\varepsilon(t) - 1$. To keep this valid in the discrete case, we define the discrete signum as

$$\text{sgn}(n) = 2\varepsilon(n) - 1 = \begin{cases} 1 & \text{if } n \geq 0 \\ -1 & \text{if } n < 0 \end{cases} \tag{1.19}$$

Table 1.2 presents the definitions of these and other functions often found as continuous signals.

1.1.4 Discrete signals vs. continuous signals

The process of obtaining a discrete signal from a continuous signal is called *discretisation*. The inverse operation is named *reconstruction*. A simple example helps to illustrate this. Let $x(t) = e^{-t}$; its McLaurin expansion

$$x(t) = e^{-t} = \sum_{n=0}^{+\infty} \frac{(-1)^n}{n!} t^n \tag{1.20}$$

converges on $\mathbb{R}$. The continuous signal $x(t)$ can be completely defined by the "coefficients" $x_n = \frac{(-1)^n}{n!}$, $n \in \mathbb{Z}_0^+$, that emerge as values of a discrete signal resulting from the discretisation of $x(t)$. This example allows us to realise that the discretisation process is not unique. In fact, if instead of powers we used, for example, orthogonal polynomials, we would obtain a different discrete signal from the very same original continuous signal. Anyway, having fixed the discretisation methodology, the reconstruction procedure gets defined as its inverse, or, if there is no inverse, as an operation that recovers the original signal as closely as possible. It is important to understand that neither the domain, nor the range of the discrete signal, are necessarily subsets of the corresponding sets in continuous-time. This is only the case in a so-called sampling process.

Definition 1.1.2. *The process of associating a discrete signal to a continuous one such that*

a) *the domain of the discrete signal is a discrete subset of the domain of the continuous signal,*

b) *the range of the discrete signal is also a discrete subset of the range of the continuous signal,*

c) *both signals assume the same values at the intersection domains,*

is named **sampling**.

The inverse operation is called *interpolation* or *reconstruction*. The number of samples per time unit is the *sampling rate*. The conversion of an analog signal into a digital one, using sampling and quantisation, is an *A/D conversion* (analog to digital conversion), and the inverse a *D/A conversion*. In general, there are infinite continuous signals that can generate a given discrete signal; in other words, it is impossible to exactly recover the original signal. However, there are situations where we can avoid this difficulty: when the signals are bandlimited, meaning that they are of bounded support in frequency domain. In mathematical terms, the signals are *analytical of exponential type*. We will not study this important problem here.

1.1.5 Operations with signals: energy and power

Definition 1.1.3. *The instantaneous power of a signal is given by*

$$p(t) = |x(t)|^2 = x(t)x^*(t) \tag{1.21}$$

Definition 1.1.4. *The energy of a continuous signal is*

$$E_{xx} = \int_{-\infty}^{+\infty} |x(t)|^2 \, dt = \int_{-\infty}^{+\infty} x(t)x^*(t) \, dt \tag{1.22}$$

and that of a discrete signal is

$$E_{xx} = \sum_{n=-\infty}^{+\infty} |x(n)|^2 = \sum_{n=-\infty}^{+\infty} x(n)x^*(n) \tag{1.23}$$

This is the well known "square integrable" mathematical concept and is useful with a rather restrict set of signals, namely those with finite duration. There are many important signals for which this quantity does not make sense because it is infinite, as are the cases of periodic signals, stationary stochastic processes, unit step, signum, ramp and so on. For some of these signals, it is possible to define instead the *average power* (or, simply, *power*) as

$$P_{xx} = \lim_{T \to \infty} \frac{1}{2T} \int_{-T}^{+T} |x(t)|^2 \, dt, \tag{1.24}$$

$$P_{xx} = \lim_{N \to \infty} \frac{1}{2N+1} \sum_{n=-N}^{+N} |x(n)|^2, \tag{1.25}$$

in the continuous and discrete cases respectively. It is a simple task to show that a signal with finite energy has null power.

In agreement with the above definitions, we can classify the signals as

Energy signals   when $0 < E_{xx} < \infty$;

Power signals   when $0 < P_{xx} < \infty$.

Some particular cases:

- All bounded finite duration signals (windows) are energy signals.
- Periodic signals are power signals.
- Stationary stochastic processes with finite variance are power signals.

Many signals do not belong to any of these classes, e.g. ramp, exponential, fractional power, etc.

1.1.6 Operations with signals: convolutions

Definition 1.1.5. *Let $x(t)$ and $y(t)$ be two square integrable (energy) continuous-time functions. We define $h(t) = f(t) * g(t)$, the convolution of $f(t)$ and $g(t)$, as*

$$z(t) = \int_{\mathbb{R}} x(t - \tau)y(\tau)\,d\tau \tag{1.26}$$

Definition 1.1.6. *Similarly, let x_n and y_n two square summable (energy) sequences. We define the convolution of x_n and y_n as the new sequence $z_n = x_n * y_n$ given by*

$$z_n = \sum_{k=-\infty}^{+\infty} x_{n-k} y_k \tag{1.27}$$

The properties of convolution are very important and are identical in both the continuous and discrete cases.

1. Linearity:
$$x * (y_1 + y_2) = x * y_1 + x * y_2 \tag{1.28}$$

2. Commutativity:
$$x * y = y * x \tag{1.29}$$

3. Differentiation: let D be a derivation operator, $Dx(t) = \frac{dx(t)}{dt}$ in the continuous case, and $Dx_n = x_n - x_{n-1}$ in the discrete case. If $x * y$ exists, then $D(x * y)$ also exists and
$$D(x * y) = Dx * y = x * Dy \tag{1.30}$$

   This is a very important property and will be generalised for fractional derivatives.

4. Associativity
$$(x * y) * z = x * (y * z) \tag{1.31}$$

   provided that $x * y$, $y * z$, $(x * y) * z$, and $x * (y * z)$ all exist.

5. Neutral element: this is the δ,
$$x * \delta = x \tag{1.32}$$

6. Inverse element: if an inverse element exists, it will be a signal x^{-1} such that
$$x * x^{-1} = \delta \tag{1.33}$$

7. Summability: if x and y are summable, $x * y$ is summable as well.

8. Duration: let the durations of x and y be, respectively, T_1 and T_2 (for continuous signals) or N and M (for discrete signals). The duration of the convolution is equal to $T_1 + T_2$ or $N + M - 1$, in the continuous or discrete case respectively.

9. Delay and lead
$$z(t) = x(t) * y(t) \Rightarrow x(t - a) * y(t - b) = z(t - a - b) \tag{1.34}$$

In particular,

$$x * \delta(t - a) = x(t - a) \qquad\qquad t, a \in \mathbb{R} \qquad (1.35)$$

$$x * \delta(n - k_0) = x(n - k_0) \qquad\qquad n, k_0 \in \mathbb{Z} \qquad (1.36)$$

10. Parity: the convolution of signals with the same parity is an even function. If they have different parity, the convolution is odd.
11. Integration and accumulation

$$\int_{\omega}^{t} x(\tau)\,d\tau = x(t) * \varepsilon(t) \qquad (1.37)$$

$$\sum_{k=-\infty}^{n} x(k) = x(n) * \varepsilon(n) \qquad (1.38)$$

This property will be generalised for fractional integration.
12. Causality: the convolution of causal signals is causal.

One of the most interesting features of convolution is the fact that it "improves the behaviour" of the functions, since the resulting function is "smoother" than the factors. As an example, the convolution of a rectangular pulse (a discontinuous function) with itself is a triangular pulse (a continuous function).

1.2 Systems

The notion of system is fundamental for the theory and applications that we will present in the following chapters. Therefore, its introduction deserves some care. For now, let us take a simple approach. The term system is used in day-to-day language to express processes of the action–reaction type. In a generic way, a system reacts to an input stimulus, producing a given behaviour or result. For example, increasing the pressure on the accelerator of an automobile changes the state of motion, incrementing the speed; the reaction of an acid with a base originates a salt and water — this is a chemical system with 2 inputs and 2 outputs; a warehouse receives orders and makes sales — this is a commercial system.

Definition 1.2.1. *Therefore, we can say that a system is a combination of components acting together to fulfill a particular goal.*

In general, a system generates outputs (finished products, desired behaviours, stimuli, etc.) given certain input stimuli (raw material, mechanical actions, desired objectives, etc.) [40, 66, 102]. Mathematically, these input stimuli and their corresponding outputs are represented by signals, and the relations among inputs and outputs are described by mathematical models. The construction of a model (modelling) has to reflect the overall behaviour of the process at hand. A theoretical model can be obtained based upon first principles, i.e. of physical laws, hypotheses about the nature of the involved processes, and, also very probably, of approximations in order to make the model manageable. The experimental data is used to estimate and validate a given model.

Notice that different systems can be described by the same mathematical model. It is well known that electrical, mechanical, chemical, or economic systems, to give only some examples, may have similar behaviours, that can be modelled with the same equations. Whatever the way a mathematical model is obtained, it is only a simplified representation of a reality. However, throughout the text, each model will be used as if, in fact, it were the real system it corresponds to. Consequently, whenever this does not cause confusion, we will designate a model simply by *system*.

Some simple engineering examples:

- A lever is a system that has one input, the force applied on one of the arms, and one output, the force exerted by the other arm. These two forces are signals which are functions of time. The relationship between input and output is linear, provided that the lever is rigid.
- A Cardan joint is a system that has one input and one output. The input can be the rotation speed of the actuated shaft, or its angular position. The output will be the rotation speed or the angular position of the other shaft. The relation between input and output will be linear only if the two shafts are exactly aligned. However, if the angle is small, a linear approximation will be perfectly acceptable.
- Consider a Wave Energy Converter (WEC) system. The sea wave that acts on the WEC is an input; it will be represented by a signal which is either the wave elevation, or the force exherted on the WEC. An eventual control action will also be an input. The output signal most likely to be studied is the energy produced by the WEC. But other outputs can be considered, important for both design and operation, like forces exherted by moving parts, their angular velocities, if they are rotating, or the oil pressure in a hydraulic circuit, if there is one involved. The relations between input and output signals are surely non-linear, though there are WECs for which linear approximations can be used.

In general, a single-input, single-output (SISO) system is defined as an application in the set of signals $\mathfrak{S}$ or, in other words, *a transformation of a signal, $x(t) \in \mathfrak{S}$, into another one, $y(t) \in \mathfrak{S}$*. Let $T[\cdot]$ be an operator that symbolically represents such a transformation. Then

$$y(t) = T[x(t)] \tag{1.39}$$

Signal $x(t)$ is the *input* or *stimulus* and $y(t)$ is the *output* or *response*. A system is said to be continuous or discrete in time if the inputs and outputs are continuous-time or discrete-time signals, respectively. A multi-input, multiple-output (MIMO) system, with n inputs and m outputs, is an application from $\mathfrak{S}^n$ to $\mathfrak{S}^m$. In this book we will consider almost exclusively SISO systems.

Example 1.2.1. *Any mobile phone network is a MIMO system.*

Systems are required to have several properties. The most interesting are now described without complete justifications.

1. Time invariance

    Definition 1.2.2. *A system is time (or, shift) invariant, if the response to a given input signal is independent of the instant at which it is applied:*

    $$T[x(t)] = y(t) \Rightarrow T[x(t - t_0)] = y(t - t_0) \tag{1.40}$$

2. Causality

    A system is causal, or not anticipatory, if it does not produce a response before input is applied, that is, before the input is non-null, the output is always null. Therefore, the output at a given time does not depend on future input values. Intuitively, causality means that, without cause, there is no effect. In a more rigourous way:

    Definition 1.2.3. *A system is causal iff, for two any inputs, $x_1(t)$ and $x_2(t)$, with $x_1(t) = x_2(t)$ for $t < t_0$, then*

    $$T[x_1(t)] = T[x_2(t)] \qquad t < t_0 \tag{1.41}$$

    Under these conditions, but with reversed inequalities, the system is said *anti-causal*. If the system is neither causal nor anti-causal, it is said *acausal*.

    Causal systems are of great importance, because any system, initially at rest, working in real time, is causal. However, causal systems are not the only ones that have meaning and/or practical usefulness. For example, the autocorrelation function of the output of a system excited by white noise can be considered as the output of an acausal system, as we will see in a later chapter. Off-line processing may be considered also as acausal. In some applications it is interesting to "flip" a stored signal to treat it as anti-causal.

3. Stability

    There are several definitions of *stability*. The most common and useful definition is the so-called *bounded-input, bounded-output* (BIBO) stability:

    Definition 1.2.4. *A system is BIBO stable if, for any bounded input, the output is bounded:*

    $$|x(t)| < A \implies \exists B : |T[x(t)]| < B, \ A, B \in \mathbb{R}^+ \tag{1.42}$$

4. Memory

    Definition 1.2.5. *A system has memory if the output at the instant t depends on the inputs and/or outputs in other instants.*

    Frequently, memory is associated with causality, due to the ability of recalling the past. However, according to this definition, an anti-causal system does have memory: a memory of future events. The concepts of causality and memory are inter-related in the following notions.

Definition 1.2.6. *If the output y(t), at any instant t, can be determined exactly from the past input and output values and from the knowledge of the input at that instant, a system is said to be* deterministic.

If, on the other hand, the output at t can only be determined with a given probability or by other statistical methods, the system is said to be stochastic.

5. Linearity

Definition 1.2.7. *A system is linear (LS) if it is represented by a linear operator, i.e. it verifies the properties of additivity and homogeneity:*

$$y(t) = T[a_1 x_1(t) + a_2 x_2(t)]$$
$$= a_1 T[x_1(t)] + a_2 T[x_2(t)]$$
$$= a_1 y_1(t) + a_2 y_2(t) \tag{1.43}$$

In simple words, we can say that a system is linear iff it verifies the *superposition principle*.

Linearity is one of the most important properties of systems. Although for many systems it is a mere simplification or approximation of an exact non-linear relation, it proves to be extremely useful in practice. We shall be interested in LS that can be expressed as associations of simple systems, described by differential or difference equations. The said simple linear systems are multiplication by a constant, time shift, and differentiation or integration. This last operation will henceforth be referred to as differintegration.

6. Interconectability

In general, systems can be considered as associations of simpler interconnected subsystems. These assume three different basic forms (see Figure 1.1):

Series   It is an association where the output of a system is the input of the next. The defining operator is given by

$$y(t) = T_2[y_1(t)] = T_2[T_1[x(t)]] = T[x(t)] \tag{1.44}$$

The equivalent operator is:

$$T = T_2[T_1[.]] = T_2 T_1 \tag{1.45}$$

Parallel   In this case the systems in parallel have the same input and the outputs are added:

$$y(t) = y_1(t) + y_2(t) = T_1[x(t)] + T_2[x(t)] = T[x(t)] \tag{1.46}$$

The equivalent operator is:

$$T = T_1 + T_2 \tag{1.47}$$

Feedback   In this case, we have two subsystems, which are often called *plant* or *direct branch*, and *feedback branch*. Together they make a *feedback loop*. The

Series

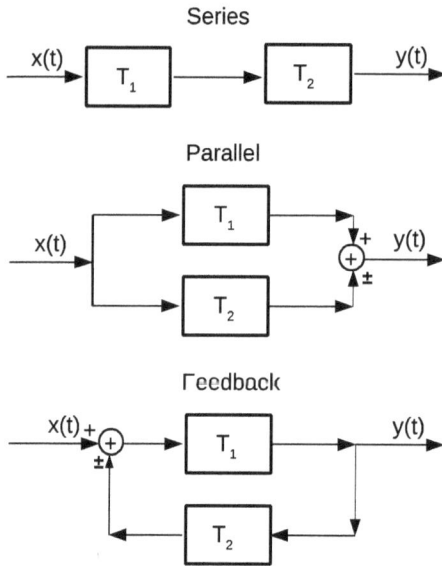

Parallel

Feedback

Fig. 1.1: Interconnection of systems. Top: series; centre: parallel; bottom: feedback.

output of the direct branch is also the output of the feedback loop and the input of the feedback branch. The output of the feedback branch is added to the input of the feedback loop to become the input of the direct branch.

$$y(t) = T_1 [x(t)] \pm T_1 T_2 [y(t)] \tag{1.48}$$

Therefore, the equivalent operator is:

$$T = [I \mp T_1 T_2]^{-1} T_1 \tag{1.49}$$

where I is the identity operator. This kind of system gives rise to the *state-space representation* of systems that we will study later.

In the case of non-linear systems, there may be other forms of association, which we will not address here.

7. Invertibility

A system is said invertible iff there exists another one that, associated in series, recovers the input. This means that there is another system with operator, T^{-1}, such that

$$T^{-1} T[x(t)] = x(t) \tag{1.50}$$

Invertibility is very important in Telecommunications and System Theory (Identification, Modelling, etc). Mathematically, and as long as $T[.]$ is a bijection, an inverse system exists. However, it may possess properties that rend it unsuitable for applications; for example, it may be unstable.

1.3 Characterisation of Linear Systems

The importance of LS in the various scientific domains cannot be overstated. While it is acknowledged that, in most current situations, linearity is only a simplified vision of reality, there are several reasons for the great use of LS:
- easy mathematical treatment,
- formal simplicity,
- easy implementation,
- good results (when conveniently used),
- ability of approximating complex systems, thereby helping its study.

This makes LS indispensable in Physics, Chemistry, all branches of Engineering, Biology, Economics, etc. For all these reasons, it is important to study them in depth. According to the characterisation of linear systems from the previous section, we will consider that LS are defined by

$$\sum_{k=0}^{N} a_k D^{\alpha_k} y(t) = \sum_{k=0}^{M} b_k D^{\beta_k} x(t) \tag{1.51}$$

where $x(t)$ is the input, $y(t)$ is the output, the a_k, $k = 0, 1, \cdots, N$ and the b_k, $k = 0, 1, \cdots, M$, are real parameters, N, M are null or positive integer orders, D is an operator defined by
- $D^a x(t) = \frac{d^a x(t)}{dt^a}$ for differential systems, not necessarily continuous, modelled using differential equations, for which $t \in \mathbb{R}$,
- $D^k x(t) = x(t - kh)$, $t = nh$, $n \in \mathbb{Z}$ for discrete systems having difference equations as modelling tools,

that we will study in the following chapters, and the α_k, $k = 0, 1, \cdots, N$ and the β_k, $k = 0, 1, \cdots, M$, parameters are derivative or delay orders, assumed to be positive reals in strictly increasing sequences.

Some results can be immediately got from (1.51):
1. If all the parameters $a_k, b_k, \alpha_k, \beta_k$ are independent of t, the system is time invariant. We will designate such systems as LTI (linear, time invariant).
2. If at least one of the orders α_k, β_k is different from zero, the system has memory.
3. If $N = M = 0$ the system has no dynamics, and its output is the input multiplied by a gain.
4. The inverse system is obtained changing the roles of the input and output.
5. The output of the system is given by

$$y(t) = h(t) * x(t) \tag{1.52}$$

where $h(t)$ is the system's output when the input is a delta — the so-called *impulse response* of the system. To prove this result we make $x(t) = \delta(t)$ and $y(t) = h(t)$ in (1.51):

$$\sum_{k=0}^{N} a_k D^{\alpha_k} h(t) = \sum_{k=0}^{M} b_k D^{\beta_k} \delta(t) \tag{1.53}$$

Convolving both sides of this equation with $x(t)$,

$$\sum_{k=0}^{N} a_k D^{\alpha_k} h(t) * x(t) = \sum_{k=0}^{M} b_k D^{\beta_k} \delta(t) * x(t) \tag{1.54}$$

Property (1.30) allows us to write

$$\sum_{k=0}^{N} a_k D^{\alpha_k} [h(t) * x(t)] = \sum_{k=0}^{M} b_k D^{\beta_k} x(t) \tag{1.55}$$

confirming (1.52). Consequently, a system can be equally described either by its impulse response or by its differential or difference equation.

In studying the behaviour of a LS with time, we use normally two kinds of working regimes:

Steady-state   In this case, it is accepted that the stimulation has started a long time ago (ideally at $t = -\infty$), so that, for a stable (in the strict sense) system, the response due to any non-null initial conditions is almost zero. This regime is usually studied employing sinusoids as the input.

Transient regime   In this case the system excitation is null until a given instant (which, without loss of generality, we can consider $t = 0$). The goal is to evaluate how the system reacts from that point on, i.e. when the excitation signal takes non null values. The most useful input for this study is the unit step, but the ramp is also used. The outputs for such cases are usually known as *time responses*.

These regimes will be studied in dedicated chapters for different systems defined using (1.51).

Exercises

1. Classify the following signals as analog, quantised, sampled, or digital:
    a) population of a country
    b) population of a country on the last day of each year
    c) blood pressure of a patient during a 24–hour period
    d) body temperature of a patient, measured daily immediately after awakening
    e) number of electricity generators connected to an isolated electrical grid
    f) power demand at an isolated electrical grid
    g) daily average power consumption of an isolated electrical grid
2. Prove that, given two continuous periodic signals $x_1(t)$ and $x_2(t)$, with periods T_1 and T_2, then $n_1 \cdot T_1 = n_2 \cdot T_2 = T$, $n_1, n_2 \in \mathbb{Z}^+$ is a sufficient condition for sum $x(t) = a_1 x_1(t) + a_2 x_2(t)$ to be a periodic signal, and that the period is T.
3. Prove that the condition above is also necessary.

4. Prove that the linear combination of two periodic discrete signals is always periodic.
5. Prove (1.10).
6. Prove that $\text{sgn}(t) = 2\varepsilon(t) - 1$ (in the continuous case) and that $\text{sgn}(n) = 2\varepsilon(n) - 1$ (in the discrete case).
7. Find the instantaneous power and the energy of each signal in Table 1.1 and in Table 1.2. If the energy is infinite, find the average power instead, for the passing T.
8. Prove that a signal with finite energy has null power.
9. Prove that all bounded finite duration signals (windows) are energy signals.
10. Prove that periodic signals are power signals.
11. Show that a ramp is neither a power signal nor an energy signal.
12. Prove each of the twelve properties of convolution in Section 1.1.6 for the case of continuous signals.
13. Do the same for the case of discrete signals.
14. Prove 1.49.
15. Consider a sinusoid with frequency equal to 25 Hz. Sample it with several sampling intervals $\{0.01, 0.02, 0.05, 0.1\}$ s. Overlap the plots of the original and sampled signals. Observe what happens with the different values. The so-called "aliasing" effect [65, 100] can be observed.

2 Continuous-time linear systems and the Laplace transform

2.1 The impulse response and the properties of linear systems

The results presented in section 1.3 will now be applied to the particular case of linear time invariant (LTI) continuous-time systems, described by differential equation (1.51) [25, 66, 102]. Let the impulse response (IR) be $h(t)$. Remember that we proved that the input/output relation is given by (1.52), which can be rewritten as

$$y(t) = \int_{-\infty}^{\infty} h(\tau)x(t-\tau)\,d\tau, \tag{2.1}$$

From this relation, we can prove immediately the following:
1. The system is (BIBO) stable iff the IR is absolutely integrable:

$$[x(t) < A \Rightarrow y(t) < B] \Leftrightarrow \int_{-\infty}^{\infty} |h(\tau)|\,d\tau < \infty \tag{2.2}$$

Here A and B are positive constants. A system verifying this requirement is said to be stable in the *strict sense or asymptotically stable*. If IR is bounded, but only absolutely integrable in any finite interval the system is *wide sense stable*.

2. Since this is an LTI system (LTIS), the output due to $\delta(t - t_0)$ (an impulse at $t = t_0$, $t_0 \in \mathbb{R}$) is $h(t - t_0)$.

3. If the system is causal (anti-causal) the IR verifies $h(t) = 0$, $t < 0$ ($h(t) = 0$, $t > 0$).

4. An LTIS is invertible if there exists a $g(t)$ such that

$$g(t) * h(t) = \delta(t). \tag{2.3}$$

Notice that, if a system is defined by a derivative D^{α}, there should exist an anti-derivative $D^{-\alpha}$ verifying $D^{-\alpha}D^{\alpha}f(t) = D^{\alpha}D^{-\alpha}f(t) = f(t)$.

2.2 Exponentials and LTIS

The (eternal) exponential, $e^{st}, t \in \mathbb{R}, s \in \mathbb{C}$, plays an important role in the study of LTIS. To see how, return to relation (2.1), that gives us a very important result: the exponential, $e^{st}, t \in \mathbb{R}$, is the eigenfunction of the LTIS. In fact, if the input is $x(t) = e^{st}$, the output is

$$y(t) = \int_{-\infty}^{\infty} h(\tau)e^{s(t-\tau)}\,d\tau = H(s)e^{st} \tag{2.4}$$

DOI 10.1515/9783110624588-002

with

$$H(s) = \int_{-\infty}^{\infty} h(\tau)e^{-s\tau}d\tau \tag{2.5}$$

The function $H(s)$ depends only on the IR of the system. Thus, it also serves to characterise the LTIS, and is called *transfer function* (TF) of the system. Relation (2.4) is valid for any $s \in \mathbb{C}$ for which $H(s)$ is finite (regular case); later we will consider singular cases.

The TF allows also computing the output of a system: we only have to express the input as a linear combination of exponentials (this was Liouville starting point for introducing fractional derivatives). Therefore, if $x(t) = \sum_{k=1}^{K_0} A_k e^{s_k t}$, $K_0 \in \mathbb{N}_0$, the output is

$$y(t) = \sum_{k=1}^{K_0} H(s_k)A_k e^{s_k t}. \tag{2.6}$$

If $H(s) = 0$ for a particular value of s, the output due to the corresponding component is identically null, and we say that the system "filters" such input component. Usually, we consider sinusoidal inputs, $s = i\omega$, and observe the output. If any component is eliminated, or if at least its amplitude is significantly reduced, we say that the LS acts as a filter for that sinusoid. This is why, by extension, linear systems are also called *filters*.

2.2.1 The Laplace transform

Let us now consider that a given "good enough" function $x(t)$, $t \in \mathbb{R}$, is the superposition of elemental exponentials given by the so-called Bromwich integral [25, 102]

$$x(t) = \frac{1}{2\pi i} \int_{\gamma} X(s)ds \cdot e^{st}, \qquad t \in \mathbb{R}. \tag{2.7}$$

The integration path γ is a vertical straight line located in a vertical strip where $X(s)$ is analytical, and we suppose that the principal value of the integral is computed. Assuming that $x(t)$ is the input to a LTIS with a TF equal to $H(s)$, the output is obtained from

$$y(t) = \frac{1}{2\pi i} \int_{\gamma} H(s)X(s) \cdot e^{st} ds, \qquad t \in \mathbb{R}. \tag{2.8}$$

Relations (2.5) and (2.7) are the analysis/synthesis pair that allows us to define the two-sided or bilateral Laplace transform (BLT), which is an efficient tool for dealing with LTIS. We will write

$$\mathcal{L}[x(t)] = X(s) = \int_{-\infty}^{\infty} x(t)e^{-st}dt, \ s \in \mathbb{C}, \tag{2.9}$$

and

$$x(t) = \mathcal{L}^{-1}[X(s)] = \frac{1}{2\pi i} \int_{\gamma} X(s) \cdot e^{st} ds, \qquad t \in \mathbb{R}. \tag{2.10}$$

Example 2.2.1 (The fractional power). *A very interesting example is the fractional power defined in (1.14), that we can rewrite as*

$$\phi(t) = \frac{t^{\alpha-1}}{\Gamma(\alpha)}\varepsilon(t), \quad \alpha > 0. \tag{2.11}$$

Its Laplace transform is

$$\mathcal{L}\left[\phi(t)\right] = \Phi(s) = \frac{1}{\Gamma(\alpha)}\int_0^\infty t^{\alpha-1}e^{-st}dt, \quad s \in \mathbb{C},$$

For simplicity, assume that s is a positive real. With a change of variable, st = τ, we get

$$\Phi(s) = \mathcal{L}\left[\phi(t)\right] = \frac{s^{-\alpha}}{\Gamma(\alpha)}\int_0^\infty \tau^{\alpha-1}e^{-\tau}d\tau,$$

The integral on the right is the Gamma function, $\Gamma(\alpha)$. We have then

$$\Phi(s) = \frac{1}{s^\alpha}, \tag{2.12}$$

Using the Cauchy theorem of complex variable function, we can prove that the above relation remains valid, even if s is complex, but Re(s) > 0.

Remark 2.2.1. *In most mathematical books on the Laplace transform, the unilateral Laplace Transform (ULT) formulation is presented:*

$$\mathcal{L}_u x(t) = \int_0^\infty x(t)e^{-st}dt, \quad s \in \mathbb{C}. \tag{2.13}$$

Obviously, whenever x(t) = 0 for all t < 0, the ULT and the BLT coincide.

However, the use of both transforms in Signals and Systems showed that the BLT has several advantages over the ULT, namely:
1. *It is more general, since the ULT can be considered as a particular case;*
2. *The class of functions to which it can be applied is much larger;*
3. *Some properties become simplified;*
4. *It offers insight into the nature of system characteristics such as stability, causality, and frequency response;*
5. *It allows the study of systems that are not causal;*
6. *It contains the Fourier transform (discussed in appendix C) as a particular case;*
7. *It does not introduce any "initial conditions". The ULT is frequently used in initial value problems, but this creates several difficulties, since they are frequently wrong.*

From now on, when mentioning the Laplace Transform (LT), the BLT definition will be presumed.

2.2.2 Existence and main properties of the Laplace transform

Let $x(t)$ be a function
- almost everywhere continuous;
- with bounded variation;
- locally integrable, in the sense that the function is absolutely integrable in any real interval $[a, b]$, so that $\int_a^b |x(t)| \, dt < \infty$;
- of exponential order, that is to say, the function does not "grow faster" than a given exponential, on each side. This means two things. First, that there are real constants A and $a > 0$ such that $|x(t)| < A.e^{at}$, when t is large and negative (say, for $t < t_1 \in \mathbb{R}$). Second, that there are real constants B and $b > 0$ such that $|x(t)| < B.e^{bt}$, when t is large (say, for $t > t_2 \in \mathbb{R}$). It also has to be true that $a > b$.

Under these conditions, the LT exists, and is analytical in a vertical strip in the complex plane defined by $b < Re(s) < a$. This strip is called *region of convergence* (ROC), and the a and b constants are the abscissas of convergence. It is possible to show that the inversion integral (2.10) converges to the half sum of the lateral values $x(t) = \frac{x(t^+)+x(t^-)}{2}$, for any $t \in \mathbb{R}$.

These conditions give some insight into the ROC of some particular functions:
- Right signal, $x(t) = 0$, $t < T \in \mathbb{R}$
  We can choose $t < t_1 < 0$, thus $a = \infty$ and the ROC is $Re(s) > b$. Therefore, the ROC corresponding to a right signal is a *right half plane*.

  Example 2.2.2. *Consider the unit step function, $\varepsilon(t)$. The computation is simple and gives*

  $$\mathcal{L}\left[\varepsilon(t)\right] = \frac{1}{s}, \quad Re(s) > 0 \tag{2.14}$$

- Left signal, $x(t) = 0$, $t > T \in \mathbb{R}$
  This is the reverse of above case: we can choose $t > t_2 > 0$, so $b = -\infty$ and the ROC is $Re(s) < a$. Therefore, the ROC corresponding to a left signal is a *left half plane*.

  Example 2.2.3. *Let the reverse unit step function, $-\varepsilon(-t)$. The computation of its LT gives*

  $$\mathcal{L}\left[-\varepsilon(-t)\right] = \frac{1}{s}, \quad Re(s) < 0 \tag{2.15}$$

- In the general case, any signal is the sum of a left and a right signals. Consequently, the ROC is the intersection of the two. If this intersection is empty, the signal has no LT.

  Example 2.2.4. *From the above results with the unit step, we can show that the constant, $c(t) = 1$, $t \in \mathbb{R}$, and signum functions do not have LT. Neither does the sinusoid $e^{i\omega_0 t}, t \in \mathbb{R}$.*
- Finite duration signal $x(t) = 0$, $t \notin [t_1, t_2]$
  It is a simple matter to conclude that the ROC is the whole complex plane, $\mathbb{C}$.

Example 2.2.5. *Consider the unit pulse function, $x(t) = 1, t \in [-T, T]$. The compu-tation is simple and gives*

$$\mathcal{L}[x(t)] = \frac{e^{sT} - e^{-sT}}{s} \tag{2.16}$$

that is analytic at $s = 0$.

Let $f(t)$ and $g(t)$ be functions with LT, $F(s)$ and $G(s)$. The main properties of the BLT are:

1. Linearity and homogeneity:

$$\mathcal{L}[a.f(t) + b.g(t)] = a.F(s) + b.G(s) \tag{2.17}$$

2. Scale change — this can be proved with the substitution of at for t, with $a \in \mathbb{R}$, into (2.4):

$$\mathcal{L}[f(at)] = \frac{1}{|a|}F(s/a) \tag{2.18}$$

3. Time reversion — this property is a particular case of the previous one with $a = -1$:

$$\mathcal{L}[f(-t)] = F(-s) \tag{2.19}$$

4. Time shift — let $a \in \mathbb{R}$; then

$$\mathcal{L}[f(t - a)] = e^{-as}F(s) \tag{2.20}$$

5. Modulation — this is the dual of the last property:

$$\mathcal{L}\left[e^{at}f(t)\right] = F(s - a) \tag{2.21}$$

6. Convolution — by defining, as seen before, the convolution between two functions $f(t)$ and $g(t)$ by

$$f(t) * g(t) = \int_{-\infty}^{\infty} f(\tau)g(t - \tau)d\tau, \tag{2.22}$$

and then inserting (2.10) in (2.22), it is possible to get

$$f(t) * g(t) = \int_{-\infty}^{\infty} f(\tau)\frac{1}{2\pi i}\int_{\gamma} G(s)e^{st}e^{-s\tau}\,ds\,d\tau = \frac{1}{2\pi i}\int_{\gamma} G(s)\left[\int_{-\infty}^{\infty} f(\tau)e^{-s\tau}d\tau\right]ds \tag{2.23}$$

from which

$$\mathcal{L}[f(t) * g(t)] = \mathcal{L}[f(t)] \cdot \mathcal{L}[g(t)] \tag{2.24}$$

7. Time derivative of the exponential — As it is well known, $D^n e^{st} = s^n e^{st}$, $t \in \mathbb{R}$. This leads from (2.10) to $\mathcal{L}[D^n f(t)] = s^n F(s)$, that is the classic form of the derivative property. Here, we want to generalise this property to fractional derivative case. If

$\alpha \in \mathbb{R}$, we want to obtain

$$D^{\alpha} f(t) = \frac{1}{2\pi i} \int_{\gamma} s^{\alpha} F(s) e^{st} \, ds, \qquad t \in \mathbb{R}. \tag{2.25}$$

from where we can deduce that

$$\mathcal{L}\left[D^{\alpha} f(t)\right] = s^{\alpha} F(s) \tag{2.26}$$

Therefore, we are assuming that we are dealing with a fractional derivative verifying

$$D^{\alpha} e^{st} = s^{\alpha} e^{st} \quad t \in \mathbb{R} \tag{2.27}$$

Remark 2.2.2. *Frequently we will work with the multivalued expressions s^{α} or $(-s)^{\alpha}$. To obtain functions, we have to fix branchcut lines. We will choose the negative or positive real half axis according to the problem at hand; we will work on the first Riemann surface. For each one we choose a branch cut line in order to define two transfer functions with regions of convergence defined by $Re(s) > 0$ and $Re(s) < 0$ – See Figure 2.1 for the s^{α} case.*

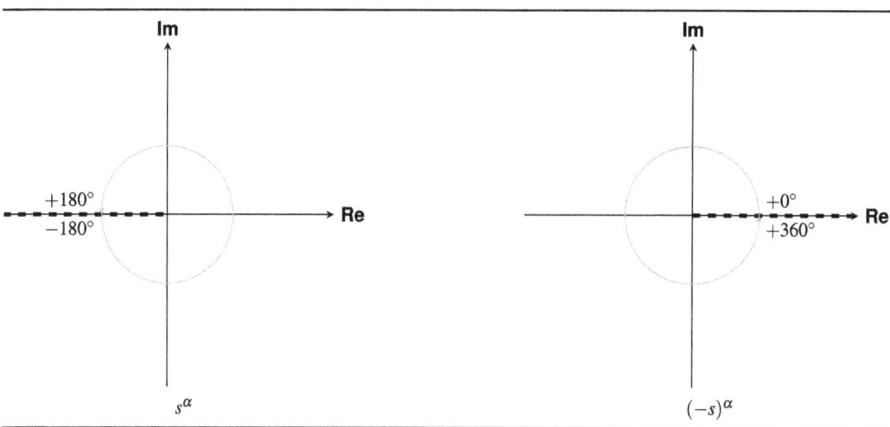

Fig. 2.1: Branchcuts and Riemann surfaces for s^{α} with regions of convergence defined by $Re(s) > 0$ and $Re(s) < 0$.

8. The Fourier transform — The BLT allows us to introduce the Fourier transform (FT). Assume that the LT of $g(t)$ has a ROC that includes the imaginary axis. Therefore, we can choose $s = i\omega$, $\omega \in \mathbb{R}$. Substituting in the analysis/synthesis integrals of the LT we obtain

$$\mathcal{F}[g(t)] = G(i\omega) = \int_{-\infty}^{\infty} g(t) e^{-i\omega t} \, dt, \quad \omega \in \mathbb{R}, \tag{2.28}$$

and

$$g(t) = \mathcal{F}^{-1}[G(i\omega)] = \frac{1}{2\pi} \int_{-\infty}^{\infty} G(i\omega) e^{i\omega t} \, d\omega, \quad t \in \mathbb{R}. \tag{2.29}$$

that are the analysis/synthesis formulae defining the Fourier transform. Frequently we write $\omega = 2\pi f$ which leads to

$$\mathcal{F}[g(t)] = G(f) = \int_{-\infty}^{\infty} g(t) e^{-i2\pi f t} \, dt, \quad f \in \mathbb{R}, \tag{2.30}$$

and

$$g(t) = \mathcal{F}^{-1}[G(f)] = \int_{-\infty}^{\infty} G(f) e^{i2\pi f t} \, df, \quad t \in \mathbb{R}. \tag{2.31}$$

We reinforce that, if a function has LT with ROC, containing the imaginary axis, the FT is obtained from the LT with the substitution of $i\omega$ for s.

To understand the involved problems, let us consider a simple example.

Example 2.2.6. Let $g(t) = e^{-at}\varepsilon(t)$, $a > 0$. Its LT is $G(s) = \frac{1}{s+a}$, $Re(s) > -a$. Notice that the ROC includes the imaginary axis. The function has FT, given by $G(i\omega) = \frac{1}{i\omega+a}$.

Now, let a decrease to 0. In the limit case, we have $g(t) = \varepsilon(t)$, that has as LT $G(s) = \frac{1}{s}$, $Re(s) > 0$, but that does not have FT given by $G(i\omega) = \frac{1}{i\omega}$. In fact, it can shown that the correct FT is $G(i\omega) = \frac{1}{i\omega} + \pi\delta(\omega)$.

This deserves a comment: we can extend the relation between both transforms even when the imaginary axis is the boundary of the ROC to all the points where the LT does not have a singularity. Singularities cause the appearance of impulses.

2.3 The differintegrator

The property of the LT stated in (2.27) tells us that there are two LS, that we will call *differintegrators*, that have

$$H(s) = s^\alpha \tag{2.32}$$

as transfer function, with $Re(s) > 0$, causal, or $Re(s) < 0$, anti-causal, as ROC. If $\alpha > 0$ we speak of "differentiator", while with $\alpha < 0$ we use the term "integrator". As we will see later, the designation "anti-differentiator" is more correct, but "integrator" is popularised by the use, and so we will keep this designation.

In the integer order cases, we have the sequence

$$\ldots s^{-n} \ldots s^{-2} \quad s^{-1} \quad 1 \quad s^1 \quad s^2 \ldots s^n \ldots . \tag{2.33}$$

in the transform domain, while in time the corresponding sequence is

$$\cdots \pm \frac{t^{n-1}}{(n-1)!}\varepsilon(\pm t) \ldots \pm \frac{t^1}{1!}\varepsilon(\pm t) \quad \pm\varepsilon(\pm t) \quad \delta(t) \quad \delta'(t) \quad \delta''(t) \ldots \delta^{(n)}(t) \ldots \tag{2.34}$$

Tab. 2.1: Examples of LT.

$g(t)$	$\mathcal{L}g(t) = G(s)$	ROC
$\delta(t)$	1	$\mathbb{C}$
$\delta^{(n)}(t)$	s^n	$\mathbb{C}$
$\varepsilon(t)$	$\frac{1}{s}$	$Re(s) > 0$
$\varepsilon(-t)$	$-\frac{1}{s}$	$Re(s) < 0$
$t\varepsilon(t)$	$\frac{1}{s^2}$	$Re(s) > 0$
$\frac{t^a}{\Gamma(a+1)}\varepsilon(t)$	$\frac{1}{s^{a+1}}$	$Re(s) > 0$
$e^{|t|}$	$\frac{?}{1-s^2}$	$|Re(s)| < 1$
$sgn(t)e^{-|t|}$	$-\frac{s}{1-s^2}$	$|Re(s)| < 1$
$e^{at}\varepsilon(t)$	$\frac{1}{s-a}$	$Re(s) > Re(a)$
$-e^{at}\varepsilon(-t)$	$\frac{1}{s-a}$	$Re(s) < Re(a)$
$\begin{cases} 1 & 0 < t \leq T \\ 0 & t < 0 \text{ and } t > T \end{cases}$	$\frac{1-e^{-sT}}{s}$	$\mathbb{C}$
$\cos(\omega_0 t)\varepsilon(t)$	$\frac{s}{s^2+\omega_0^2}$	$Re(s) > 0$
$\sin(\omega_0 t)\varepsilon(t)$	$\frac{\omega_0}{s^2+\omega_0^2}$	$Re(s) > 0$
$e^{-at}\cos(\omega_0 t)\varepsilon(t)$	$\frac{(s+a)}{(s+a)^2+\omega_0^2}$	$Re(s) > -Re(a)$
$e^{-at}\sin(\omega_0 t)\varepsilon(t)$	$\frac{\omega_0}{(s+a)^2+\omega_0^2}$	$Re(s) > -Re(a)$
$\frac{t^N}{N!}e^{at}\varepsilon(t)$	$\frac{1}{(s-a)^{N+1}}$	$Re(s) > Re(a)$
$-\frac{t^N}{N!}e^{at}\varepsilon(-t)$	$\frac{1}{(s-a)^{N+1}}$	$Re(s) < Re(a)$
$x(t)\cos(\omega_0 t)$	$\frac{1}{2}X(s - i\omega_0) + \frac{1}{2}X(s + i\omega_0)$	$\mathbb{C}$
$x(t)\sin(\omega_0 t)$	$\frac{1}{2i}X(s - i\omega_0) - \frac{1}{2i}X(s + i\omega_0)$	$\mathbb{C}$
$x^{(n)}(t)\varepsilon(t)$	$s^n X(s) - \sum_{j=0}^{n-1} x^{(j)}(0)s^j$	$\mathbb{C}$

Now, we want to fill in the gaps with fractional or irrational powers, such as

$$\ldots s^{-n} \ldots s^{-2} \quad s^{-\sqrt{2}} \quad s^{-1} \quad s^{-0.4} \quad 1 \quad s^{0.5} \quad s^1 \quad s^{\sqrt{3}} \quad s^2 \ldots s^n \ldots \tag{2.35}$$

to obtain a time sequence given by

$$\ldots \pm \frac{t^{n-1}}{\Gamma(n)}\varepsilon(\pm t) \ldots \pm \frac{t^1}{1!}\varepsilon(\pm t) \quad \pm \frac{t^{\sqrt{2}-1}}{\Gamma(\sqrt{2})}\varepsilon(\pm t) \quad \pm \varepsilon(\pm t) \quad \pm \frac{t^{0.4-1}}{\Gamma(\sqrt{0.4})}\varepsilon(\pm t) \tag{2.36}$$

$$\delta(t) \quad \pm \frac{t^{-0.5-1}}{\Gamma(\sqrt{-0.5})}\varepsilon(\pm t) \quad \delta'(t) \quad \pm \frac{t^{-\sqrt{3}-1}}{\Gamma(-\sqrt{3})}\varepsilon(\pm t) \quad \delta''(t) \ldots \delta^{(n)}(t) \ldots$$

The left part of the sequence is proved in the following.

In (2.11) to (2.12), we showed that the impulse response corresponding to the causal integrator is the causal fractional power

$$\mathcal{L}\left[\frac{t^{\alpha-1}}{\Gamma(\alpha)}\varepsilon(t)\right] = \frac{1}{s^\alpha}, Re(s) > 0 \tag{2.37}$$

with $\alpha > 0$. Now, let us compute the LT of the anti-causal power:

$$\mathcal{L}\left[-\frac{t^{\alpha-1}}{\Gamma(\alpha)}\varepsilon(-t)\right] = -\frac{1}{\Gamma(\alpha)}\int\limits_\infty^0 t^{\alpha-1}e^{-st}\,dt = \frac{(-1)^\alpha}{\Gamma(\alpha)}\int\limits_0^\infty t^{\alpha-1}e^{-(-s)t}\,dt, = \frac{1}{s^\alpha}, Re(s) < 0. \tag{2.38}$$

Therefore,

$$\mathcal{L}\left[\pm\frac{t^{\alpha-1}}{\Gamma(\alpha)}\varepsilon(\pm t)\right] = \frac{1}{s^\alpha}, \pm Re(s) > 0. \tag{2.39}$$

The left hand side represents the impulse responses of fractional integrators.

Now, we are in condition of computing the fractional derivative of the fractional power. We are going to consider the causal case, leaving the anti-causal as exercise:

$$\mathcal{L}\left[D^\beta \frac{t^{\alpha-1}}{\Gamma(\alpha)}\varepsilon(t)\right] = s^\beta \frac{1}{s^\alpha} = \frac{1}{s^{\alpha-\beta}}, Re(s) > 0. \tag{2.40}$$

If $\beta \le \alpha$, the solution is immediate:

$$D^\beta \frac{t^{\alpha-1}}{\Gamma(\alpha)}\varepsilon(t) = \frac{t^{\alpha-\beta-1}}{\Gamma(\alpha-\beta)}\varepsilon(t), \tag{2.41}$$

where $\frac{t^{-1}}{\Gamma(0)}\varepsilon(t) = \delta(t)$. To justify this, in an informal way, notice that when $t \ne 0$ the numerator is finite and the denominator is infinite, and so it is reasonable to say $\frac{t^{-1}}{\Gamma(0)}\varepsilon(t) = 0$; while for $t = 0$ both numerator and denominator are infinite, but it can be shown that as $t \to 0$ an infinite value is obtained. A rigorous proof can be found in [20]. The case $\beta > \alpha$ corresponds to the derivatives of the impulse: since $s^\alpha = s^N \frac{1}{s^{-\alpha+N}}$, then $D^\alpha \delta(t) = D^N \frac{t^{-\alpha+N-1}}{\Gamma(-\alpha+N)}\varepsilon(t)$. The successive computation of integer order derivatives originates the appearance of singular infinite terms. If they are rejected, we verify that (2.41) remains valid for any order. In particular,

$$D^\alpha \delta(t) = \frac{t^{-\alpha-1}}{\Gamma(-\alpha)}\varepsilon(t). \tag{2.42}$$

A very curious result can be obtained from (2.41) by an integer order differentiation relatively to α. Let $\psi(.) = D\Gamma(.)$. With some simple manipulations, we get

$$D^\beta t^\alpha \log(t)\varepsilon(t) = \frac{\Gamma(\alpha+1)}{\Gamma(\alpha-\beta+1)}t^{\alpha-\beta}\varepsilon(t)\left[\log(t) + \psi(\alpha+1) - \psi(\alpha-\beta+1)\right]. \tag{2.43}$$

In particular,

$$D^\beta \log(t)\varepsilon(t) = \frac{1}{\Gamma(-\beta+1)}t^{-\beta}\varepsilon(t)\left[\log(t) - \gamma - \psi(-\beta+1)\right] \tag{2.44}$$

where $\gamma = -\psi(1) \approx 0.5772156649\ldots$ is the Euler–Mascheroni constant.

2.4 Impulse and step responses of causal LS

Return back to system (1.51), and use relations (2.4) and (2.27) to obtain the transfer function

$$H(s) = \frac{B(s)}{A(s)} = \frac{\sum\limits_{k=0}^{M} b_k s^{\beta_k}}{\sum\limits_{k=0}^{N} a_k s^{\alpha_k}} \tag{2.45}$$

with a suitable choice of the ROC. Here we will consider the causal case only. According to (2.5), the transfer function is the LT of the impulse response. Therefore, we have three equivalent ways of representing a given LS:
– Differential equation (DE)
– Impulse response (IR)
– Transfer function (TF)

Probably, the most important is the transfer function, since it gives insights into the structure of the system, which are difficult to obtain from the others. In the following, we shall be concerned with the computation of the IR from the TF. To do it, we will present a theorem that states a relation between the behaviours of both.

2.4.1 The general initial value theorem

The Abelian initial value theorem (IVT) is a very important result in dealing with the Laplace transform [124]. This theorem relates the asymptotic behaviour of a causal signal, $\phi(t)$, as $t \to 0^+$, to the asymptotic behaviour of its Laplace transform, $\Phi(s) = \mathcal{L}[\phi(t)]$, as $\sigma = Re(s) \to \infty$.

Theorem 2.4.1 (The initial-value theorem). *Assume that $\phi(t)$ is a causal signal such that:*
– *in some neighbourhood of the origin, it is a regular distribution, corresponding to an integrable function;*
– *its Laplace transform is $\Phi(s)$, with ROC $Re(s) > 0$.*
Also assume that there is a real number $\beta > -1$ such that $\lim\limits_{t \to 0^+} \phi(t) t^{\beta}$ exists and is a finite complex value. Then

$$\lim_{t \to 0^+} \frac{\phi(t)}{t^{\beta}} = \lim_{\sigma \to \infty} \frac{\sigma^{\beta+1} \Phi(\sigma)}{\Gamma(\beta + 1)} \tag{2.46}$$

For proof see [124] (section 8.6, pages 243–248).

As referred above, $\mathcal{L}[\phi^{(\alpha)}(t)] = s^{\alpha} \Phi(s)$ for $Re(s) > 0$. On the other hand, attending to the IVT for $\beta = 0$,

$$\phi(0^+) = \lim_{\sigma \to \infty} \sigma \Phi(\sigma) \tag{2.47}$$

we can write

$$\lim_{\sigma \to \infty} \sigma^{\beta+1} \Phi(\sigma) = \lim_{t \to 0^+} \frac{\phi(t)}{t^{\beta}} \tag{2.48}$$

which is given by (2.46).

Corollary 2.4.1. 1. *If $-1 < \alpha < \beta$, then*

$$\lim_{t \to 0^+} \frac{\phi(t)}{t^\alpha} = 0 \tag{2.49}$$

2. *If $\alpha > \beta$, then*

$$\lim_{t \to 0^+} \frac{\phi(t)}{t^\alpha} = \infty \tag{2.50}$$

To prove this we only have to write

$$\lim_{t \to 0^+} \frac{\phi(t)}{t^\alpha} = \lim_{t \to 0^+} \frac{\phi(t)}{t^\beta} \frac{t^\beta}{t^\alpha} = 0 \tag{2.51}$$

and note that the first fraction has a finite limit given by (2.48), while the limit of the second fraction is zero in case 1 and infinite in case 2. This means that function $\phi(t)$ has the behaviour $\phi^{(\beta)}(0+)t^\beta$ near the origin.

2.4.2 The general final value theorem

The Abelian final value theorem (FVT) is the dual of the IVT [124].

Theorem 2.4.2 (The Final-value theorem). *Under the same asumptions of the IVT, we can state*

$$\lim_{t \to \infty} \frac{\phi(t)}{t^\beta} = \lim_{\sigma \to 0^+} \frac{\sigma^{\beta+1} \Phi(\sigma)}{\Gamma(\beta+1)} \tag{2.52}$$

For proof see [124] (section 8.7, pages 249–251). This result is useful in studying transient responses of systems. Notice, in particular, that $\lim_{t \to \infty} \phi(t) = \lim_{\sigma \to 0^+} \sigma \Phi(\sigma)$ that expresses the classical FVT.

2.4.3 Transfer function series representation

Let $H(s)$ be a transfer function with an associated ROC $Re(s) > 0$. Based on the IVT, it is possible to obtain a decomposition of $H(s)$ into a sum of negative power functions plus an error term. The procedure is as follows [92]:
1. Define $R_0(s) = H(s)$, and let the corresponding IR be $r_0(t)$.
2. Let γ_0 be the real value such that

$$\lim_{\sigma \to \infty} \sigma^{\gamma_0} H(\sigma) = A_0 \tag{2.53}$$

where A_0 is finite and non null. Then let

$$R_1(s) = H(s) - A_0 s^{-\gamma_0} \tag{2.54}$$

It is clear that $\lim_{\sigma \to \infty} \sigma^{\gamma_0} R_1(\sigma) = 0$ and $A_0 = \lim_{t \to 0^+} r_0(t) = h^{(\gamma_0-1)}(0^+)$.

3. Now repeat the process. Let γ_1 be the real value such that

$$\lim_{\sigma \to \infty} \sigma^{\gamma_1} R_1(\sigma) = A_1 \tag{2.55}$$

where A_1 is finite and non null. Again, introduce

$$R_2(s) = H(s) - A_0 s^{-\gamma_0} - A_1 s^{-\gamma_1} \tag{2.56}$$

with $\lim_{\sigma \to \infty} \sigma^{\gamma_1} R_2(\sigma) = 0$ and $A_1 = \lim_{t \to 0^+} r_1^{(\gamma_1 - 1)}(t)$, with $r_1(t)$ as the inverse of $R_1(s)$.

4. In general, let γ_n be the real value for which

$$\lim_{\sigma \to \infty} \sigma^{\gamma_n} R_n(\sigma) = A_n \tag{2.57}$$

where A_n is finite and non null. We arrive at function

$$R_n(s) = H(s) - \sum_{k=0}^{n-1} A_k s^{-\gamma_k} \tag{2.58}$$

and $\lim_{\sigma \to \infty} \sigma^{\gamma_n} R_n(\sigma) = 0$. As above, $A_n = \lim_{t \to 0^+} r_{n-1}^{(\gamma_n - 1)}(t) = r_{n-1}^{(\alpha_n)}(0^+)$, to be coherent with the initial value theorem: $\gamma_n = \alpha_n + 1$, for $n \in \mathbb{Z}_0^+$.

We can write:

$$H(s) = \sum_{k=0}^{n-1} A_k s^{-\gamma_k} + R_n(s) \tag{2.59}$$

leading us to conclude that $H(s)$ can be expanded in a Laurent like power series.

Theorem 2.4.3 (Generalised Laurent series). *Let*

$$H(s) = \sum_{0}^{\infty} r_k^{(\alpha_k)}(0^+) s^{-\gamma_k} \tag{2.60}$$

If the series in (2.60) is approximated by a truncated summation as in (2.59), then the error is bounded by $|R_n(s)| = o\left(\left|\frac{1}{s^\gamma}\right|\right)$, with $\gamma > \gamma_n$, and consequently the error in time is less than $\frac{t^{\alpha_n}}{\Gamma(\alpha_n + 1)}$.

It is not easy to establish the ROC of the series in (2.60). If $\alpha_k = k\alpha$, we can apply the theory of Z transform [102]. In this case, if the sequence $A_k = r_k^{(k\alpha)}(0^+)$ is of exponential order, the series converges in the region that is the exterior of a circle with centre at the origin.

According to the procedure above, the successive functions $r_k^{(\alpha_k)}(t)$ are obtained from $h(t)$ by removing the inverses of $r_m^{(\alpha_m)}(0^+) s^{-\gamma_m}$ for $m < k$. It can be seen that the derivatives of order α_k of such terms are singular at the origin. Therefore, the terms $r_k^{(\alpha_k)}(t)$ are equal to the analytic part of $h^{(\alpha_k)}(t)$. However, when the derivative orders are positive integers, this does not happen, because the derivatives of the removed terms are either zero or derivatives of the impulse $\delta(t)$, null for $t = 0^+$. In the following we will represent them by $h_a^{(\alpha_k)}(t)$.

Theorem 2.4.4 (The generalised MacLaurin series). *Let $h(t)$ be the impulse response corresponding to $H(s)$. Then*

$$h(t) = \sum_{k=0}^{\infty} h_a^{(\gamma_k-1)}(0^+)\frac{t^{\gamma_k-1}}{\Gamma(\gamma_k)}\varepsilon(t) \tag{2.61}$$

that can also be written as

$$h(t) = \sum_{k=0}^{\infty} h_a^{(\alpha_k)}(0^+)\frac{t^{\alpha_k}}{\Gamma(\alpha_k+1)}\varepsilon(t) \tag{2.62}$$

Remark 2.4.1. *These relations remain valid even with $\gamma_0 = 0$ ($\alpha_0 = -1$), provided that we accept that*

$$\frac{t^{-1}}{\Gamma(0)}\varepsilon(t) = \delta(t)$$

Corollary 2.4.2 (The generalised step response). *From (2.60), the LT of the (causal) step response is*

$$R_\varepsilon(s) = \sum_{k=0}^{\infty} r_k^{(\alpha_k)}(0^+)s^{-\gamma_k-1} \tag{2.63}$$

which gives

$$r_\varepsilon(t) = \sum_{0}^{\infty} h_a^{(\gamma_k)}(0^+)\frac{t^{\gamma_k}}{\Gamma(\gamma_k+1)}\varepsilon(t) \tag{2.64}$$

Example 2.4.1. $H(s) = \frac{s^{\alpha-\beta}}{s^\alpha+1}$ *is the LT of the so-called two-parameter Mittag-Leffler function. (See more about the MLF in Appendix F.) Proceeding as pointed above,*

$$A_0 = 1 \text{ and } \gamma_0 = \beta$$
$$A_1 = -1 \text{ and } \gamma_1 = \alpha + \beta$$

Repeating the process, we obtain

$$R_n(s) = (-1)^n\frac{1}{s^{\beta+n.\alpha}(s^\alpha+1)} \tag{2.65}$$

with

$$A_n = (-1)^n \text{ and } \gamma_n = n.\alpha + \beta \tag{2.66}$$

This leads to

$$H(s) = s^{-\beta}\sum_{n=0}^{\infty}(-1)^n s^{-n\alpha} \tag{2.67}$$

Its inverse is easily obtained:

$$h(t) = \sum_{n=0}^{\infty} (-1)^n \frac{t^{n\alpha+\beta-1}}{\Gamma(n\alpha+\beta)} \cdot \varepsilon(t) = t^{\beta-1} E_{\alpha,\beta}(-t) \tag{2.68}$$

where

$$E_{\alpha,\beta}(z) = \sum_{n=0}^{\infty} \frac{z^{n\alpha}}{\Gamma(n\alpha+\beta)} \tag{2.69}$$

is the two-parameter Mittag-Leffler function (MLF) [22].

Example 2.4.2. *Let* $H(s) = \frac{1}{s^{\sqrt{3}}+s^{\sqrt{2}}+1}$. *The coefficients of the generalised Laurent series are*

$$A_n = (-1)^n \tag{2.70}$$

and the powers are

$$\gamma_n = (n+1)\sqrt{3} - n\sqrt{2} \tag{2.71}$$

for $n = 0, 1, \ldots$

Remark 2.4.2. *The procedure can be applied to TF useful in some applications, such as*

$$H(s) = \frac{1}{\sqrt{1+s^2}} \tag{2.72}$$

$$H(s) = \frac{1}{s\sqrt{1+s}} \tag{2.73}$$

or the one in the following example, that include non-integer powers of s but do not follow the format in (2.45). The corresponding systems are called "implicit systems".

Example 2.4.3. *Let* $H(s) = \arctan(1/s)$. *This TF is an example of an implicit system. We will use the procedure described above, but, since we now deal with a more involved function, the different steps will be carefully described.*

1.

$$\lim_{\sigma\to\infty} \sigma \cdot \arctan(1/\sigma) = \lim_{\sigma\to\infty} \frac{\arctan(1/\sigma)}{1/\sigma} = \lim_{v\to 0} \frac{\arctan(v)}{v} = \lim_{v\to 0} \frac{\frac{1}{1+v^2}}{1} = 1$$

Thus $A_0 = 1$, $\gamma_0 = 1$, *and* $R_0(s) = \arctan(1/s) - 1/s$.

2.

$$\lim_{\sigma\to\infty} \sigma^2 [\sigma \cdot \arctan(1/\sigma) - 1/\sigma] = 0$$

We have to try the next power:

$$\lim_{\sigma\to\infty} \sigma^3 [\arctan(1/\sigma) - 1/\sigma] = \lim_{v\to 0} \frac{\arctan(v) - v}{v^3} = \lim_{v\to 0} \frac{D^3 [\arctan(v)]}{3!} =$$

$$= \lim_{v\to 0} \frac{D^2 \left[\frac{1}{1+v^2}\right]}{3!} = -\frac{2!}{3!} = -\frac{1}{3}$$

We obtain $A_1 = -1/3$, $\gamma_1 = 3$, *and* $R_1(s) = \arctan(1/s) - 1/s + 1/(3s^3)$.

3.

$$\lim_{\sigma \to \infty} \sigma^4 \left[\arctan(1/\sigma) - 1/\sigma + 1/(3\sigma^3) \right] = 0$$

Again we have to try the next power:

$$\lim_{\sigma \to \infty} \sigma^5 \left[\arctan(1/\sigma) - 1/\sigma + 1/(3\sigma^3) \right] = \lim_{v \to 0} \frac{\arctan(v) - v + (v^3)/3}{v^5} =$$

$$= \lim_{v \to 0} \frac{D^5 \left[\arctan(v) \right]}{5!} =$$

$$= \lim_{v \to 0} \frac{D^4 \left[\frac{1}{1+v^2} \right]}{5!} = -\frac{4!}{5!} = -\frac{1}{5}$$

leading to $A_2 = -1/5$, $\gamma_2 = 5$, *and* $R_2(s) = \arctan(1/s) - 1/s + 1/(3s^3) - 1/(5s^5)$.

4. *By inference:*

$$H(s) = \sum_{n=0}^{\infty} (-1)^n \frac{1}{2n+1} s^{-2n-1} \tag{2.74}$$

5. *The inverse Laplace transform gives:*

$$h(t) = \sum_{n=0}^{\infty} (-1)^n \frac{1}{2n+1} \frac{t^{2n}}{(2n)!} \cdot \varepsilon(t) = \frac{\sum_{n=0}^{\infty} (-1)^n \frac{t^{2n+1}}{(2n+1)!}}{t} \cdot \varepsilon(t) = \frac{\sin(t)}{t} \cdot \varepsilon(t) \tag{2.75}$$

Here is another example, not easy to interpret in terms of a differential equation:

Example 2.4.4. *Consider the fractional power of a zero or a pole:*

$$G(s) = \left(1 + \frac{s}{a} \right)^{\pm a} \tag{2.76}$$

This case can be treated in a different manner. We can write

$$G(s) = \left(\frac{s}{a} \right)^{\pm a} \left(1 + \frac{a}{s} \right)^{\pm a} = \left(\frac{s}{a} \right)^{\pm a} \sum_{k=0}^{\infty} \binom{\alpha}{k} \left(\frac{a}{s} \right)^k \tag{2.77}$$

and

$$G(s) = \sum_{k=0}^{\infty} \binom{\alpha}{k} \left(\frac{a}{s} \right)^{\mp a + k} \tag{2.78}$$

This case is studied in the frequency domain in chapter 4.

These last two examples show how to treat fractional implicit transfer functions, which do not correspond to a fractional differential equation defined by (1.51). In the last example, (2.77) shows that they can rather be said to be "infinite order" integer order differential equations.

2.5 Transfer function series representation for commensurate LS

As an anticipation of the development presented in the next chapter, we are going to consider the special case of transfer functions corresponding to commensurate fractional linear equations with constant coefficients. These are given by

$$G(s) = \frac{\sum\limits_{k=0}^{M} b_k s^{\alpha k}}{\sum\limits_{k=0}^{N} a_k s^{\alpha k}} \tag{2.79}$$

where α can be any positive real. They have this name because all orders α_k and β_k in (2.45) are now commensurate; α is the commensurable order.

To particularise the above algorithm for this situation, introduce a sequence got from the numerator coefficients:

$$b_{N-i}^0 = \begin{cases} b_{M-i} & i \leq M \\ 0 & i > M \end{cases} \quad i = 0, \dots, N \tag{2.80}$$

The denominator in (2.79) will be represented by $D(s)$. It is not difficult task to conclude that

$$A_0 = b_M \quad \text{and} \quad \gamma_0 = (N - M)\alpha \tag{2.81}$$

On the other hand, the repetition of the above described algorithm leads to

$$R_n(s) = \frac{\sum\limits_{k=0}^{N-1} b_k^n s^{\alpha k}}{s^{(N+n-M)\alpha} D(s)} \tag{2.82}$$

where (notice that the upper letters in b_k^n are superscripts, not powers)

$$b_{N-k}^n = b_{N-1-k}^{n-1} - A_{n-1} \cdot a_{N-1-k}, \quad k = 0, \dots, N-1 \tag{2.83}$$

and

$$A_n = b_N^n \quad \text{and} \quad \gamma_n = (N + n - M)\alpha \tag{2.84}$$

To simplify computations, we can arrange a tabular form. In the first line of the table, we put the symmetric of the denominator coefficients. In the second, we put the coefficients of numerator, inserting zeros, if necessary, to have the same number of coefficients in the first row.

$-a_N = -1$	$-a_{N-1}$	...	...	...	$-a_1$	$-a_0$	
b_N^0	b_{N-1}^0	...	...	...	b_1^0	b_0^0	$A_0 = b_N^0$

To compute line n, proceed as follows:
1. Multiply the first line of the table (i.e. the symmetric of the denominator coefficients) by $A_{n-1} = b_N^{n-1}$.
2. Add the result to line $n-1$ of the table.
3. Discard the first coefficient, shift the result to the left, and insert a zero to the right.

If more than one zero appear in the beginning of the line, we slide the values to the left as well. The corresponding increase in the power of s^α is equal to the number of slides.

Example 2.5.1. *For TF $H(s) = \frac{1}{s^{2\alpha}+3s^{\alpha}+2}$, we have the following table:*

−1	−3	−2			
0	0	1	$A_0 = 0$		we must slide the line 2 columns to the left
1	0	0	$A_0 = 1$	$\gamma_0 = 2\alpha$	and that is why γ_0 has this value
−3	−2	0	$A_1 = -3$	$\gamma_1 = 3\alpha$	
7	6	0	$A_2 = 7$	$\gamma_2 = 4\alpha$	
−15	−14	0	$A_3 = -15$	$\gamma_3 = 5\alpha$	
31	30	0	$A_4 = 31$	$\gamma_4 = 6\alpha$	

The line with coefficient A_1 is obtained using $1 \times [-1 \; -3 \; -2] + [1 \; 0 \; 0]$, and shifting the result one column to the left (thus $\gamma_1 = \gamma_0 + \alpha$).

The line with coefficient A_2 is obtained using $-3 \times [-1 \; -3 \; -2] + [-3 \; -2 \; 0]$, and shifting the result one column to the left (thus $\gamma_2 = \gamma_1 + \alpha$).

The line with coefficient A_3 is obtained using $7 \times [-1 \; -3 \; -2] + [7 \; 6 \; 0]$, and shifting the result one column to the left. It should be clear by now how to proceed further. Consequently,

$$H(s) = \frac{1}{s^{2\alpha} + 3s^{\alpha} + 2} = s^{2\alpha} - 3s^{3\alpha} + 7s^{4\alpha} - 15s^{5\alpha} + 31s^{6\alpha} + \ldots \qquad (2.85)$$

Example 2.5.2. *Consider TF $H(s) = \frac{1}{s^3 + s^{\frac{1}{3}} + 1}$. This example is more difficult than the previous one, as the fraction decomposition involves a larger number of complex terms. $H(s) = \frac{1}{s^3 + s^{1/3} + 1} = \frac{1}{s^{9\alpha} + s^{\alpha} + 1}\big|_{\alpha = \frac{1}{3}}$ as series of the form $H(s) = \sum_0^{n-1} A_k s^{-\gamma_k} + R_n(s)$. The next table presents the computation of the series coefficients of $H(s)$ for $n = 10$:*

−1	0	0	0	0	0	0	0	−1	−1			
1	0	0	0	0	0	0	0	0	0	$\gamma_0 = 9\alpha$	$A_0 = 1$	line shifted 9 columns to the left
−1	−1	0	0	0	0	0	0	0	0	$\gamma_1 = 1/\alpha$	$A_1 = -1$	line shifted 8 columns to the left
−1	0	0	0	0	0	0	1	1	0	$\gamma_2 = 18\alpha$	$A_2 = -1$	line shifted 1 column to the left
1	2	1	0	0	0	0	0	0	0	$\gamma_3 = 25\alpha$	$A_3 = 1$	line shifted 7 columns to the left
2	1	0	0	0	0	0	−1	−1	0	$\gamma_4 = 26\alpha$	$A_4 = 2$	line shifted 1 column to the left
1	0	0	0	0	0	−1	−3	−2	0	$\gamma_5 = 27\alpha$	$A_5 = 1$	line shifted 1 column to the left
−1	−3	−3	−1	0	0	0	0	0	0	$\gamma_6 = 33\alpha$	$A_6 = -1$	line shifted 6 columns to the left
−3	−3	−1	0	0	0	0	1	1	0	$\gamma_7 = 34\alpha$	$A_7 = -3$	line shifted 1 column to the left
−3	−1	0	0	0	0	1	4	3	0	$\gamma_8 = 35\alpha$	$A_8 = -3$	line shifted 1 column to the left
−1	0	0	0	0	1	4	6	3	0	$\gamma_9 = 36\alpha$	$A_9 = -1$	line shifted 1 column to the left

Consequently,

$$H(s) = \frac{1}{s^3 + s^{\frac{1}{3}} + 1} = s^3 - s^{\frac{17}{3}} - s^6 + s^{\frac{25}{3}} + 2s^{\frac{26}{3}} + s^9 - s^{11} - 3s^{\frac{34}{3}} - 3s^{\frac{35}{3}} - s^{12} + \dots$$

$$(2.86)$$

Exercises

1. Use the Cauchy theorem of complex variable function to prove that (2.12) remains valid when $Re(s) > 0$.
2. Prove the LT of the unit step (2.14).
3. Prove the LT of the pulse (2.16).
4. Prove property (2.17) of the LT from its definition (2.9).
5. Prove property (2.20) of the LT from its definition (2.9).
6. Prove property (2.21) of the LT from (2.20).
7. Prove the Fourier transforms of example 2.2.6 from definition (2.28).
8. Prove (2.43) from (2.41) using the $\alpha - 1$ for α substitution followed by a derivative (order 1) relatively to α.
9. Show from the error boundary $|R_n(s)| = o\left(\left|\frac{1}{s^\gamma}\right|\right)$ of the approximated Laurent series (2.59) that the error in time is in fact less than $\frac{t^{\alpha n}}{\Gamma(\alpha_n + 1)}$.
10. Plot the IR of (2.76) for some fixed value of a, and several values of $-1 < \alpha < 0$. Then compare this with the IR of (2.79) with no zeros ($M = 0$ and $b_0 = 1$) and the same pole, for the same values of α.
11. Apply the algorithm in section 2.5 to the following commensurate transfer functions, and obtain the corresponding series representations:
    a) $G_1(s) = \dfrac{1}{s^{0.5} + 2s^{0.25} + 3}$
    b) $G_2(s) = \dfrac{s^{0.2} - 1}{s^{0.4} - 2s^{0.2} + 17}$
    c) $G_3(s) = \dfrac{s^{0.7} + s^{0.35}}{s^{0.7} + s^{0.35} + 3}$

3 Fractional commensurate linear systems: time responses

3.1 Impulse and step responses: the Mittag-Leffler function

In the previous chapter, we showed how to compute the impulse and step responses of systems described by the general fractional differential equation (1.51). The results we got there are useful for input/output computations, but not suitable for a deep study of the system. However, this is a difficult task, unless the system assumes a simpler form. This is the case of commensurate transfer functions given by (2.79) that we reproduce here [53, 87]:

$$H(s) = \frac{B(s^\alpha)}{A(s^\alpha)} = \frac{\sum\limits_{k=0}^{M} b_k s^{\alpha k}}{\sum\limits_{k=0}^{N} a_k s^{\alpha k}} \tag{3.1}$$

The corresponding differential equation is

$$\sum_{k=0}^{N} a_k D^{\alpha k} y(t) = \sum_{k=0}^{M} b_k D^{\alpha k} x(t) \tag{3.2}$$

where all differentiation orders are multiples of some α, the commensurate order.

Example 3.1.1. *Equation $[aD^{2/3} + bD^{1/2}]y(t) = x(t)$ is commensurable: notice that it can be written as $[aD^{4\times\frac{1}{6}} + bD^{3\times\frac{1}{6}} + 0D^{2\times\frac{1}{6}} + 0D^{\frac{1}{6}}]y(t) = x(t)$. The commensurate order is $\frac{1}{6}$, and the corresponding TF in*

$$\frac{Y(s)}{X(s)} = \frac{1}{as^{4\times\frac{1}{6}} + bs^{3\times\frac{1}{6}}} \tag{3.3}$$

Remark 3.1.1. *If all differentiation orders are rational, a differential equation is always commensurate. Of course, in many cases one or more coefficients have to be zero, as seen in the example above.*

The roots of the numerator polynomial $B(s^\alpha)$ are called **pseudo-zeroes**, and those of the denominator polynomial $A(s^\alpha)$ **pseudo-poles**.

Example 3.1.2. *Transfer function (3.3) does not have pseudo-zeros. Let $u = s^\alpha$. The denominator polynomial is $au^4 + bu^3$. Thus, the pseudo-poles are $u = 0$ (double) and $u = -\frac{b}{a}$.*

The use of the term "pseudo" is related to the fact that a root of polynomial $B(u)$ or polynomial $A(u)$, where $u = s^\alpha$, may be, or may not be, a zero or pole of the system. To understand this, let a be a root of $B(u)$ of multiplicity 1. Then $B(s) = (s^\alpha - a) \cdot B_1(s)$.

DOI 10.1515/9783110624588-003

The zero exists if $s^\alpha - a = 0$. This only happens if $|\arg(a)| < \alpha\pi$. To understand the problem, remember that s^α is a multivalued expression. Fix the negative real axis as branchcut line and assume that the function is continuous from above. Also assume that we work on the first Riemann sheet. Thus, the corresponding function is defined by

$$s^\alpha = |s|^\alpha e^{i\alpha\theta} \quad -\pi < \theta \le \pi \tag{3.4}$$

This means that the solution of $s^\alpha - a = 0$ is obtained from

$$|s|^\alpha e^{i\alpha\theta} = |a|e^{i\phi} \tag{3.5}$$

We get

$$|s| = |a|^{1/\alpha} \text{ and } \phi = \alpha\theta \tag{3.6}$$

So, there is only a root if the pseudo-zero is in the angular region of the complex plane defined by $-\pi\alpha < \theta \le \pi\alpha$.

For the roots of $A(s^\alpha)$, the situation is similar. This allows us to decompose the impulse response in two components, one integer and another fractional, as we will do below in subsection 3.1.2.

For comparisons with the integer order case, we perform a substitution $s \to s^\alpha$. This implies that the interval $[0, \pi)$ is transformed into the interval $[0, \alpha\pi)$. Without losing generality, we will assume in the following that $0 < \alpha \le 1$. The inversion of TF (3.1) follows quite closely the classic procedure:

1. Transform $H(s)$ into $H(u)$, by making $u = s^\alpha$.
2. Expand $H(u)$ in partial fractions.
3. Substitute back u for s^α, to obtain the partial fraction decomposition (Appendix E)

$$H(s) = \sum_{k=0}^{N_p} \frac{R_k}{(s^\alpha - p_k)^{n_k}}, \tag{3.7}$$

where N_p is the number of parcels, p_k the pseudo-poles, and the n_k are the corresponding multiplicities.

Now we will have to invert a generic partial fraction with a pseudo-pole p:

$$F(s) = \frac{1}{s^\alpha - p}. \tag{3.8}$$

Observe that, having solved the problem of this partial fraction with a denominator of order one, other powers are very easy to tackle. In fact,

$$\frac{1}{(s^\alpha - p)^2} = D_p \left[\frac{1}{s^\alpha - p} \right], \tag{3.9}$$

where D_p means the derivative in order to p. The generalisation is simple but rarely found in practice.

There are two ways of expressing the inverse of (3.8).

3.1.1 In terms of the Mittag-Leffler function

The inversion is easily performed using the geometric series

$$F(s) = \frac{1}{s^\alpha - p} = s^{-\alpha} \sum_{n=0}^{\infty} p^n s^{-an} = \sum_{n=1}^{\infty} p^{n-1} s^{-an}, \tag{3.10}$$

valid for $|ps^{-\alpha}| < 1$. Choosing the region of convergence $Re(s) > |p|^{1/\alpha}$ we will arrive at the causal inverse of $F(s)$:

$$f(t) = \sum_{n=1}^{\infty} p^{n-1} \frac{t^{n\alpha-1}}{\Gamma(n\alpha)} \varepsilon(t). \tag{3.11}$$

This function is called alpha-exponential and is normally expressed in terms of the Mittag-Leffler function (MLF) as

$$f(t) = t^{\alpha-1} E_{\alpha,\alpha}(pt^\alpha)\varepsilon(t). \tag{3.12}$$

Function $f(t)$ is the impulse response corresponding to partial fraction (3.8). If $\alpha = 1$, we obtain the classic result: $f(t) = e^{pt}\varepsilon(t)$.

The step response $r_\varepsilon(t)$ of (3.10) can be obtained from (3.8) as

$$\mathcal{L}[r_\varepsilon(t)] = \frac{1}{s^\alpha - p} \cdot \frac{1}{s} = \sum_{1}^{\infty} p^{n-1} s^{-an-1}, \tag{3.13}$$

and

$$r_\varepsilon(t) = t^\alpha E_{\alpha,\alpha+1}(pt^\alpha) \cdot \varepsilon(t) = \frac{1}{p}[E_{\alpha,1}(pt^\alpha) - 1] \cdot \varepsilon(t). \tag{3.14}$$

With expressions (3.12) and (3.14) we are able to compute the impulse and step responses of any LTI systems defined by the transfer function (3.8).

Remark 3.1.2. *Defining the ramp function by $r(t) = t \cdot \varepsilon(t)$ and knowing that its LT is $R(s) = \frac{1}{s^2}$ for $Re(s) < 0$, it is a simple task to show that the corresponding response is*

$$r_r(t) = t^{\alpha+1} E_{\alpha,\alpha+2}(pt^\alpha) \cdot \varepsilon(t). \tag{3.15}$$

Remark 3.1.3. *As seen in chapter 2,*

$$\mathcal{L}\left[-\frac{t^{\alpha-1}}{\Gamma(\alpha)}\varepsilon(-t)\right] = \frac{1}{s^\alpha}, \quad Re(s) < 0 \tag{3.16}$$

which allows us to obtain the impulse and step responses of the anti-causal system described by (3.8).

3.1.2 By integer/fractional decomposition

The inversion of a transfer function based on the MLF has an important drawback: the solutions rely on one, or several, series, that create severe computational problems.

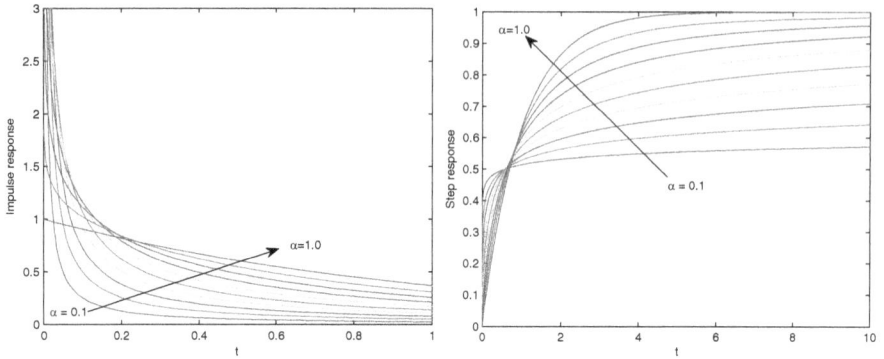

Fig. 3.1: Impulse (left) and step (right) responses of (3.8) for $\alpha = 0.1k, \ \ k = 1, 2, \cdots, 10$ and $p = -1$.

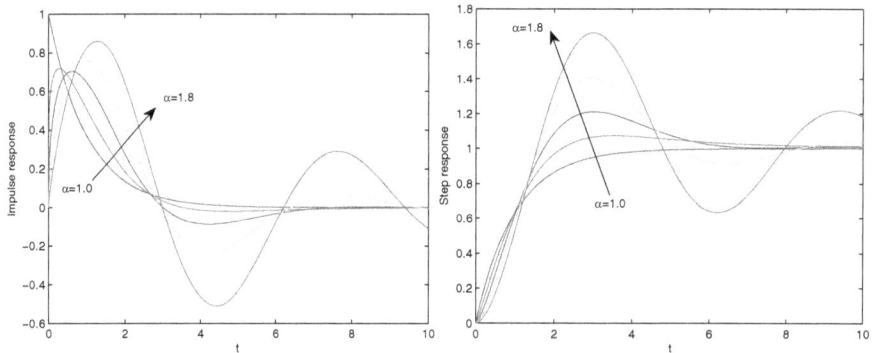

Fig. 3.2: Impulse (left) and step (right) responses of (3.8) for $\alpha = 1 + 0.2k, \ \ k = 0, 1, \cdots, 4$ and $p = -1$.

Furthermore, such solution masks the underlying structure of the system, in the sense that it does not highlight the presence of two different terms:

- One component of integer order that inherits the classical behaviour, mainly oscillations and (un)stability;
- Another component of fractional order responsible for the long range behaviour of the fractional linear systems, that is intrinsically stable as we will demonstrate in the sequel.

To obtain these two components, we perform the inversion by using the Bromwich integral for the inverse LT [78]. Although in this section we are considering the commensurate case, for this result we are going to assume the general case (2.45), and we assume that the poles are known. As the transform must be analytic on the right half complex plane, we choose the left half real axis and assume that $H(s)$ is continuous from above on the branchcut line and verifies $\lim_{s \to \infty} H(s) = 0, \ |\arg(s)| < \pi$. We will assume also that $\lim_{s \to 0} sH(s) = 0$.

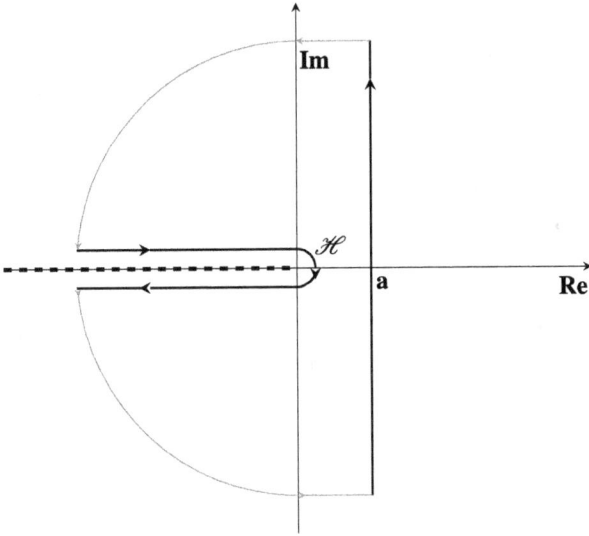

Fig. 3.3: Hankel $\mathcal{H}$ path, together with the closing arcs

Let (γ_k, p_k), $k = 1, 2, \ldots, K_0$, be the pairs (order, pole) such that

$$H(s) = \frac{B(s)}{\bar{A}(s) \prod_1^{K_0} (s^{\gamma_k} - p_k)}$$

where $\bar{A}(s)$ represents the part of the denominator corresponding to pseudo-poles that are not poles.

In these conditions we can use the integration path $\mathcal{H}$ [26, 87] and we apply the residue theorem. Let $u \in \mathbb{R}^+$ and consider $H(e^{i\pi}u)$ and $H(e^{-i\pi}u)$, the values of $H(s)$ immediately above and below the branch cut line. Decomposing the integral parcells according to each path section [26], we get

$$h(t) = \sum_{k=1}^{K_0} A_k e^{p^{1/\gamma_k} t} \varepsilon(t) + \frac{1}{2\pi i} \int_0^\infty \left[H(e^{-i\pi}u) - H(e^{i\pi}u) \right] e^{-\sigma t} du \cdot \varepsilon(t) \qquad (3.17)$$

where the constants A_k, $k = 1, 2, \ldots, K_0$, are the residues of $H(s)$ at p^{1/γ_k}.

Computing the LT of both sides in (3.17), we obtain

$$H(s) = H_i(s) + H_f(s) \qquad (3.18)$$

where the integer order part is

$$H_i(s) = \sum_{k=1}^{K_0} \frac{A_k}{s - p^{1/\gamma_k}}, \qquad Re(s) > \max(Re(p^{1/\gamma_k})) \qquad (3.19)$$

and the fractional part is

$$H_f(s) = \frac{1}{2\pi i} \int_0^\infty \left[H(e^{-i\pi} u) - H(e^{i\pi} u) \right] \frac{1}{s+u} du \qquad (3.20)$$

valid for $Re(s) > 0$. The integer order part of the impulse response (3.19) is the classical sum of exponentials (or sinusoids), eventually multiplied by integer powers. The possibly sinusoidal behaviour comes from this term. Its numerical computation does not put any severe problem.

The fractional part is expressed as an integral of the product of an exponential, with negative real exponent, by a bounded function, that is zero at the origin and at infinite. This means that the integral is easily computed by means of a simple numerical procedure. Using an uniform sampling interval σ_0, we can write

$$h(t) = \sum_{k=1}^{K_0} A_k e^{p_k^{1/\gamma_k} t} \epsilon(t) + \frac{1}{2\pi i} \sum_{n=0}^{L} \left[H(e^{-i\pi} \sigma_n) - H(e^{i\pi} \sigma_n) \right] e^{-\sigma_n t} \sigma_0 \qquad (3.21)$$

with $\sigma_n = n\sigma_0$, $n = 0, 1, \ldots, L$. The sampling interval σ_0 is chosen to guaranty that the fraction in (3.21) is small for $\sigma = L\sigma_0$.

Remark 3.1.4. *Consider the TF in (3.7) with $n_k = 1$, $k = 1, 2, \ldots$*

$$H(s) = \sum_{k=1}^N \frac{R_k}{s^\alpha - p_k} \qquad (3.22)$$

where R_k and p_k, $k = 1, 2, \ldots, N$ are the residues (at p_k, $k = 1, 2, \ldots$) and poles obtained by substituting u for s^α in (3.1).

Consider the set $Q = \{p_k : |\arg p_k| \leq \pi/\alpha, \ k = 1, 2, \ldots, N\}$. The integer part $h_i(t)$ of the impulse response is given by:

$$h_i(t) = \sum_{k=1; p_k \in Q}^N R_k \cdot B_k e^{p_k^{1/\alpha} t} \epsilon(t) \qquad (3.23)$$

where R_k, $k = 1, 2, \ldots, N$ represent the coefficients obtained from the partial fraction decomposition and $B_K = \frac{1}{\alpha p_k^{1-1/\alpha}}$ is the residue of $\frac{1}{s^\alpha - p_k}$ at $p_k^{1/\alpha}$. The corresponding LT becomes

$$H_i(s) = \sum_{k=1; p_k \in Q}^N \frac{R_k \cdot B_k}{s - p_k^{1/\alpha}} \qquad (3.24)$$

for $Re(s) > \max(Re(p_k^{1/\alpha}))$, $p_k \in Q$ defined above.

With generality, consider the situation where we have K pseudo-poles. The result expressed in (3.21) allows us to state that:

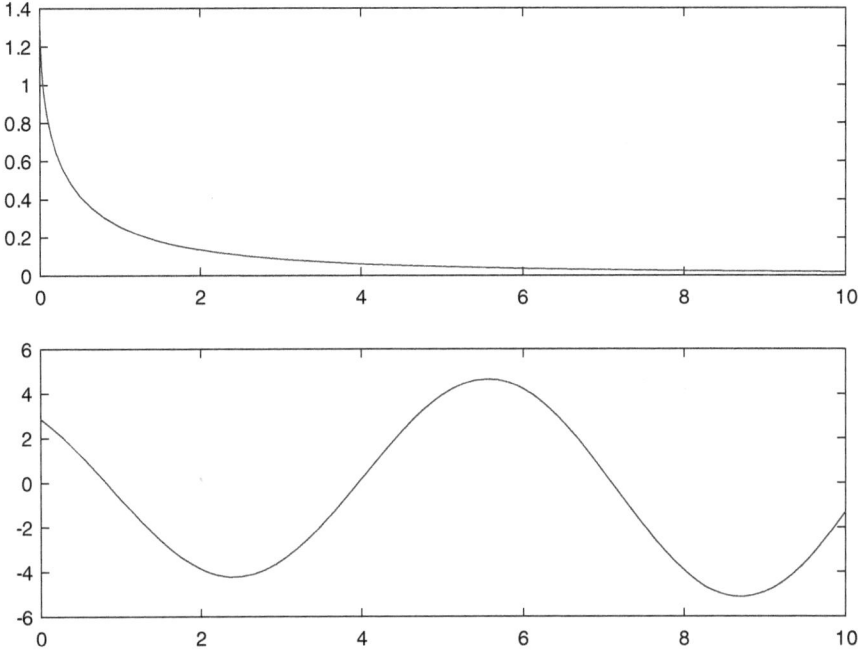

Fig. 3.4: Fractional and integer parts of the sum of impulse responses of (3.21) with $\alpha = 0.51$ and pseudo-poles $p = e^{\pm i\pi/4}$.

- For $\gamma_k = 1$, $k = 1, 2, \ldots, K$, we have no fractional component. The TF is a sum of partial fractions and each one has an exponential for solution. The corresponding TF is the quotient of two polynomials in s.
- For $\gamma_k < 1$, $k = 1, 2, \ldots, K$, we may have two components depending on the location of p_k in the complex plane:
  - If $|\arg(p_k)| > \pi\gamma_k$, $k = 1, 2, \ldots, K$, then we do not have the integer order component: it is a purely fractional system;
  - If $|\arg(p_k)| \leq \pi\gamma_k$, for some $k = 1, 2, \ldots, K$, then the system is of mixed character, in the sense that we have both components;
  - If $|\arg(p_k)| = \frac{\pi}{2}\gamma_k$, for some $k = 1, 2, \ldots, K$, the integer order component is sinusoidal, but the fractional component exists as well.

3.2 Stability

3.2.1 Knowing the pseudo-poles

Consider (3.17) and particularise for the simplest case, $G(s) = \frac{1}{s^\gamma - p}$, where we have only one pole $p^{1/\gamma}$:

$$g(t) = \frac{1}{\gamma} p^{1/\gamma - 1} e^{p^{1/\gamma} t} \varepsilon(t) + \frac{\sin(\gamma\pi)}{\pi} \int_0^\infty \frac{u^\gamma}{u^{2\gamma} - 2p \cos(\gamma\pi) + p^2} e^{-\sigma t} du \cdot \varepsilon(t) \qquad (3.25)$$

We can extract some conclusions:

1. The fractional part is always bounded, for $\gamma > 0$ and any $p \in \mathbb{C}$. In fact, it is a simple matter to verify that function $\dfrac{u^\gamma}{u^{2\gamma} - 2p\cos(\gamma\pi) + p^2}$ is bounded. Therefore,

$$\int_0^\infty \left| \frac{u^\gamma}{u^{2\gamma} - 2p\cos(\gamma\pi) + p^2} \right| e^{-\sigma t}\, du < \frac{A}{t}, \quad t > 0 \qquad (3.26)$$

2. As said, the integer part only exists if $-\pi < \arg(p) \leq \pi$. In this case, we have three situations:
   - $|\arg(p)| < \gamma\frac{\pi}{2}$ — the exponential increases without bound – unstable system;
   - $|\arg(p)| > \gamma\frac{\pi}{2}$ — the exponential decreases to zero – stable system;
   - $|\arg(p)| = \gamma\frac{\pi}{2}$ — the exponential oscillates sinusoidally – critically stable system.

Remark 3.2.1. *Frequently, we find affirmations like "the region $|\arg(p)| > \gamma\pi$ does not have any physical significance". This is a wrong statement. In fact, consider the low-pass RC circuit, with TF $H(s) = \frac{1}{1+RCs}$, as is well known. It has a pole with $\arg(p) = \pi$. Now consider a fractional capacitor with order $\gamma < 1$ (see e.g. the excellent paper [123]). The TF is*

$$H(s) = \frac{1}{1 + RC_\gamma s^\gamma} \qquad (3.27)$$

Therefore, the pseudo-pole is $p = -\frac{1}{RC_\gamma}$ and $\arg(p) = \pi > \pi\gamma$, showing that the affirmation is incorrect. Moreover, this shows that we can conceive a LS without poles, since pseudo-pole p is not a pole of (3.27).

The above considerations allow us to conclude that the behaviour of stable systems can be integer, fractional, or mixed:

- Classical integer order systems have impulse responses corresponding to linear combinations of exponentials that, in general, go to zero very fast. They are short memory systems.
- In fractional systems without poles, the exponential component disappears. These are long memory systems. All pseudo-poles have arguments with absolute values greater than π/γ, where $\gamma < 1$.
- Mixed systems have both components. Some pseudo-poles have arguments with absolute values larger than $\frac{\pi}{2}\gamma$.

Remark 3.2.2. *The three rules above are usually known as* Matignon's theorem, *that can be put in the following way: transfer function (3.7) is stable iff all pseudo-poles p_k verify $|\arg(p_k)| > \gamma\frac{\pi}{2}$, $\forall k$.*

3.2.2 Routh-Hurwitz criterion for an integer TF

The above procedure for studying stability demands knowing the pseudo-poles. In the integer order case, there are several criteria to evaluate the stability of a given linear system without the knowledge of the poles. One of the most important is the Routh-Hurwitz criterion, that gives information on the number of poles on the right hand half complex plane [11]. With suitable modification this can be used for fractional commensurate systems as well, as seen below in section 3.2.4.

We will recursively construct a set of alternately even and odd polynomials. With the coefficients of these polynomials, we will construct the *Routh Table*, which will allow us to study the distribution of the zeros of the original polynomial. Because polynomial coefficients are associated with a given order of recursion, we represent it by an upper script (which is not a power). So our original polynomial will be written

$$P_N(s) = \sum_{k=0}^{N} a_k^N s^k \qquad (3.28)$$

If N is even, we will denote by $F_N(s)$ and $F_{N-1}(s)$ its even and odd parts, respectively. If N is odd, we will denote by $F_N(s)$ and $F_{N-1}(s)$ its odd and even parts, respectively.

If N is even, we write:

$$F_N(s) = \sum_{k=0}^{N/2} a_{2k}^N s^{2k} \qquad (3.29)$$

and

$$F_{N-1}(s) = \sum_{k=0}^{N/2} a_{2k-1}^{N-1} s^{2k-1} \qquad (3.30)$$

where we set $a_{-1} = 0$. Conversely, if N is odd, we set:

$$F_N(s) = \sum_{k=0}^{(N-1)/2} a_{2k+1}^N s^{2k+1} \qquad (3.31)$$

and

$$F_{N-1}(s) = \sum_{k=0}^{(N-1)/2} a_{2k}^{N-1} s^{2k} \qquad (3.32)$$

Let us construct the sequence of polynomials $F_i(s)$, $i = N-2, ..., 0$, using the recursive relation

$$F_{i-2}(s) = F_i(s) - \eta_i s F_{i-1}(s) \qquad i = N, N-1, ..., 2 \qquad (3.33)$$

where η_i is obtained so that the higher power coefficient of $F_{i-2}(s)$ is null. This algorithm is presented in tabular form. Let us consider the case where N is even. The case where N is odd is similar. The algorithm begins with the coefficients of polynomials $F_N(s)$ and $F_{N-1}(s)$ written in the first two lines. The third line is constructed as follows:

1. Compute coefficient $\eta_N = \dfrac{a_N^N}{a_{N-1}^{N-1}}$;
2. Multiply the second line by η_N, and subtract the result from the first line;
3. Remove the first coefficient (which is of course zero, by construction) and put the obtained values in the third line.

The fourth line is constructed similarly, using lines 2 and 3; in general, the line k is built using lines $k-1$ and $k-2$.

Routh table for an integer TF

s^N	a_N^N	a_{N-2}^N	...	...	...	a_3^N	a_1^N	
s^{N-1}	a_{N-1}^{N-1}	a_{N-3}^{N-1}	...	...	...	a_2^{N-1}	a_0^{N-1}	
s^{N-2}	a_{N-2}^{N-2}	a_{N-4}^{N-2}	...	...	...	a_1^{N-2}		$\eta_N = \frac{a_N^N}{a_{N-1}^{N-1}}$
s^{N-3}	a_{N-3}^{N-3}	a_{N-5}^{N-3}	...	...	a_2^{N-3}	a_0^{N-3}		$\eta_{N-1} = \frac{a_{N-1}^{N-1}}{a_{N-2}^{N-2}}$
s^{N-4}	a_{N-4}^{N-4}	a_{N-6}^{N-4}	...	...	a_1^{N-4}			$\eta_{N-2} = \frac{a_{N-2}^{N-2}}{a_{N-3}^{N-3}}$
$\vdots$	$\vdots$	$\vdots$					$\vdots$	$\vdots$
s	a_1^1							$\eta_3 = \frac{a_4^3}{a_3^2}$
s^0	a_0^0							$\eta_2 = \frac{a_3^2}{a_1^1}$

This table is used together with the *Routh-Hurwitz criterion*: **an N-degree polynomial has all its roots in the left half plane (Hurwitz polynomial) iff the $N+1$ values of the first column of the Routh matrix have the same sign** [11].

If there is a sign variation, or a zero coefficient, we can immediately conclude that there are roots that do not lie in the left complex half plane. If all elements of the first column are nonzero, **the number of sign changes gives the number of roots in the right hand half plane.**

Example 3.2.1. $P(s) = 2s^4 + s^3 + 3s^2 + 5s + 10$

s^4	2	3	10	η
s^3	1	5	0	
s^2	-7	10		2
s	6.43	0		$-1/7$
s^0	10			

The sign sequence is $+ + - + +$, so there are two sign variations. Therefore, there are two roots in the right half plane. In fact, the polynomial roots are: $-1.0055 \pm i0.93311$ and $0.7555 \pm i1.4444$

Example 3.2.2. $P(s) = 6s^5 + 5s^4 + 4s^3 + 3s^2 + 2s + 1$

s^5	6	4	2	η
s^4	5	3	1	
s^3	0.4	0.8	0	6/5
s^2	-7	1		$-0.4/7$
s	6/7	0		$-1/7$
s^0	1			

The sign sequence is $+++-++$, so there are two sign variations. Therefore, there are two roots on the right half plane.

In constructing the table we can always multiply or divide all the elements in a line by a positive number. In the following table we repeat the previous example, applying this rule to make calculations more convenient.

s^5	3	2	1	η	*division by 2*
s^4	5	3	1		
s^3	1	2	0	3/5	*multiplication by 5*
s^2	−7	1		5	
s	15	0		−1/7	*multiplication by 7*
s^0	1				

The conclusions must be the same.

3.2.3 Special cases of the Routh-Hurwitz criterion for an integer TF

When filling in the Routh Table, it is impossible to apply the algorithm above in two circumstances that lead to $\eta = \infty$:

- line of zeroes;
- null leading elements.

A line of zeroes occurs when the even and odd parts of the polynomial are not prime, and the common factor is a polynomial with roots symmetrically disposed relatively to the origin (it is an even polynomial). It is not very difficult to verify that the presented algorithm destroys the remainder of both polynomials until two consecutive lines are proportional; hence the next line will be null (this will always happen in an odd line). The solution consists of replacing the line of zeros by the coefficients of the polynomial obtained by differentiating the previous polynomial. An example helps to make this clear.

Example 3.2.3. $P(s) = s^8 + 36s^7 + 13s^6 + 252s^5 + 73s^4 + 1116s^3 + 211s^2 + 900s + 150$

s^8	1	13	73	211	150	η	
s^7	1	7	31	25			*division by 36*
s^6	6	42	186	150		1	*proportional to the line above*
s^5	0	0	0	0			*line with zeroes*

The solution of the problem is the substitution of the line with zeroes by the one we obtain using the coefficients of the derivative of the polynomial in the immediately preceding line: $P_6(s) = 6s^6 + 42s^4 + 186s^2 + 150$. This line can be divided by a suitable constant, in this case 6, to make $P_6(s) = s^6 + 7s^4 + 31s^2 + 25$. Its derivative is $P_5(s) = 6s^5 + 28s^3 + 62s$, and this can again be divided by 2. The Routh table is as follows:

s^8	1	13	73	211	150		η	
s^7	1	7	31	25				*division by* 36
s^6	1	7	31	25			1	*division by* 6
s^5	3	14	31				*differentiation*	*division by* 2
s^4	7	62	75				1/3	*multiplication by* 3
s^3	−11	−1	0				3/7	*multiplication by* 7/8
s^2	675	825					−7/11	*multiplication by* 11
s	8400	0					−11/675	
s^0	825							

There are two sign variations; so it is concluded that there are two roots in the right half plane. We can also conclude that there are at least two pure imaginary roots. (Why?) In fact, the polynomial roots are: −35.8326, ±1 ± 2i, ±i and −0.1674.

If a leading zero appears, it is replaced by an infinitesimal quantity ϵ, which may be either positive or negative. Calculations proceed and the number of sign changes is found both for a positive and for a negative ϵ. Again, an example makes this clearer.

Example 3.2.4. $P(s) = s^7 + s^6 + s^5 + s^4 + s^3 + s^2 + s + 1$

s^7	1	1	1	1	α	
s^6	1	1	1	1	0	
s^5	0	0	0		1	*line of zeroes*
s^5	3	2	1	0		*division by* 2
s^4	1	2	3		1/3	
s^3	−1	−2	0		3	*division by* 4
s^2	0	3			−1	*leading zero*
s^2	ϵ	3				*replace zero with* ϵ
s	$\frac{3-2\epsilon}{\epsilon}$	0			$-\frac{1}{\epsilon}$	
s^0	3				$-\frac{\epsilon}{3-2\epsilon}$	

The elements in the first column are 1, 1, 3, 1, −1, ϵ, $\frac{3-2\epsilon}{\epsilon}$ *and* 3. *Suppose that the infinitesimally small ϵ is positive. Then $\frac{3-2\epsilon}{\epsilon} \approx \frac{3}{\epsilon}$ is also positive, and there will be two sign changes. Suppose that ϵ is negative instead. Then $\frac{3-2\epsilon}{\epsilon} \approx \frac{3}{\epsilon}$ is also positive, and there will be again two sign changes. Consequently, there are always two sign changes, which means that there are two poles in the right half plane (plus two pure imaginary poles). In fact, the polynomial has the following roots: $\pm\frac{\sqrt{2}}{2} \pm \frac{\sqrt{2}}{2}i$, ±i and −1.*

There is an alternative procedure that we will present in the next sub-section.

3.2.4 Routh-Hurwitz criterion for a fractional commensurate TF

The generalisation of the Routh table for the fractional case was done by [44] and is very similar to the integer order case. In fact, the starting "even and odd" polynomials are substituted by the "real and imaginary" pseudo-polynomials. Let our original pseudo-polynomial be written as

$$P_N(s) = \sum_{k=0}^{N} a_k^N s^{\alpha k} \tag{3.34}$$

and let $s = \omega$, $\omega > 0$. We obtain

$$P_N(\omega) = \sum_{k=0}^{N} a_k^N \cos\left(\frac{k\pi\alpha}{2}\right)\omega^{\alpha k} + i \sum_{k=0}^{N} a_k^N \sin\left(\frac{k\pi\alpha}{2}\right)\omega^{\alpha k}. \tag{3.35}$$

We will recursively construct a set of alternately "real" and "imaginary" pseudo-polynomials. With the coefficients of these pseudo-polynomials, we will construct a *Routh Table*, which will allow us to study the distribution of the zeros of the original pseudo-polynomial. If $N\alpha$ is not an odd positive integer, we define the pseudo-polynomials

$$F_N(s) = \sum_{k=0}^{N} a_k^N \cos\left(\frac{k\pi\alpha}{2}\right)\omega^{\alpha k} \tag{3.36}$$

and

$$F_{N-1}(s) = \sum_{k=0}^{N} a_k^N \sin\left(\frac{k\pi\alpha}{2}\right)\omega^{\alpha k} \tag{3.37}$$

If $N\alpha$ is an odd positive integer, we set:

$$F_N(s) = \sum_{k=0}^{N} a_k^N \sin\left(\frac{k\pi\alpha}{2}\right)\omega^{\alpha k} \tag{3.38}$$

and

$$F_{N-1}(s) = \sum_{k=0}^{N} a_k^N \cos\left(\frac{k\pi\alpha}{2}\right)\omega^{\alpha k} \tag{3.39}$$

As above, we construct the sequence of polynomials $F_i(s)$, $i = N - 2, \ldots, 0$ using the recursive relation

$$F_{i-2}(s) = -F_i(s) + \eta_i s F_{i-1}(s) \qquad i = N, N-1, \ldots, 2 \tag{3.40}$$

where η_i is obtained so that the higher power coefficient of $F_{i-2}(s)$ is null. Again, the algorithm begins with the coefficients of polynomials $F_N(s)$ and $F_{N-1}(s)$ written in the first two lines. The third line is constructed as follows:

1. Compute coefficient $\eta_N = \frac{a_N^N}{a_{N-1}^{N-1}}$;
2. Multiply the second line by η_N, and subtract it the first line – note that there is a sign change when compared with the classic integer order procedure above described;
3. Remove the first coefficient (which is of course zero, by construction) and put the obtained values in the third line.

The fourth line is constructed similarly, using lines 2 and 3; in general, the line k is built using lines $k-1$ and $k-2$.

Routh table for a fractional, commensurate TF

a_N^N	a_{N-1}^N	...	...	...	a_1^N	a_0^N
a_N^{N-1}	a_{N-1}^{N-1}	...	...	...	a_1^{N-1}	
a_N^{N-2}	a_{N-1}^{N-2}	...	...	...	a_1^{N-2}	$\eta_N = \frac{a_N^N}{a_N^{N-1}}$
a_N^{N-3}	a_{N-1}^{N-3}	...	...	...		$\eta_{N-1} = \frac{a_N^{N-1}}{a_N^{N-2}}$
$\vdots$	$\vdots$				$\vdots$	$\vdots$
a_1^1						$\eta_3 = \frac{a_N^3}{a_N^2}$
a_0^0						$\alpha_2 = \frac{a_2^2}{a_1^1}$

Also the stability criterion is different: let V_l be the number of sign changes in the leftmost column, and V_r the number of sign changes in the rightmost non-null elements of each row. Then, the number of unstable roots (pseudo-poles of a transfer function) is $\left[\frac{Na}{2}\right] - V_l + V_r$, where $[x]$ is a rounding operator defined by $[x] = \lfloor x + 0.5 \rfloor$.

Example 3.2.5. $P(s) = s^{\frac{4}{3}} - 4s^{\frac{2}{3}} + 8$

$$a_0^2 = 8\cos 0, \ a_1^2 = -4\cos\frac{\pi\frac{2}{3}}{2}, \ a_2^2 = 1\cos\frac{2\pi\frac{2}{3}}{2},$$

and

$$a_0^1 = 8\sin 0, \ a_1^1 = -4\sin\frac{\pi\frac{2}{3}}{2}, \ a_2^1 = 1\sin\frac{2\pi\frac{2}{3}}{2}.$$

-0.5	-2	8	η
0.866	-3.464		
4	-8	-0.577	
1.732		0.217	
8		2.309	

The leftmost column is $-0.5, 0.866, 4, 1.732, 8,$ *and thus* $V_l = 1$. *The rightmost non-null elements of each row are* $8, -3.464, -8, 1.732, 8,$ *and thus* $V_r = 2$. *So there are* $\left[\frac{2\frac{2}{3}}{2}\right] - 1 + 2 = 2$ *unstable pseudo-poles. In fact, they are* $2 \pm 2i$, *and since* $\alpha = \frac{2}{3}$ *they are unstable.*

Example 3.2.6. $P(s) = s^{\frac{2}{3}} - 4s^{\frac{1}{3}} + 8$

$$a_0^2 = 8\cos 0, \ a_1^2 = -4\cos\frac{\pi\frac{1}{3}}{2}, \ a_2^2 = 1\cos\frac{2\pi\frac{1}{3}}{2},$$

and

$$a_0^1 - = 8\sin 0, \ b_1^1 = -4\sin\frac{\pi\frac{1}{3}}{2}, \ a_2^1 = 1\sin\frac{2\pi\frac{1}{3}}{2}.$$

0.5	-3.464	8	η
0.866	-2		
2.309	-8	0.577	
-1		0.375	
8		-2.309	

The leftmost column is $0.5, 0.866, 2.309, -1, 8,$ *and thus* $V_l = 2$. *The rightmost non-null elements of each row are* $8, -2, -8, -1, 8,$ *and thus* $V_r = 2$. *So there are* $\left[\frac{2\frac{1}{3}}{2}\right] - 2 + 2 = 0$ *unstable pseudo-poles. In fact, since now* $\alpha = \frac{1}{3}$, *the pseudo-poles are unstable.*

3.2.5 Special cases of the Routh-Hurwitz criterion for a fractional commensurate TF

Special cases in the table are handled as follows:
- Zero in the first column: slide to left until a nonzero appears. The continuation of the table is the same as well as the rules for root distribution.
- Line of zeroes:
  1. Compute the derivative (order 1) of the previous pseudo-polynomial;
  2. Divide the resulting coefficients by α;
  3. Substitute the line of zeroes by these coefficients.

  The rule for the number of unstable roots is different. V_l will now be the number of sign changes in the leftmost column, from the top until the last line before a special case; V_r will be the number of sign changes in the rightmost non-null elements of each row, also from the top until the last line before a special case. Let V_{l2} be the number of sign changes in the leftmost column, from the last line before a special case until the bottom, and let V_{r2} be the number of sign changes in the rightmost non-null elements of each row, also from the last line before a special case until the bottom. Then, the number of unstable roots (pseudo-poles of a transfer function) is $\left[\frac{N\alpha}{2}\right] - V_{r2} + V_{l2} - V_l + V_r$, and the number of critically stable roots is $2(V_{r2} - V_{l2})$.

Example 3.2.7. $P(s) = s - 4s^{\frac{1}{2}} + 8$

This time, $a_0^2 = 8\cos 0$, $a_1^2 = -4\cos\frac{\pi\frac{1}{2}}{2}$, $a_2^2 = 1\cos\frac{2\pi\frac{1}{2}}{2} = 0$, and so the line must be shifted to the left. Also, $a_0^1- = 8\sin 0$, $b_1^1 = -4\sin\frac{\pi\frac{1}{2}}{2}$, $a_2^1 = 1\sin\frac{2\pi\frac{1}{2}}{2}$.

-2.828	8	η
1	-2.828	
2		-2.828
2.828		0.5

$V_l = 1$ *is found from* $-2.828, 1$. $V_r = 1$ *is found from* $8, -2.828$. $V_{l2} = 0$ *is found from* $1, 2, 2.828$. $V_{r2} = 1$ *is found from* $-2.828, 2, 2.828$. *So there are* $\left[\frac{2\frac{1}{2}}{2}\right] - 1 + 0 - 1 + 1 = 0$ *unstable pseudo-poles, and* $2(1 - 0) = 2$ *critical ones.*

3.3 Causal periodic impulse response

In causal linear system applications, oscillatory behaviour is very important, because we are often interested in removing or, at least, attenuating it. The transient behaviour of such systems is embedded in the impulse or step responses. In the present case we are interested in knowing under which conditions the impulse response is oscillatory, and, in particular, periodic [115, 117]. To study such problem we consider a TF given by (3.7). It is straightforward to conclude that an oscillatory response will appear if [78, 91]:

- $\arg(p_k) = a\pi/2$ — condition for a pure oscillation;
- $n_k = 1$ — required for stability reasons.

But these conditions are not sufficient to guaranty a causal periodic output. The impulse response of a single pseudo-pole, given by (3.25), shows that:

1. The integer part (the first term) is sinusoidal for any value of γ such that $\arg(p) = \gamma\frac{\pi}{2}$;
2. The fractional part (the second term) is not periodic for any value of γ such that $\arg(p) = \gamma\frac{\pi}{2}$;
3. The fractional part is null iff $\gamma \in \mathbb{Z}$. Therefore, only with derivatives of order 1 we have pure sinusoidal responses.

Example 3.3.1. *Figure 10.1 shows the integer and fractional parts of the impulse response for a systems with $\alpha = 0.51$ and 2 pseudo-poles $p = e^{\pm i\pi/4}$.*

3.4 Initial conditions

Consider the above referred fractional RC circuit described by the differential equation

$$RC_a v_o^{(\alpha)}(t) + v_o(t) = v_i(t) \tag{3.41}$$

where $v_i(t)$ and $v_o(t)$ are the input and output voltages. If the input signal is identically null, the ouput is also null, unless there is any energy stored in the system (capacitor). This stored energy defines what is normally called initial-conditions (IC).

Traditionally, the IC problem is solved with the unilateral LT, that may not give the correct solution. One of the reasons is that it uses the instant $(t = 0^+)$ instead of $t = 0^-$. In fractional systems study, the Caputo derivative is frequently used, because it uses some initial values that depend on integer order derivatives. This is frequently a poor solution. The IC depend on the structure of the system, not on the tool we use to analyse it [67, 75].

A suitable solution will be described next and is based on the following assumptions:

1. Equation (3.2) is defined for any $t \in \mathbb{R}$;
2. Our observation window is the unit step $\varepsilon(t)$;
3. The IC depend on the structure of the system and are independent of the tools that we adopt for the analysis;

4. The IC are the values assumed by the variables at the instant where the observation window opens.

Therefore, let us return to the above example. The IC is physically represented by the charge in the capacitor and translated outside by the voltage at the terminals of the device. To search for the solution, multiply both sides of the above equation by the unit step:

$$RC_\alpha v_0^{(\alpha)}(t)\varepsilon(t) + v_0(t)\varepsilon(t) = 0 \tag{3.42}$$

Then we should relate $v_0^{(\alpha)}(t)\varepsilon(t)$ with $[v_0(t)\varepsilon(t)]^{(\alpha)}$. In the integer order case, this is a simple task with the *jump formula*. In the fractional case, the situation may be more involved.

First note that, in general, function $v_0(t)\varepsilon(t)$ will have a discontinuity, or jump, at the origin. Therefore, before computing its derivative, it is convenient to remove the discontinuity. Let $y(t) = v_0(t)\varepsilon(t) - v_0(0)\varepsilon(t)$. This is a continuous function, so we can compute its derivative. We want

$$RC_\alpha y^{(\alpha)}(t) = -y(t) \ t \in \mathbb{R} \tag{3.43}$$

or

$$RC_\alpha [v_0(t)\varepsilon(t) - v_0(0)\varepsilon(t)]^{(\alpha)} = -v_0(t)\varepsilon(t) \tag{3.44}$$

giving

$$RC_\alpha [v_0(t)\varepsilon(t)]^{(\alpha)} - v_0(0)\delta^{(\alpha-1)}(t) = -\frac{1}{RC_\alpha}v_0(t)\varepsilon(t) \tag{3.45}$$

Comparing this equation with (3.42) we see that we substituted $[v_0(t)\varepsilon(t)]^{(\alpha)} - v_0(0)\delta^{(\alpha-1)}(t)$ for $v_0^{(\alpha)}(t)\varepsilon(t)$. Setting $V_0(s) = \mathcal{L}[v_0(t)\varepsilon(t)]$, we obtain

$$s^\alpha V_0(s) - s^{\alpha-1}v_0(0) = -\frac{V_0(s)}{RC_\alpha} \tag{3.46}$$

making clear the appearance of the IC. If $\alpha = 1$ we obtain the classic formula. This procedure was generalised in [75] leading to the *fractional jump formula*:

$$y^{\alpha_N}(t)\varepsilon(t) = [y(t)\varepsilon(t)]^{\alpha_N} - \sum_{0}^{N-1} y^{(\alpha_m)}(0)\delta^{(\alpha_N-\alpha_m-1)}(t) \tag{3.47}$$

In the commensurate case it assumes the form

$$y^{\alpha N}(t)\varepsilon(t) = [y(t)\varepsilon(t)]^{\alpha N} - \sum_{0}^{N-1} y^{(\alpha m)}(0)\delta^{(\alpha N-\alpha m-1)}(t) \tag{3.48}$$

In the integer order case, it simplifies to

$$y^{N}(t)\varepsilon(t) = [y(t)\varepsilon(t)]^{N} - \sum_{0}^{N-1} y^{(m)}(0)\delta^{(N-m-1)}(t) \tag{3.49}$$

For $N = 1$, we get $y'(t)\varepsilon(t) = [y(t)\varepsilon(t)]' - y(0)\delta(t)$, as expected. Also $y''(t)\varepsilon(t) = [y(t)\varepsilon(t)]'' - y(0)\delta'(t) - y'(0)\delta(t)$.

These formulae are substituted in both input and output signals of differential equation (3.2) to obtain the complete solution, defined for $t \geq 0$:

$$\sum_{k=0}^{N} a_k y(t)^{(\alpha_k)} = \sum_{k=0}^{M} b_k x(t)^{(\beta_k)} +$$

$$+ \sum_{k=1}^{N} a_k \sum_{m=0}^{k-1} y^{(\alpha_m)}(0)\delta^{(\alpha_k - \alpha_m)}(t) - \sum_{k=1}^{M} b_k \sum_{m=0}^{k-1} x^{(\beta_m)}(0)\delta^{(\beta_k - \beta_m)}(t). \quad (3.50)$$

From this general formulation we obtain a version for the commensurate case:

$$\sum_{k=0}^{N} a_k y(t)^{(\alpha k)} = \sum_{k=0}^{M} b_k x(t)^{(\alpha k)} +$$

$$+ \sum_{k=1}^{N} a_k \sum_{m=0}^{k-1} y^{(\alpha m)}(0)\delta^{(\alpha k - \alpha m)}(t) - \sum_{k=1}^{M} b_k \sum_{m=0}^{k-1} x^{(\alpha m)}(0)\delta^{(\alpha k - \alpha m)}(t). \quad (3.51)$$

The above formulation is fully compatible with classic results. Obviously, we can apply the LT. For this last situation, we obtain

$$Y(s) = \frac{B(s)}{A(s)} X(s) + \frac{C(s)}{A(s)} \quad (3.52)$$

where

$$A(s) = \sum_{k=0}^{N} a_k s^{\alpha k} \quad (3.53)$$

$$B(s) = \sum_{k=0}^{M} b_k s^{\alpha k} \quad (3.54)$$

$$C(s) = \sum_{k=1}^{N} a_k \sum_{m=0}^{k-1} y^{(\alpha m)}(0) s^{\alpha(k-m)} - \sum_{k=1}^{M} b_k \sum_{m=0}^{k-1} x^{(\alpha m)}(0) s^{\alpha(k-m)} \quad (3.55)$$

If $\alpha = 1$, we recover the classic formula:

$$C(s) = \sum_{k=1}^{N} a_k \sum_{m=0}^{k-1} y^{(m)}(0) s^{(k-m)} - \sum_{k=1}^{M} b_k \sum_{m=0}^{k-1} x^{(m)}(0) s^{(k-m)} \quad (3.56)$$

3.4.1 Special singular system cases

Let us face now the problem we have when the *characteristic pseudo-polynomial* in the denominator has root of order m for $s = p$ and the input is the exponential $e^{pt}, t \in \mathbb{R}$. To solve the problem we start by factorising $A(s)$ by putting in evidence the presence of the pole:

$$A(s) = (s^\alpha - p^\alpha)^m \bar{A}(s) \quad (3.57)$$

This implies that we can decouple the original equation into two sub-equations:

$$\sum_{k=0}^{N-m} \bar{a}_k D^{\alpha k} u(t) = \sum_{k=0}^{M} b_k D^{\beta k} x(t) \tag{3.58}$$

$$\sum_{k=0}^{m} (-1)^{1-k} \binom{m}{k} p^{(m-k)\alpha} D^{k\alpha} y(t) = u(t) \tag{3.59}$$

Before going into the solution, we introduce a useful result [88]:

Lemma 1. *Let* $x(t) = t^K e^{pt}$, $t \in \mathbb{R}$, $K \in \mathbb{N}_0$ *be the input of (3.57). The output* $y(t)$ *is given by*

$$y(t) = \sum_{j=0}^{K} \binom{K}{j} \bar{H}^{(j)}(p) t^{K-j} e^{pt} \tag{3.60}$$

provided that p *is not a pole of* $\bar{H}(s)$ *that is the TF of the system defined by (3.58)*

$$\bar{H}(s) = \frac{B(s)}{\bar{A}(s)} \tag{3.61}$$

Therefore, the solution $u(t)$ of (3.58) is given by (3.60)

$$u(t) = \sum_{j=0}^{K} \binom{K}{j} \bar{H}^{(j)}(p) t^{K-j} e^{pt} = u_0(t) e^{pt} \tag{3.62}$$

According to (3.60), $u(t) = u_0(t)e^{pt}$, and $u_0(t)$ is a polynomial of degree K. Similarly, $y(t) = y_0(t)e^{pt}$, but now the degree of $y_0(t)$ is $J = K + m$. Using the generalised Leibniz rule [87], for the product the γ order fractional derivative of $y(t)$ is given by

$$D^{\gamma}[t^J e^{pt}] = \sum_{i=0}^{J} \binom{\gamma}{i} D^i y_0(t) p^{\gamma-i} e^{pt} \tag{3.63}$$

Insert this result into (3.59) to obtain

$$\sum_{k=0}^{m} (-1)^{1-k} \binom{m}{k} p^{(m-k)\alpha} \sum_{i=0}^{J} \binom{k\alpha}{i} D^i y_0(t) p^{k\alpha-i} = u_0(t) \tag{3.64}$$

After some manipulation, we get an ordinary integer order differential equation:

$$\sum_{i=0}^{J} p^{\alpha-i} \left[\sum_{k=0}^{m} (-1)^{1-k} \binom{m}{k} \binom{k\alpha}{i} \right] y_0^{(i)}(t) = u_0(t) \tag{3.65}$$

Conversely,

$$\sum_{k=0}^{m} (-1)^{m-k} \binom{m}{k} \binom{k\alpha}{i} = \sum_{k=0}^{m} (-1)^{m-k} \binom{m}{k} \frac{(-k\alpha)_i}{i!}.$$

The Pochhammer symbol $(-k\alpha)_i$ is a polynomial in k with degree equal to i. Therefore, attending to

$$\sum_{k=0}^{1}(-1)^{1-k}\binom{1}{k}k^i = 0$$

for $i < m$ [33] , we can rewrite the above equation as

$$\sum_{i=0}^{K}p^{\alpha-i-1}\left[\sum_{k=0}^{m}(-1)^{1-k}\binom{m}{k}\binom{k\alpha}{i+m}\right]y_0^{(i+m)}(t) = u_0(t) \tag{3.66}$$

Introduce a new function $v(t) = y_0^{(m)}(t)$, solution of the equation

$$\sum_{i=0}^{K}p^{\alpha-i-m}\left[\sum_{k=0}^{m}(-1)^{m-k}\binom{m}{k}\binom{k\alpha}{i+m}\right]v^{(i)}(t) = u_0(t) \tag{3.67}$$

This new differential equation has the transfer function

$$G(s) = \frac{1}{\sum_{i=0}^{K}A_i s^i} \tag{3.68}$$

with

$$A_i = p^{\alpha-i-m}\left[\sum_{k=0}^{m}(-1)^{m-k}\binom{m}{k}\binom{k\alpha}{i+m}\right], \quad i = 0, 1, \ldots, K. \tag{3.69}$$

To obtain the solution of equation (3.59), we use the result in (3.60), showing that $u_0(t)$ is a polynomial with order equal to K: $u_0(t) = \sum_{i=0}^{K}U_i t^i$. Then

$$y_0(t) = \sum_{i=0}^{K}U_i\sum_{j=0}^{i}G^{(j)}(0)\frac{(i-j)!}{(i-j+m)!}t^{i-j+m} \tag{3.70}$$

Theorem 3.4.1. *Singular case*
Let $x(t) = e^{pt}$ be the input to a given singular system having a pole with multiplicity m at $s = p$. The output is [88]

$$y(t) = U_0 G(0)\frac{t^m}{m!}e^{pt} \tag{3.71}$$

Here the U_0 comes from in (3.60), and $G(s)$ is given by (3.68) and (3.69).

Example 3.4.1. *Take the equation*

$$y^{(3\alpha)} + y^{(2\alpha)} - 4y^{(\alpha)} + 2y = e^t \tag{3.72}$$

with $x(t) = e^t$. The point $s = 1$ is a pole of the transfer function of order $m = 1$. On the other hand, $\bar{H}(s) = \frac{1}{s^{2\alpha}+2s^{\alpha}-2}$. So

$$u(t) = \bar{H}(1)e^t = e^t \tag{3.73}$$

So we must look for the solution of $y^{(\alpha)}(t) - y(t) = u(t)$. Assume that $y(t) = [At]e^t$. Using the Leibniz rule $y^{(\alpha)}(t) = [At + \alpha A]e^t$ that inserted in the equation gives $A = 1/\alpha$.

Exercises

1. Find, if possible, analytical expressions for the impulse, step and ramp responses of the following transfer functions, and compute and plot all responses for a suitable range of values of t:

   a) $G_1(s) = \dfrac{1}{s^\alpha + 1}$, with $\alpha = \frac{1}{4}, \frac{1}{2}, \frac{3}{4}, 1, \frac{5}{4}$

   b) $G_2(s) = \dfrac{1}{s^{\frac{1}{3}} + a}$, with $a = 10, 1, 0.1, -0.1$

   c) $G_3(s) = \dfrac{1}{(s^{\frac{1}{5}} + p_1)(s^{\frac{1}{5}} + p_2)}$, with $p_1, p_2 = 1 \pm 2i, \pm 2i, -1 \pm 2i, -6 \pm 2i, -10 \pm 2i$

   d) $G_4(s) = \dfrac{s^\alpha + 1}{s^{0.5} + 1}$, with $\alpha = 0.1, 0.25, 1$

2. Apply the Routh-Hurwitz criterion to study the stability of TF with the following denominator polynomials:
   (a) $P(s) = 4s^4 + 3s^3 + s^2 + 5s + 1$
   (b) $P(s) = 5s^5 + 2s^4 + 2s^3 + s^2 + 3s + 1$
   (c) $P(s) = s^5 + 4s^4 + 8s^3 + 8s^2 + 7s + 4$
   (d) $P(s) = 2s^4 + s^3 + 3s^2 + 5s + 10$
   (e) $P(s) = s^8 + s^7 + s^6 + s^5 + s^4 + s^3 + s^2 + s + 1$

3. Apply the Routh-Hurwitz criterion to study the stability of TF with the following denominator polynomials:
   (a) $P(s) = s - 2s^{\frac{1}{2}} + 4$
   (b) $P(s) = s^{\frac{4}{3}} - 2s^{\frac{1}{3}} + 4$
   (c) $P(s) = s^{\frac{8}{5}} - 2s^{\frac{4}{5}} + 4$
   (d) $P(s) = s^{\frac{12}{5}} - 2s^{\frac{6}{5}} + 4$
   (e) $P(s) = s^{\frac{8}{3}} + 4s^{\frac{4}{3}} + 8$

4. Prove the impulse response in the last line of Table 3.1, showing that the series representation of the transfer function has (F.2) as inverse.

Tab. 3.1: Time responses of important transfer functions, valid for $t > 0$.

Function	Unit impulse response	Unit step response	Unit slope ramp response	See
s^α	$\dfrac{t^{-\alpha-1}}{\Gamma(-\alpha)}$	$\dfrac{t^{-\alpha}}{\Gamma(1-\alpha)}$	$\dfrac{t^{-\alpha+1}}{\Gamma(2-\alpha)}$	Equation (2.39)
$\dfrac{1}{s^\alpha \pm p}$	$t^{\alpha-1}E_{\alpha,\alpha}(\mp pt^\alpha)$	$t^\alpha E_{\alpha,\alpha+1}(\mp pt^\alpha)$	$t^{\alpha+1}E_{\alpha,\alpha+2}(\mp pt^\alpha)$	Equations (3.12), (3.13)–(3.14)
$\dfrac{s^{\alpha-\beta}}{s^\alpha \pm p}$	$t^{\beta-1}E_{\alpha,\beta}(\mp pt^\alpha)$	$t^\beta E_{\alpha,\beta+1}(\mp pt^\alpha)$	$t^{\beta+1}E_{\alpha,\beta+2}(\mp pt^\alpha)$	Exercise 3.1

4 The fractional commensurate linear systems. Frequency responses

4.1 Steady-state behaviour: the frequency response

In this chapter, we continue the study of commensurate linear systems described by transfer function

$$H(s) = \frac{B(s)}{A(s)} = \frac{\sum\limits_{k=0}^{M} b_k s^{k\alpha}}{\sum\limits_{k=0}^{N} a_k s^{k\alpha}}. \tag{4.1}$$

We can rewrite it as

$$H(s) = \frac{B(s)}{A(s)} = b_M \frac{\prod\limits_{k=1}^{M_z}(s^\alpha - z_k)^{m_k}}{\prod\limits_{n=1}^{N_p}(s^\alpha - p_n)^{m_n}}, \tag{4.2}$$

where M_z and N_z are the number of distinct numerator and denominator roots, respectively, and the integers m_k and m_n denote the corresponding multiplicities.

For our objectives, it is preferable to give another form to (4.2):

$$H(s) = K_0 \frac{\prod\limits_{k=1}^{M_z}\left[\left(\frac{s}{v_k}\right)^\alpha + 1\right]^{m_k}}{\prod\limits_{n=1}^{N_p}\left[\left(\frac{s}{\vartheta_n}\right)^\alpha + 1\right]^{m_n}}, \tag{4.3}$$

Here, K_0 is called *static gain*. As it is easy to verify, $\vartheta_k = (-p_k)^{1/\alpha}$ and $v_k = (-z_k)^{1/\alpha}$. With these changes, (4.2) assumes a more classical form [87, 116, 119].

Since all the coefficients a_k and b_k are real, all values of z_k and p_k are either real, or, being complex, appear in conjugate pairs. In this case, it is usual to join the corresponding terms. Let $\vartheta = (-p)^{1/\alpha}$ with p being a generic pseudo-pole

$$\left[\left(\frac{s}{\vartheta}\right)^\alpha + 1\right]\left[\left(\frac{s}{\vartheta^*}\right)^\alpha + 1\right] = \left(\frac{s^\alpha}{|\vartheta|}\right)^2 + \left(\frac{s}{|\vartheta|}\right)^\alpha \frac{2Re(\vartheta^\alpha)}{|\vartheta|^\alpha} + 1 \tag{4.4}$$

that can be written as

$$\left[\left(\frac{s}{\vartheta}\right)^\alpha + 1\right]\left[\left(\frac{s}{\vartheta^*}\right)^\alpha + 1\right] = \left(\frac{s}{\omega_n}\right)^{2\alpha} + 2\zeta\left(\frac{s}{\omega_n}\right)^\alpha + 1 \tag{4.5}$$

with $\omega_n = |\vartheta| = |p|^{\frac{1}{\alpha}}$ and $\zeta = \frac{Re(\vartheta^\alpha)}{|\vartheta|^\alpha} = \cos[\arg(\vartheta)\alpha] = -\cos[\arg(p)]$. When $\alpha = 1$, ω_n is called *natural frequency* of the system, while ζ is the *damping ratio*. As $\arg(p) > \alpha\frac{\pi}{2}$

DOI 10.1515/9783110624588-004

for a stable causal system, ζ is not necessarily positive. We have

$$-\cos\frac{\alpha\pi}{2} < \zeta < 1. \tag{4.6}$$

This formulation is suitable for introducing the notion of *frequency response* (FR) as follows.

As seen in previous chapter, the exponentials are the eigenfunctions of the LTIS. In particular, the complex sinusoidal case is very important; when $x(t) = e^{i\omega_0 t}$ $\omega_0 \in \mathbb{R}$, the output is

$$y(t) = H(i\omega_0)e^{i\omega_0 t}, \ t \in \mathbb{R} \tag{4.7}$$

provided that $H(i\omega_0)$ exists. This is the frequency response of the system and is given by the Fourier transform of $h(t)$:

$$H(i\omega) = \int_{-\infty}^{\infty} h(\tau)e^{-i\omega\tau}d\tau, \ \omega \in \mathbb{R} \tag{4.8}$$

Therefore, if $h(t)$ has LT with ROC including the imaginary axis, we can make $s = i\omega$, obtaining the Fourier transform as a particular case of the LT.

If the coefficients in (4.1) are real, then

1. $|H(\omega)|$ is the *amplitude spectrum*, or *gain*, and is an even function,
2. $\varphi(\omega) = \arg H(\omega)$ is the *phase spectrum*, or simply *phase*, and is an odd function.

Example 4.1.1. *Consider a system defined by the differential equation*

$$y'''(t) + y''(t) - 4y'(t) + 2y(t) = x''(t) - 4x(t) \tag{4.9}$$

and assume that $x(t) = e^{i\pi t}$*. Then*

$$y(t) = \frac{\pi^2 + 4}{i\pi^3 + \pi^2 + 4i\pi - 2}e^{i\pi t}. \tag{4.10}$$

Remark that, if $x(t) = e^{\pm i2t}$*, then* $y(t) = 0$*.*

Let the input be $x(t) = \cos(\omega_0 t) = \frac{1}{2}e^{i\omega_0 t} + \frac{1}{2}e^{-i\omega_0 t}$. The output is

$$y(t) = H(i\omega_0)\frac{1}{2}e^{i\omega_0 t} + H(-i\omega_0)\frac{1}{2}e^{-i\omega_0 t}. \tag{4.11}$$

Therefore, we have the following theorem:

Theorem 4.1.1. *Sinusoidal case*
   The output of the system defined by (4.1) when $x(t) = \cos(\omega_0 t)$ *is given by*

$$y(t) = |H(\omega_0)|\cos(\omega_0 t + \varphi(\omega_0)). \tag{4.12}$$

where $H(i\omega_0) = |H(\omega_0)|e^{\varphi(\omega_0)}$*.*

This is also true for implicit fractional TF as well. From this result it would be immediate to compute the solution for $x(t) = \sin(\omega_0 t)$.

Example 4.1.2. *Consider again the system defined by*

$$y'''(t) + y''(t) - 4y'(t) + 2y(t) = x''(t) - 4x(t)$$

and assume that $x(t) = \cos(\pi t)$. Then

$$y(t) = \frac{\pi^2 + 4}{\sqrt{(\pi^2 - 2)^2 + \pi^2(\pi^2 + 4)^2}} \cos(\pi t + \phi). \tag{4.13}$$

with $\phi = \arctan\left[\frac{\pi(\pi^2 + 4)}{\pi^2 - 2}\right]$.

Problem 4.1.1. *"Fractionalise" the previous example and compute the output for the same input.*

$$y^{(3\alpha)}(t) + y^{(2\alpha)}(t) - 4y^{(\alpha)}(t) + 2y(t) = x^{(2\alpha)}(t) - 4x(t)$$

Remark 4.1.1. *When $H(i\omega_0) = 0$, $y(t)$ is identically null. This is the reason why systems described by linear differential equations are called filters: a filter amplifies or attenuates the amplitude, and from 4.12 it is seen that when doing so it also introduces a change in the phase of the input sinusoid.*

Relation (4.12) reveals the important role of $|H(\omega)|$ and $\varphi(\omega)$, and shows that it is enough to know them for $\omega \geq 0$, since $H(-i\omega) = H^*(i\omega)$. The gain is normally expressed in decibel (dB),

$$A(\omega) = 20\log_{10}|H(\omega)| = 10\log_{10}|H(\omega)|^2. \tag{4.14}$$

For an explicit fractional TF, and using (4.3), gain and phase assume a general form given by

$$A(\omega) = 20\log_{10}(K_0) + \sum_{k=1}^{M_z} m_k 20\log_{10}\left|\left(\frac{i\omega}{v_k}\right)^\alpha + 1\right| - \sum_{k=1}^{N_p} n_k 20\log_{10}\left|\left(\frac{i\omega}{\theta_k}\right)^\alpha + 1\right|, \tag{4.15}$$

$$\varphi(\omega) = \arg(K_0) + \sum_{k=1}^{M_z} m_k \arg\left[\left(\frac{i\omega}{v_k}\right)^\alpha + 1\right] - \sum_{k=1}^{N_p} n_k \arg\left[\left(\frac{i\omega}{\theta_k}\right)^\alpha + 1\right]. \tag{4.16}$$

The most common graphical representations of frequency responses are:
1. Bode diagrams
   show $A(\omega)$ and $\varphi(\omega)$ in two separate plots (often reduced to their asymptotic straight lines), with frequencies in a logarithmic scale in the abscissas, and $A(\omega)$ in dB.
2. Polar plots
   express $Im[H(\omega)]$ as a function of $Re[H(\omega)]$ with $\omega > 0$ as a parameter. The

Nyquist diagram is similar to the polar plot, but shows the response for $\omega \in \mathbb{R}$ and not only for $\omega > 0$. Since $H(-i\omega)$ is the complex conjugate of $H(i\omega)$, the Nyquist diagram has the same curve of the polar plot, and then another curve, symmetrical in relation to the real axis.

3. Nichols diagrams

show $A(\omega)$, usually in dB, as a function of $\varphi(\omega)$, in degrees. These are the less useful frequency plots.

In this chapter we will consider Bode diagrams. The others will be treated when we present feedback systems.

4.2 Bode diagrams of typical explicit systems

Relation (4.12) is of a major importance in designing, modelling, and testing systems using the frequency response. Frequently, simple representations of the amplitude and phase — typically the asymptotes of the Bode diagram — are enough [11]. In what follows, and for the reasons explained in Section 4.1, frequency responses are given only for $\omega > 0$, due to the hermitian symmetry property of the FT [98, 119].

4.2.1 The differintegrator

The frequency response of $G_1(s) = s^\alpha$, $\alpha \in \mathbb{R}$, is

$$G_1(j\omega) = (i\omega)^\alpha, \tag{4.17}$$

$$A_1(\omega) = 20\log_{10} \omega^\alpha = 20\alpha \log_{10} \omega, \tag{4.18}$$

$$\varphi_1(\omega) = \frac{\alpha\pi}{2}. \tag{4.19}$$

So, the amplitude Bode diagrams are straight lines with slope 20α dB per decade. The phase diagrams are horizontal straight lines with ordinate $\frac{\alpha\pi}{2}$. They are represented in Figure 4.1.

4.2.2 One pseudo-pole or one pseudo-zero

The frequency response of $G_2(s) = \left[\left(\dfrac{s}{a} \right)^\alpha + 1 \right]^{\pm 1}$, $\alpha, a \in \mathbb{R}^+$, is

$$G_2(i\omega) = \left[\left(\frac{i\omega}{a} \right)^\alpha + 1 \right]^{\pm 1}$$

$$A_2(\omega) = 20\log_{10} \left[\sqrt{\left(\frac{\omega}{a} \right)^{2\alpha} + 2\left(\frac{\omega}{a} \right)^\alpha \cos \frac{\alpha\pi}{2} + 1} \right]^{\pm 1}$$

$$= \pm 10\log_{10} \left[\left(\frac{\omega}{a} \right)^{2\alpha} + 2\left(\frac{\omega}{a} \right)^\alpha \cos \frac{\alpha\pi}{2} + 1 \right], \tag{4.20}$$

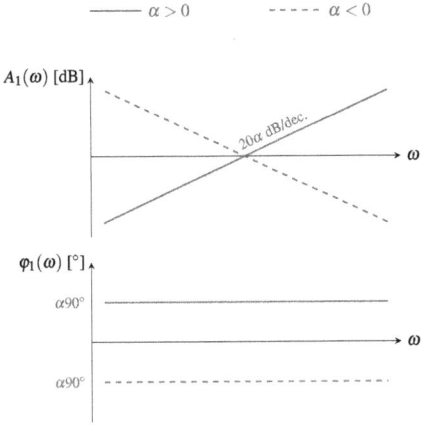

Fig. 4.1: Bode diagrams of $G_1(s) = s^\alpha$, $\alpha \in \mathbb{R}$.

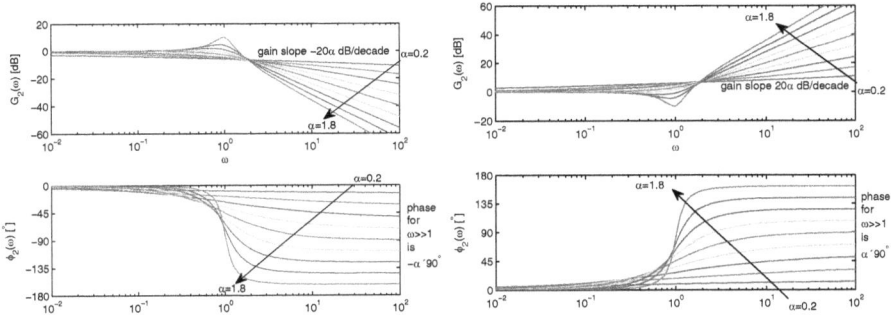

Fig. 4.2: Frequency responses of $G_2(s) = \left(\left(\frac{s}{a}\right)^\alpha + 1\right)^{\pm 1}$, for several values of $\alpha \leq 1$; a was set to 1.

$$\varphi_2(\omega) = \pm \arctan \frac{\left(\frac{\omega}{a}\right)^\alpha \sin \frac{\alpha\pi}{2}}{\left(\frac{\omega}{a}\right)^\alpha \cos \frac{\alpha\pi}{2} + 1}. \tag{4.21}$$

The asymptotic expressions are easily obtained.
- For low frequencies $\omega \ll a$, we have $G_2(i\omega) \approx 1$,
- For high frequencies $\omega \gg a$, we have $G_2(i\omega) \approx \left(\frac{j\omega}{a}\right)^{\pm \alpha}$,
- At the corner frequency $\omega = a$ rad·s^{-1},

$$G_2(i\omega) = \left(1 + e^{i\frac{\alpha\pi}{2}}\right)^{\pm 1}. \tag{4.22}$$

Therefore $A_2(a) = \pm 20 \log_{10} \sqrt{[2 + 2\cos \frac{\alpha\pi}{2}]} = \pm 20 \log_{10} \sqrt{2}\sqrt{1 + \cos \frac{\alpha\pi}{2}}$, and $\phi_2 = \frac{\alpha\pi}{4}$.
- For the particular case, when $\alpha = 1$, at the corner frequency $\omega = a$ rad·s^{-1}, $A_2(a) = \mp 20 \log_{10} \sqrt{2} \approx \mp 3$ dB. Because of this value for the gain at the corner frequency of this particular simple plant, a usual criterion to describe filters is to state the lowest frequency above which the gain is always ∓ 3 dB above/below the static gain. This frequency is the 3 dB point that serves to define the *filter bandwidth*.

– The 3 dB point for the general case $\alpha \in \mathbb{R}^+$ is obtained when

$$\left(\frac{\omega}{a}\right)^{2\alpha} + 2\left(\frac{\omega}{a}\right)^{\alpha} \cos\frac{\alpha\pi}{2} + 1 = 2.$$

Consequently, the Bode amplitude asymptotes of $G_2(s)$ are given by

$$A_2(\omega) = \begin{cases} 0 & \omega \le a \\ \pm 20\alpha\log_{10}\omega \mp 20\alpha\log_{10}a & \omega > a \end{cases}, \tag{4.23}$$

i.e. by two half straight lines that join at $\omega = a$, where the amplitude is $A_2(a)_{|dB} = \pm 3 \mp 10\log_{10}\left[1 + \cos\frac{\alpha\pi}{2}\right]$. To draw the phase Bode spectrum the procedure is a bit different. Asymptotically, we have.

$$\varphi_2(\omega) = \begin{cases} 0 & \omega \ll a \\ \pm\frac{\alpha\pi}{2} & \omega \gg a \end{cases}. \tag{4.24}$$

The Bode phase spectrum is drawn as follows:

1. For $\omega \le a/10$, $\phi_2(\omega) = 0$ (low frequency asymptote).
2. For $\omega \ge 10a$, $\phi_2(\omega) = \frac{\alpha\pi}{2}$ (high frequency asymptote).
3. In the two decade interval $\omega \in (a/10, 10a)$ the spectrum increases (for a pseudo-zero) or decreases (for a pseudo-pole) $\frac{\alpha\pi}{4}$ radians per decade (rad/dec).
4. At $\omega = a$, the phase is exactly $\phi_2(a) = \pm\frac{\alpha\pi}{4}$.

Remark 4.2.1. *Strictly speaking there is no need to study the case $\alpha > 1$, but it may be useful to do so. Calculating the derivative of $A_2(\omega)$ when $\alpha > 1$, it can be seen that, if $1 < \alpha < 3 \vee 5 < \alpha < 7 \vee 9 < \alpha < 11\ldots$, the gain will have a peak at*

$$\omega_{peak} = a\left(-\cos\frac{\alpha\pi}{2}\right)^{\frac{1}{\alpha}}, \tag{4.25}$$

$$A_2(\omega_{peak}) = \pm 20\log_{10}\left|\sin\frac{\alpha\pi}{2}\right|. \tag{4.26}$$

Otherwise, the gain increases or decreases monotonously. From the derivative of the phase $\phi_2(\omega)$, we verify that there are four possible cases. For the a pseudo-pole, they are as follows:

– *If $0 < \alpha < 2 \vee 4 < \alpha < 6 \vee 8 < \alpha < 10 \vee \ldots$, the phase will decrease from 0 to $\lfloor\frac{\alpha}{4}\rfloor 2\pi - \alpha\frac{\pi}{2}$.*
– *If $2 < \alpha < 4 \vee 6 < \alpha < 8 \vee 10 < \alpha < 12 \vee \ldots$, the phase will increase from 0 to $\left(1 + \lfloor\frac{\alpha}{4}\rfloor\right)2\pi - \alpha\frac{\pi}{2}$.*
– *If $\alpha \in \{4, 8, 12, 16\ldots\}$, the phase is constant and equal to 0.*
– *If $\alpha \in \{2, 6, 10, 14\ldots\}$ the phase will have a discontinuity at $\omega = a$ rad·s^{-1} and jumps from 0 to $\pm\pi$.*

In the case of a pseudo-zero, the behavior is symmetrical.

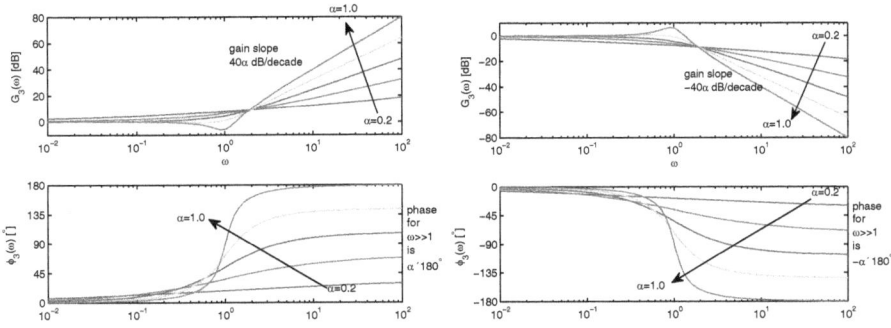

Fig. 4.3: Amplitude and phase spectra of $G_2(s) = \left[(s/\omega_n)^{2\alpha} + 2\zeta(s/\omega_n)^{\alpha} + 1\right]^{\pm}$, for several values of $\alpha = 0.2k$, $k = 1, 2, \cdots, 5$, with $\zeta = 0.25$.

4.2.3 Complex conjugate pair of pseudo-poles or pseudo-zeroes

We are going to study the frequency response of the pair of complex conjugate pseudo-poles or pseudo-zeroes

$$G_3(s) = \left[(s/\omega_n)^{2\alpha} + 2\zeta(s/\omega_n)^{\alpha} + 1\right]^{\pm 1}, \quad \alpha, a, \zeta \in \mathbb{R}^+, \tag{4.27}$$

To do it, we begin by studying the behaviour of the function

$$D(\omega) = (i\omega/\omega_n)^{2\alpha} + 2\zeta(i\omega/\omega_n)^{\alpha} + 1 = (\omega/\omega_n)^{2\alpha}e^{i\alpha\pi} + 2\zeta(\omega/\omega_n)^{\alpha}e^{i\alpha\pi/2} + 1 \tag{4.28}$$

We have
$$D(\omega)\bar{D}(\omega)^* =$$

$$= \left[\left(\frac{\omega}{\omega_n}\right)^{2\alpha}e^{i\alpha\pi} + 2\zeta\left(\frac{\omega}{\omega_n}\right)^{\alpha}e^{i\alpha\pi/2} + 1\right]\left[\left(\frac{\omega}{\omega_n}\right)^{2\alpha}e^{-i\alpha\pi} + 2\zeta\left(\frac{\omega}{\omega_n}\right)^{\alpha}e^{-i\alpha\pi/2} + 1\right]$$

$$= \left(\frac{\omega}{\omega_n}\right)^{4\alpha} + 4\zeta\left(\frac{\omega}{\omega_n}\right)^{3\alpha}\cos\frac{\alpha\pi}{2} + 2\left(\frac{\omega}{\omega_n}\right)^{2\alpha}(2\zeta^2 + \cos\alpha\pi) + 4\zeta\left(\frac{\omega}{\omega_n}\right)^{\alpha}\cos\frac{\alpha\pi}{2} + 1$$

an analytical expression which is not simple. However, it is useful to obtain the asymptotic behaviour and the corresponding Bode diagrams.

- For low frequencies $\omega \ll \omega_n$, we have $D(\omega)D(\omega)^* \approx 1$,
- For high frequencies $\omega \gg \omega_n$, we have $D(\omega)D(\omega)^* \approx \left(\frac{\omega}{\omega_n}\right)^{4\alpha}$,
- At the corner frequency $\omega = \omega_n$ rad·s^{-1}, $D(\omega_n) = e^{i\alpha\pi} + 2\zeta e^{i\alpha\pi/2} + 1$. Therefore $|D(\omega_n)| = \sqrt{2}\sqrt{1 + 4\zeta\cos\frac{\alpha\pi}{2} + \zeta^2 + \cos\alpha\pi}$.

The Bode amplitude spectrum corresponding to $G_3(s)$ is given by

$$A_3(\omega)_{|dB} = \begin{cases} 0 & \omega \le \omega_n \\ -40\alpha\log_{10}\omega + 40\alpha\log_{10}\omega_n & \omega > \omega_n \end{cases}. \tag{4.29}$$

In a log-plot, it is constituted by two half straight lines that join together at $\omega = \omega_n$, where the amplitude is

$$A_3(\omega_n)|_{dB} = -6 - 10\log_{10}\left[1 + 4\zeta\cos\frac{\alpha\pi}{2} + \zeta^2 + \cos\alpha\pi\right]. \tag{4.30}$$

Although the asymptotic plot is enough for many problems, it is important to have information about the possible existence of an overshoot. To do it, we need to look for the extrema of $D(\omega)D(\omega)^*$ that must be obtained from the positive solutions of the equation:

$$x^3 + 3\zeta\cos\frac{\alpha\pi}{2}x^2 + (2\zeta^2 + \cos\alpha\pi)x + \zeta\cos\frac{\alpha\pi}{2} = 0. \tag{4.31}$$

obtained by differentiation of $D(x)D(x)^*$, with $x = \omega/\omega_n$. We will perform this study in several steps:

1. If $0 < \alpha \leq \frac{1}{2}$, we remark that the coefficients of the polynomial in (4.31) are all positive. By the Descartes rule of signs, there are no positive roots. Consequently, the amplitude spectrum has no overshoot.
2. For $\alpha = 1$, the equation has a solution given by $x^2 = 1 - 2\zeta^2$. Thus, there exists an overshoot at $\omega = \omega_n\sqrt{1 - 2\zeta^2}$ if $\zeta < \frac{1}{\sqrt{2}}$.
3. For $\frac{1}{2} < \alpha < 1$, the situation is similar to the previous one. In this case $\cos\alpha\pi < 0$, and coefficient $(2\zeta^2 + \cos\alpha\pi)$ can be null or negative, if $\zeta^2 < \frac{-\cos\alpha\pi}{2}$. However, this does not ensure the existence of an overshoot. This only happens if (4.31) has 3 real roots.
4. Simulations show that one of the 3 real roots is near the origin, for small $|\zeta|$. This means that $\zeta\cos\frac{\alpha\pi}{2} \approx 0$. Then the searched root is given approximately by

$$x = -\frac{3}{2}\zeta\cos\frac{\alpha\pi}{2} + \sqrt{\frac{9}{4}\zeta^2\cos^2\frac{\alpha\pi}{2} - 2\zeta^2 - \cos\alpha\pi} \tag{4.32}$$

provide that

$$\zeta^2 < \frac{-\cos\alpha\pi}{2} \tag{4.33}$$

Then, there may exist an overshoot at $\omega \approx \omega_n\sqrt{1 - 2\zeta^2}$ if $\zeta < \frac{1}{\sqrt{2}}$.

Now we are going to study the phase behaviour of the function

$$D(\omega) = (\omega/\omega_n)^{2\alpha}e^{i\alpha\pi} + 2\zeta(\omega/\omega_n)^\alpha e^{i\alpha\pi/2} + 1. \tag{4.34}$$

Denote its phase by $\phi_D(\omega)$. We deduce immediately:
-   For low frequencies $\omega \ll \omega_n$, we have $\phi_D(\omega) \approx 0$;
-   For high frequencies $\omega \gg \omega_n$, we have $\phi_D(\omega) \approx \alpha\pi$;
-   At the corner frequency $\omega = \omega_n$ rad·s^{-1}, $D(\omega_n) = e^{i\alpha\pi} + 2\zeta e^{i\alpha\frac{\pi}{2}} + 1$. Therefore

$$\phi_D(\omega_n) = \arctan\left(\frac{\sin(\alpha\pi) + 2\zeta\sin(\alpha\frac{\pi}{2})}{1 + \cos(\alpha\pi) + 2\zeta\cos(\alpha\frac{\pi}{2})}\right). \tag{4.35}$$

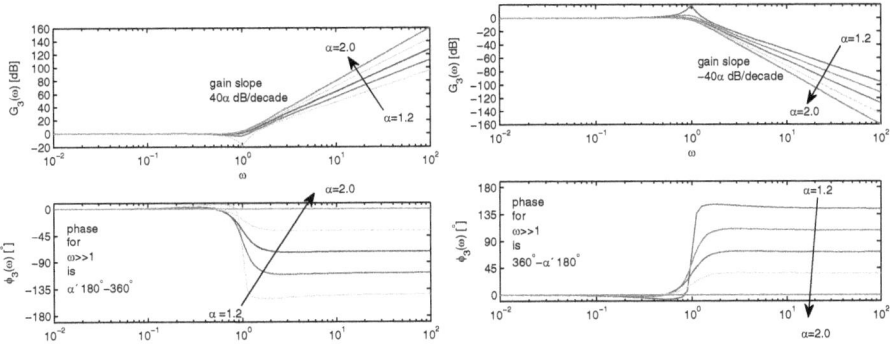

Fig. 4.4: Amplitude and phase spectra of $G_2(s) = \left[(s/\omega_n)^{2\alpha} + 2\zeta(s/\omega_n)^{\alpha} + 1\right]^{\pm}$, for several values of $\alpha = 1 + 0.2k$, $k = 1, 2, \cdots, 5$, with $\zeta = 0.25$.

The Bode phase spectrum is drawn as follows:

1. For $\omega \leq \omega_n/10$, we have $\phi_2(\omega) = 0$ (low frequency asymptote);
2. For $\omega \geq 10\omega_n$, we have $\phi_2(\omega) = \alpha\pi$;
3. In the two decade interval $\omega \in (\omega_n/10, 10\omega_n)$, the spectrum increases/decreases $(\pm 1) \frac{\alpha\pi}{2}$ per decade;
4. At $\omega = \omega_n$, the phase is exactly $\phi_2(a) = \pm\frac{\alpha\pi}{2}$ only if $\zeta = 0$.

Remark 4.2.2. *In the above study, we could verify that the phase corresponding to a zero in the right half complex plane is identical to the one of a pole in the left half-plane. This lead us to classify the systems according to the phase. An LTI system is*

- *minimum phase, if all zeros are in the left complex half plane;*
- *maximum phase, if all the zeros are in the right complex half-plane;*
- *mixed phase if there are at least one zero in each half-plane.*

There is a relationship between the type of phase and the way how the step response evolves.

4.3 Implicit systems

Till now, we studied systems defined by linear differential equations of finite order (1.51) and showed that they can be characterised by a transfer function involving pseudo-polynomials in positive powers of s. However, this does not cover all the interesting transfer functions found in practice. There are many others that we call *implicit systems* and that can be described by other kinds of transfer functions. For example [98, 119],

- Fractions with implicit fractional order poles or zeroes, given by $(s + p)^{\pm\alpha}$;
- Irrational functions, such as $\frac{\arctan(\sqrt{s})}{\sqrt{s}}$.

4.3.1 Implicit fractional order zero or pole

The frequency response of the implicit fractional-order TF

$$G_4(s) = \left(\frac{s}{a} + 1\right)^{\pm a}, \quad \alpha, a \in \mathbb{R}^+, \tag{4.36}$$

is

$$G_4(i\omega) = \left(\frac{i\omega}{a} + 1\right)^{\pm a}, \tag{4.37}$$

$$A_4(\omega) = 20\log_{10}\left(1 + \frac{\omega^2}{a^2}\right)^{\pm\frac{\alpha}{2}} = \pm 10\alpha\log_{10}\left(1 + \frac{\omega^2}{a^2}\right), \tag{4.38}$$

$$\varphi_4(\omega) = \pm\alpha\arctan\frac{\omega}{a}. \tag{4.39}$$

There are no peaks in the gain, and the phase always increases/decreases monotonously from 0 to $\pm\alpha\frac{\pi}{2}$. For low and high frequencies, the asymptotic approximations for $G_2(i\omega)$ are valid for $G_4(i\omega)$ as well. Therefore, the asymptotic frequency response of $G_4(s)$ is given by

$$A_4(\omega) = \begin{cases} 0 & \omega \ll a \\ \pm 10\alpha\log_{10}2 & \omega = a \\ \pm 20\alpha\log_{10}\omega \mp 20\alpha\log_{10}a & \omega \gg a \end{cases}, \tag{4.40}$$

$$\varphi_4(\omega) = \begin{cases} 0 & \omega \ll a \\ \pm\frac{\alpha\pi}{4} & \omega = a \\ \pm\frac{\alpha\pi}{2} & \omega \gg a \end{cases}. \tag{4.41}$$

The asymptotic diagrams of Figure 4.2, considering only the expressions for low and high frequencies, apply to $G_4(s)$ as well. Figure 4.5 has the frequency responses of TF similar to those of Figure 4.2, so that the differences between them may be more clear.

4.3.2 Fractional PID

There are two different TF called fractional proportional+integral+derivative (PID) controller, which are given by

$$G_5(s) = K_p\left(1 + \frac{1}{(T_i s)^\lambda} + (T_d s)^\mu\right), \quad K_p, T_i, T_d, \lambda, \mu > 0 \tag{4.42}$$

$$G_6(s) = K_p\left(1 + \frac{1}{T_i s}\right)^\lambda (1 + T_d s)^\mu, \quad K_p, T_i, T_d, \lambda, \mu > 0 \tag{4.43}$$

where K_p is the proportional gain, T_i is the integral time, and T_d the derivative time. Form (4.42) is an explicit fractional transfer function, while (4.43) is implicit.

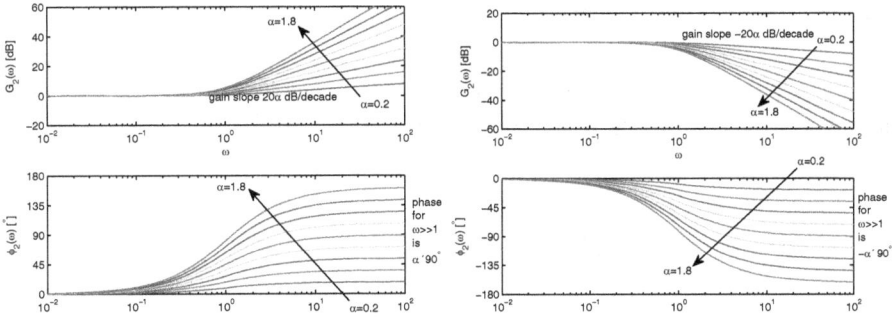

Fig. 4.5: Amplitude and phase spectra of $G_2(s) = \left[(s/\omega_n)^{2\alpha} + 2\zeta(s/\omega_n)^\alpha + 1\right]^{\pm 1}$, for several values of $\alpha = 1 + 0.2k$, $k = 1, 2, \cdots, 5$.

The frequency responses are

$$G_5(i\omega) = K_p \left(1 + \frac{1}{(T_i i\omega)^\lambda} + (T_d i\omega)^\mu \right),$$ (4.44)

$$G_6(i\omega) = K_p \left(1 + \frac{1}{T_i i\omega} \right)^\lambda (1 + T_d i\omega)^\mu.$$ (4.45)

The asymptotic behaviours of both responses are the same. At low frequencies,

$$G_5(i\omega) \approx \frac{K_p}{i^\lambda T_i \omega^\lambda} \approx G_6(i\omega),$$ (4.46)

$$A_5(\omega) \approx 20 \log_{10} \frac{K_p}{T_i} - 20\lambda \log_{10} \omega \approx A_6(\omega),$$ (4.47)

$$\varphi_5(\omega) \approx \arg[i^{-\lambda}] \approx \varphi_6(\omega).$$ (4.48)

At high frequencies,

$$G_5(i\omega) \approx i^\mu K_p T_d \omega^\mu \approx G_6(i\omega),$$ (4.49)

$$A_5(\omega) \approx 20 \log_{10}(K_p T_d) + 20\mu \log_{10} \omega \approx A_6(\omega),$$ (4.50)

$$\varphi_5(\omega) \approx \arg[i^\mu] \approx \varphi_6(\omega).$$ (4.51)

In summary, the gain begins with a slope of -20λ dB per decade, and ends with a slope of 20μ dB per decade. The phase goes from $-\lambda\frac{\pi}{2}$ to $\mu\frac{\pi}{2}$ (or, in the case of $G_5(s)$, an angle with the same locus in the complex plane). This asymptotic behaviour is shown in Figure 4.6. Moreover, Figures 4.7 and 4.8 show the actual frequency responses for two particular cases of $G_5(s)$ and two of $G_6(s)$; in each Figure, one case has the zeros close to each other, and the other has the zeros clearly separated.

4.3.3 Fractional lead compensator

The fractional lead compensator is used in control to increase the phase of a system around a chosen frequency, which may improve its stability in closed loop, as shall be seen below in chapter 6. It is defined by the TF

$$G_7(s) = \left(\frac{\tau s + a}{s + a}\right)^\alpha, \quad \alpha, a \in \mathbb{R}^+, \ \tau > 1, \tag{4.52}$$

and its frequency response is

$$G_7(i\omega) = \left(\frac{\tau i\omega + a}{i\omega + a}\right)^\alpha, \tag{4.53}$$

$$A_7(\omega) = 20 \log_{10}\left(\frac{a^2 + \tau^2\omega^2}{a^2 + \omega^2}\right)^{\frac{\alpha}{2}}$$

$$= 10\alpha \log_{10}\left(\frac{a^2 + \tau^2\omega^2}{a^2 + \omega^2}\right), \tag{4.54}$$

$$\varphi_7(\omega) = \alpha\left(\arctan\frac{\tau\omega}{a} - \arctan\frac{\omega}{a}\right). \tag{4.55}$$

For low frequencies $\omega \ll a$, we have $G_7(j\omega) \approx 1$, while for high frequencies $\omega \gg a$, we have $G_7(i\omega) \approx \tau^\alpha$. Joining this with the values for corner frequency $\omega = \frac{a}{\sqrt{\tau}}$ rad·$^{-1}$, the following asymptotic frequency response is obtained:

$$A_7(\omega) = \begin{cases} 0 & \omega \ll \frac{a}{\sqrt{\tau}} \\ 10\,\alpha\log_{10}\tau & \omega = \frac{a}{\sqrt{\tau}} \\ 20\,\alpha\log_{10}\tau & \omega \gg \frac{a}{\sqrt{\tau}} \end{cases}, \tag{4.56}$$

$$\varphi_7(\omega) = \begin{cases} 0 & \omega \ll \frac{a}{\sqrt{\tau}} \\ \alpha\arctan\sqrt{\tau} - \alpha\arctan\frac{1}{\sqrt{\tau}} & \omega = \frac{a}{\sqrt{\tau}} \\ 0 & \omega \gg \frac{a}{\sqrt{\tau}} \end{cases}. \tag{4.57}$$

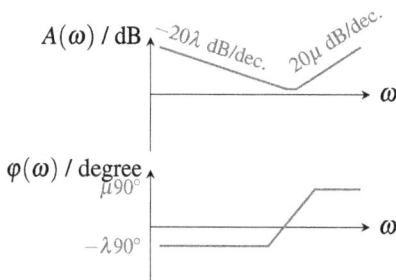

Fig. 4.6: Asymptotic Bode diagrams of $G_5(s) = K_p\left(1 + \frac{1}{(T_i s)^\lambda} + (T_d s)^\mu\right)$, $K_p, T_i, T_d, \lambda, \mu > 0$ and of $G_6(s) = K_p\left(1 + \frac{1}{T_i s}\right)^\lambda(1 + T_d s)^\mu$, $K_p, T_i, T_d, \lambda, \mu > 0$.

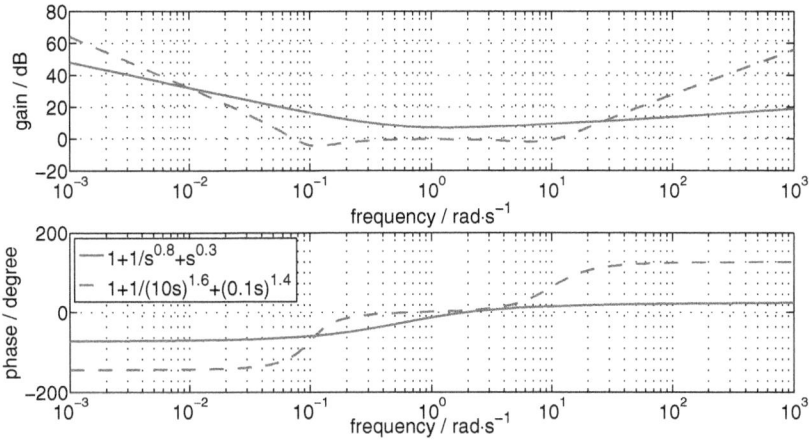

Fig. 4.7: Amplitude and phase spectra of $1 + \frac{1}{s^{0.8}} + s^{0.3}$ and $1 + \frac{1}{(10s)^{1.6}} + (0.1s)^{1.4}$ (particular cases of $G_5(s)$).

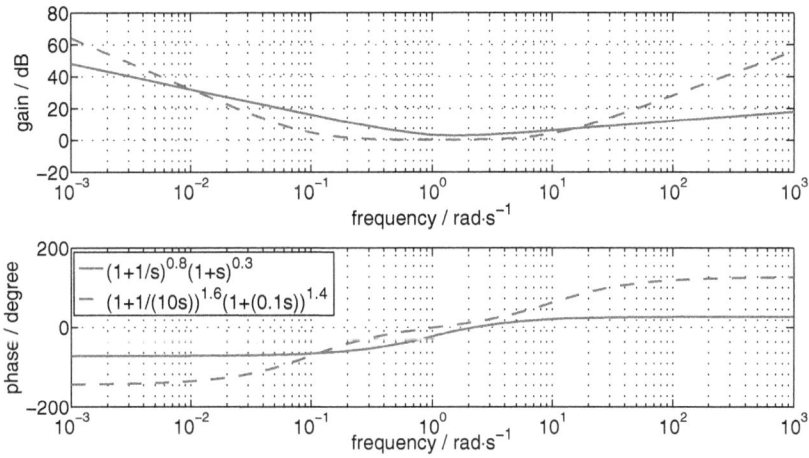

Fig. 4.8: Amplitude and phase spectra of $\left(1 + \frac{1}{s}\right)^{0.8}(1 + s)^{0.3}$ and $\left(1 + \frac{1}{10s}\right)^{1.6}(1 + 0.1s)^{1.4}$ (particular cases of $G_6(s)$).

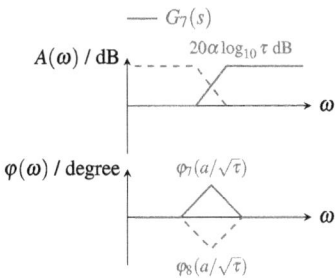

Fig. 4.9: Asymptotic Bode diagrams of $G_7(s) = \left(\frac{\tau s + a}{s + a}\right)^\alpha$, $a \in \mathbb{R}^+$, $\tau > 1$ and $G_8(s) = \left(\frac{s + a}{s + \frac{a}{\tau}}\right)^\alpha$, $\alpha, a \in \mathbb{R}^+$, $\tau > 1$.

See Figure 4.9 for the asymptotic frequency response, that illustrates the purpose for which this TF is used in control. It also shows that the price to pay is a positive gain of $G_7(s)$ at higher frequencies. Since in practical applications there often is high frequency noise, this will mean that such noise will be filtered less efficiently.

4.3.4 Fractional lag compensator

The fractional lag compensator is used in control to increase the static gain of a plant, which may improve its precision in closed loop, as shall be seen below in chapter 6. It is defined by the TF

$$G_8(s) = \left(\frac{s+a}{s+\frac{a}{\tau}} \right)^\alpha, \quad \alpha, a \in \mathbb{R}^+, \; \tau > 1, \tag{4.58}$$

The corresponding frequency response is

$$G_8(i\omega) = \left(\frac{i\omega + a}{i\omega + \frac{a}{\tau}} \right)^\alpha, \tag{4.59}$$

$$A_8(\omega) = 20 \log_{10} \left(\frac{a^2 + \omega^2}{\frac{a^2}{\tau^2} + \omega^2} \right)^{\frac{\alpha}{2}}$$

$$= 10\alpha \log_{10} \left(\frac{a^2 + \omega^2}{\frac{a^2}{\tau^2} + \omega^2} \right), \tag{4.60}$$

$$\varphi_8(\omega) = \alpha \arctan \frac{\omega}{a} - \alpha \arctan \frac{\tau\omega}{a}. \tag{4.61}$$

For low frequencies $\omega \ll a$, we have $G_8(i\omega) \approx \tau^\alpha$, while for high frequencies $\omega \gg a$, we have $G_8(i\omega) \approx 1$. Joining this with the values for corner frequency $\omega = \frac{a}{\sqrt{\tau}}$ rad·$^{-1}$, the following asymptotic frequency response is obtained:

$$A_8(\omega) = \begin{cases} 20\alpha \log_{10} \tau & \omega \ll \frac{a}{\sqrt{\tau}} \\ 10\alpha \log_{10} \tau & \omega = \frac{a}{\sqrt{\tau}} \\ 0 & \omega \gg \frac{a}{\sqrt{\tau}} \end{cases}, \tag{4.62}$$

$$\varphi_8(\omega) = \begin{cases} 0 & \omega \ll \frac{a}{\sqrt{\tau}} \\ \alpha \arctan \frac{1}{\sqrt{\tau}} - \alpha \arctan \sqrt{\tau} & \omega = \frac{a}{\sqrt{\tau}} \\ 0 & \omega \gg \frac{a}{\sqrt{\tau}} \end{cases}. \tag{4.63}$$

See Figure 4.9 for the asymptotic frequency response, that illustrates the purpose for which this TF is used in control. It also shows that the price to pay is a negative phase of $G_8(s)$ at some frequency. This must be carefully managed, as it may decrease the stability of the closed loop, as shall be seen below in chapter 6.

4.4 Approximations to transfer functions based upon a frequency response

For practical purposes, such as computational simulations or physical implementations (electrical, or mechanical) using integer order components, fractional-order TF
are often approximated by some integer-order TF with a frequency response that is in
some sense similar, in a given frequency band. The three more interesting methods to
do so are here presented (see also the numerical methods in [42, 43]).

4.4.1 Oustaloup's CRONE approximation

The CRONE approximation (French acronym of Non-Integer Order Robust Control) of
s^α, due to Alain Oustaloup, is essentially an integer order pole/zero stable model, with
all the N poles and N zeroes being real and negative (it is a minimum phase system).
The poles and zeroes are recursively and alternatively placed within a frequency range
$[\omega_l, \omega_h]$ [94, 96]:

$$s^\alpha \approx C \prod_{m=1}^{N} \frac{1 + \frac{s}{\omega_{z,m}}}{1 + \frac{s}{\omega_{p,m}}}, \tag{4.64}$$

$$\omega_{z,m} = \omega_l \left(\frac{\omega_h}{\omega_l} \right)^{\frac{2m-1-\alpha}{2N}}, \tag{4.65}$$

$$\omega_{p,m} = \omega_l \left(\frac{\omega_h}{\omega_l} \right)^{\frac{2m-1+\alpha}{2N}}. \tag{4.66}$$

The gain C is set to obtain the correct gain at $\omega = 1$ rad·s^{-1}, which is $|(i\omega)^\alpha| = 1$, $\forall \alpha \in \mathbb{R}$
(i.e. 0 dB). If the frequency 1 rad·s^{-1} is outside the bandwidth $[\omega_l, \omega_h]$, the correct gain
at any other suitable frequency will have to be adjusted instead.
    The recursion relation between poles and zeros is

$$\frac{\omega_{z,m+1}}{\omega_{z,m}} = \frac{\omega_{p,m+1}}{\omega_{p,m}} = \left(\frac{\omega_h}{\omega_l} \right)^{\frac{1}{N}}, \tag{4.67}$$

and causes them to be equidistant in a logarithmic scale of frequencies. For $\alpha \in]-1, 1[$
this approximation is good in the frequency range $[10\omega_l, \frac{\omega_h}{10}]$. Both the gain and the
phase have some ripples that decrease as N increases. It is a good practice to set N to
at least the number of decades in $[\omega_l, \omega_h]$ for achieving good results.
    For $|\alpha| > 1$, the approximation is poorer. Since $s^\alpha = s^{\lceil \alpha \rceil} s^{\alpha - \lceil \alpha \rceil}$ or $s^\alpha = s^{\lfloor \alpha \rfloor} s^{\alpha - \lfloor \alpha \rfloor}$,
this can be solved approximating only the last term $s^{\alpha - \lceil \alpha \rceil}$ or $s^{\alpha - \lfloor \alpha \rfloor}$, which is then
multiplied by the required integer power of s.
    If $\alpha \in]-1, 0[$, the decomposition $s^\alpha = s^{\lfloor \alpha \rfloor} s^{\alpha - \lfloor \alpha \rfloor}$ may be used as well, to ensure
that there is the effect of an integer integrator.
    When using this method to approximate a fractional TF, first each fractional
power of s in the numerator and in the denominator is approximated, and then the
approximations are replaced in the desired TF.

4.4.2 The Carlson's approximation

The Carlson's approximation of s^{α}, $\alpha = \frac{1}{n}$, $n \in \mathbb{N}$, is based on the Newton–Raphson iterative numerical method to find a root of an algebraic equation. After k iterations, the approximation $G_k(s)$ of s^{α} is given by [8]

$$G_{k+1}(s) = G_k(s) \frac{\left(\frac{1}{\alpha} - 1\right) G_k^{\frac{1}{a}}(s) + \left(\frac{1}{\alpha} + 1\right) s}{\left(\frac{1}{\alpha} + 1\right) G_k^{\frac{1}{a}}(s) + \left(\frac{1}{\alpha} - 1\right) s}. \tag{4.68}$$

The first approximation $G_1(s)$ is usually set to $G_1(s) = 1$, which is a very poor approximation, that will be improved by subsequent iterations. The order of the TF $G_k(s)$ increases with k and depends on the order α being approximated. The range of frequencies where $G_k(s)$ is a good approximation also increases with k, but is always centred on $\omega = 1$ rad·s^{-1}. This poses no problem, since multiplying all zeros and poles by a constant ω_c results in an approximation around frequency $\omega = \omega_c$ rad·s^{-1} (the gain must be corrected accordingly). Poles and zeros are are always stable but not necessarily real.

While the CRONE approximation fixes directly the number of zeros and poles, and also the frequency range of interest, in the Carlson's approximation both these things depend on the number iterations performed k, and there is no clear expression to find them from k. For a similar performance, the Carlson's approximation often has to be of higher order than a CRONE approximation. But the major limitation of this method is that the fractional order must be the inverse of an integer.

Fractional TF are approximated as combinations of approximations, as with the CRONE method.

4.4.3 Matsuda's approximation

The Matsuda method [59] provides a TF that approximates any given frequency behaviour $G(i\omega)$, known at frequencies $\omega_0, \omega_1, \ldots \omega_N$ (which do not need to be ordered). In fact, it only requires the gain; no information on the phase is required. It can be used to approximate any fractional-order TF by providing the corresponding gain. It works equally for any commensurate, non-commensurate, or implicit TF. (It is not necessary to approximate fractional powers of s separately to combine them.)

$$G(s) \approx d_0(\omega_0) + \frac{s - \omega_0}{d_1(\omega_1)+} \frac{s - \omega_1}{d_2(\omega_2)+} \frac{s - \omega_2}{d_3(\omega_3)+} \cdots = \left[d_0(\omega_0); \frac{s - \omega_{k_1}}{d_k(\omega_k)} \right]_{k=1}^{N}, \tag{4.69}$$

$$d_0(\omega) = |G(i\omega)|, \tag{4.70}$$

$$d_k(\omega) = \frac{\omega - \omega_{k-1}}{d_{k-1}(\omega) - d_{k-1}(\omega_{k-1})}, \quad k = 1, 2, \ldots N. \tag{4.71}$$

This is a continued fraction interpolation with $\left\lceil \frac{N}{2} \right\rceil$ zeros and $\left\lfloor \frac{N}{2} \right\rfloor$ poles. Hence $G(s)$ is causal only if N is even (i.e. $N + 1$ is odd). The range of frequencies where the approximation is valid depends on how they are distributed. A logarithmic distribution of the frequencies is the simplest case, but better results may be obtained concentrating frequencies around those values for which the response varies more abruptly.

4.5 System modelling and identification from a frequency response

In many daily life applications we deal with systems, eventually nonlinear, that can be studied and characterised through the frequency response. For them, we assume that their frequency response $G(i\omega)$ is approximated by

$$G(i\omega) \approx \frac{\sum\limits_{k=0}^{M} b_k(i\omega)^{\alpha_k}}{\sum\limits_{k=0}^{N} a_k(i\omega)^{\alpha_k}} \tag{4.72}$$

The quality of approximation can be measured by any optimising criterion, for example, least squares. In this case, we have to perform a minimisation on all the involved parameters, coefficients and orders, assumed to be time invariant. Eventually, a criterion involving also time responses can be formulated.

As an example of an application of this procedure, a model for electrochemical capacitors was obtained and validated by experimental tests [57, 58]. The resulting TF model assumed the form

$$Z(s) = R_0 + \frac{1}{C_1 s^{\gamma_1}} + \frac{1}{C_2 s^{\gamma_2}} + \frac{1}{C_3 s^{\gamma_1+\gamma_2}} \tag{4.73}$$

where R_0 is a resistance; C_1, C_2, and C_3 are coefficients which define the equivalent capacitance of electrochemical capacitors model; γ_1 and γ_2 are the fractional orders. Similar model creation can be seen in other applictions [6, 15, 16, 32, 113].

If the assumed model has commensurate orders, we have

$$G(i\omega) \approx \frac{\sum\limits_{k=0}^{m} b_k(i\omega)^{k\alpha}}{1 + \sum\limits_{k=1}^{n} a_k(i\omega)^{k\alpha}}, \tag{4.74}$$

or

$$G(i\omega) \approx K_0 \frac{\prod\limits_{k=1}^{M_z} \left[\left(\frac{i\omega}{\nu_k} \right)^{\alpha} + 1 \right]^{m_k}}{\prod\limits_{n=1}^{N_p} \left[\left(\frac{i\omega}{\vartheta_n} \right)^{\alpha} + 1 \right]^{m_n}}, \tag{4.75}$$

In the case of (4.74), the identification problem can be solved by Levy's identification method, that represents a slight variation of the least-squares fit. The form that this method takes for a commensurate fractional-order TF with a known, fixed α is described in which follows [119, 121].

Let us consider a model such as (4.74), with known values of N_0, M_0, and α, so that the unknowns are the coefficients a_k and b_k. We will set $a_0 = 1$ (to make each TF unique) and let

$$N(i\omega) = \sum_{k=0}^{M_0} b_k(i\omega)^{k\alpha}, \tag{4.76}$$

$$D(i\omega) = 1 + \sum_{k=1}^{N_0} a_k(i\omega)^{k\alpha}, \tag{4.77}$$

so that the frequency response of the desired model is $\frac{N(i\omega)}{D(i\omega)}$. If the frequency response to be fitted is $G(i\omega_p)$, $p = 1, 2 \ldots N_f$, (where N_f is the number of frequencies where the frequency response is known), then, instead of minimising the energy of the error defined by

$$\epsilon(i\omega_p) = G(i\omega_p) - \frac{N(i\omega_p)}{D(i\omega_p)}, \quad p = 1, 2, \cdots .N_f \tag{4.78}$$

(which would be a nonlinear problem), Levy's method minimises the square of the norm (energy) of

$$E(i\omega_p) = \epsilon(i\omega_p)D(i\omega_p) = G(i\omega_p)D(i\omega_p) - N(i\omega_p). \tag{4.79}$$

It is possible to give higher weight to the measurements of some frequencies (e.g. those with higher reliability) with a frequency-dependent weight $w(\omega_p)$ (if this is not desired, it is sufficient to set $w(\omega_p) = 1$, $\forall p = 1, 2 \ldots N_f$). Therefore, we minimise the total weighted energy error

$$E_N = \sum_{p=1}^{N_f} w(\omega_p)|E(i\omega_p)|^2 \tag{4.80}$$

by calculating the derivative of E_N relatively to all the coefficients, and then equating the derivative to zero, obtaining in this wise a set of linear equations. Solving this system of $N_0 + M_0 + 1$ equations gives the desired coefficients.

The best way to do so is to use complex numbers; otherwise, the algorithm introduces a frequency dependence that originates poor results for some frequencies (mainly low frequencies). Since $|E|^2 = EE^*$, the derivatives of E_N in order to coefficients a_k and b_k are given by

$$\begin{cases} \sum_{p=1}^{N_f} w(\omega_p) \left[\dfrac{\partial E}{\partial a_k} E^* + E \dfrac{\partial E^*}{\partial a_k} \right] = 0, & k = 1 \ldots n \\[4mm] \sum_{p=1}^{N_f} w(\omega_p) \left[\dfrac{\partial E}{\partial b_k} E^* + E \dfrac{\partial E^*}{\partial b_k} \right] = 0, & k = 0 \ldots m \end{cases} \tag{4.81}$$

From (4.76)–(4.77) and (4.79) it can be seen that

$$\begin{array}{llll} \frac{\partial E}{\partial a_k} = G(\omega)(i\omega)^{kq}, & \frac{\partial E^*}{\partial a_k} = G^*(\omega)(-i\omega)^{kq}, & k = 1\ldots n \\[2mm] \frac{\partial E}{\partial b_k} = -(i\omega)^{kq}, & \frac{\partial E^*}{\partial b_k} = -(-i\omega)^{kq}, & k = 0\ldots m \end{array} \tag{4.82}$$

Let us also define, so as to simplify notation,

$$\left\{ \begin{array}{l} A(\omega) \overset{\text{def}}{=} |G(\omega)| \\[2mm] \varphi(\omega) \overset{\text{def}}{=} \arg[G(\omega)] \end{array} \right. \Rightarrow \left\{ \begin{array}{l} G(\omega) = A(\omega)e^{i\varphi(\omega)} \\[2mm] G^*(\omega) = A(\omega)e^{-i\varphi(\omega)} \end{array} \right. \tag{4.83}$$

Taking into account that

$$E^*(\omega) = G^*(i\omega)D^*(i\omega) - N^*(i\omega), \tag{4.84}$$

after a tedious, but not difficult manipulation, we get

$$\left\{ \begin{array}{l} \sum_{p=1}^{N_f} w(\omega_p)\left[A(\omega_p)\sum_{j=0}^{N_0}\left\{a_j C_{k-j}\omega_p^{j\alpha}\right\} - \sum_{j=0}^{M_0}\left\{b_j D_{k-j}\omega_p^{j\alpha}\right\}\right] = 0, \\[2mm] \quad k = 1\ldots N_0 \\[4mm] \sum_{p=1}^{N_f} w(\omega_p)\left[A(\omega_p)\sum_{j=0}^{N_0}\left\{a_j D_{k-j}\omega_p^{j\alpha}\right\} - \sum_{j=0}^{M_0}\left\{b_j C_{k-j}\omega_p^{j\alpha}\right\}\right] = 0, \\[2mm] \quad k = 0\ldots M_0 \end{array} \right. \tag{4.85}$$

where $C_{k-j} = \cos\left[\alpha(k-j)\frac{\pi}{2}\right]$ and $D_{k-j} = \cos\left[\varphi(\omega_p) + \alpha(j-k)\frac{\pi}{2}\right]$. As to the number of frequencies N_f that should be used, while in principle data from only one frequency should suffice to find a model, in practice (due to noise and unavoidable inaccuracies) N_f has to be larger, or the identified model will be poor. It is a good idea to spread frequencies ω_p over the range where the model has to be valid, and to concentrate them around those where the behaviour of the system changes.

A drawback of this method is that, as mentioned, N_0, M_0, and α are fixed in advance. Finding them along with the coefficients is not a linear problem. So, if they are not known, the best solution is to try several values and keep the best compromise between goodness of fit and complexity. Nelder-Mead's simplex search method and heuristic optimisation methods have been used for this.

An expression alternative to (4.85) can be found stacking, rather than summing, the contributions of the different frequencies, and then computing the pseudo-inverse to find the best possible solution for the resulting over-defined equation system [119, 121].

4.6 Filters

4.6.1 Generalities. Ideal filters

The designation of filter is derived from the ability of certain systems (not necessarily linear) to eliminate or attenuate strongly certain harmonics or frequency bands. In this case, the system acts as a frequency selector, letting some pass and retaining others. This is, for example, the case of the filter used for the tuning of radio emitters or receivers. To this type of system we give the name of *selectivity filter* or *selector*. The band where the filter practically does not change the signal is called *passband*, in opposition to the rejection band or *stopband*, where the signal is eliminated or strongly attenuated. In the case of non-ideal filters, there is a band between passband and stopband, called *transition band*.

However, there are other linear systems, also called filters, which are intended not to filter in the preceding sense, but to change the shape of the spectrum of a given signal: they are called *form filters*. The concept of form filter is extremely important in the modelling of certain signals found in practice. However, their study goes beyond the scope of this text, so we will not return to this topic.

We will pay attention exclusively to linear systems as selectors, that will be referred to only as filters. To understand the action of a filter we introduce the notion of ideal filter. This concept is extremely important in Signal Processing, even though an ideal filter has no physical realizability, and is always an acausal system. In the following figures the spectral diagrams of three ideal filter types are depicted. Usually only the amplitude diagram is shown. The phase is always assumed to be a linear function of the frequency. Since this type of phase only causes a delay in time, we will often assume that it is zero [61, 100].

The ideal filters are as follows:

–   Lowpass filter

Low frequency signals are passed, other frequencies are eliminated:

$$H_{LP}(\omega) = \begin{cases} 1 & |\omega| \leq \Omega_c \\ 0 & |\omega| > \Omega_c \end{cases} \tag{4.86}$$

Parameter Ω_c is called cut-off frequency. In real filters, it is the frequency at which the amplitude decreases 3 dB, for the reason stated in section 4.2.2. The impulse response is a sinc function:

$$h_{LP}((t) = \frac{\sin \Omega_c t}{\pi t} \tag{4.87}$$

This confirms that the ideal lowpass filter is an acausal filter, obviously not physically realisable.

–   Bandpass filter

Only signals with frequencies in a band are passed; the band is an interval that does not contain the origin:

$$H_{BP}(\omega) = \begin{cases} 1 & \Omega_L \leq |\omega| \leq \Omega_H \\ 0 & |\omega| < \Omega_L \wedge |\omega| > \Omega_H \end{cases} \tag{4.88}$$

The centre frequency is $\omega_c = \frac{\Omega_L + \Omega_H}{2}$. The corresponding impulse response is easily obtained attending to the properties of the Fourier transform. In fact, it is enough to note that a translation in frequency is obtained by multiplying the time function by a sinusoid. So, multiplying $h_{LP}(t)$ by $\cos(\omega_c t)$ we obtain the impulse response of a bandpass filter.

– Highpass filter
It is the reverse of a lowpass filter:

$$H_{HP}(\omega) = 1 - H_{LP}(\omega) \tag{4.89}$$

– Bandstop filter
It is the reverse of a bandpass filter:

$$H_{SB}(\omega) = 1 - H_{BP}(\omega) \tag{4.90}$$

In applications some further filters deserve to be considered:

– Notch filter
This is an extreme case of the bandstop filter, that rejects just one specific frequency:

$$H_{NF}(\omega) = (\omega^2 + \omega_0^2)G(\omega), \tag{4.91}$$

where $G(\omega)$ is the frequency response of any system.

– Comb filter
This filter has multiple regularly spaced narrow passbands. It can be considered as a sequence of notches.

– Allpass filter
It is a filter with constant amplitude spectrum. It only changes the phase. Every allpass filter consists in the product of pole-zero systems of the type

$$H_A(\omega) = \frac{s^\alpha - (-1)^\alpha p^*}{s^\alpha - p}. \tag{4.92}$$

In figure 4.10 we depict an overlap of the amplitude responses of different ideal and realisable filters. As seen, the difference is significative; the ideal serves as goal for designing practical filters. The design procedure is an optimisation problem with several constraints.

In figure 4.11, we represent the various parameters we take in consideration when designing a filter. We present the specifications for a low pass filter, but for the other kinds of filters they are similar.

– δ_p – band pass ripple
– $A_p = 20\log_{10}\left(\frac{1+\delta_p}{1-\delta_p}\right)$ – bandpass attenuation
– δ_s – amplitude of stop band ripple
– ω_c – pass band edge frequency
– ω_s – stop band edge frequency
– $\Delta\omega_i = \omega_s - \omega_c$ – transition band
– $A_s = -20\log_{10}(\delta_s)$ – stop band attenuation
– Roll-off—the rate at which attenuation increases beyond the cut-off frequency
– Order of the filter – degree of the denominator polynomial

Usually we fix the band pass (BP) amplitude equal to 1. The designed filter will have an amplitude that can have oscillations in the interval $[1 - \delta_p, 1 + \delta_p]$, where δ_p is the band pass ripple, which must be a small value.

4.6.2 Prototypes of integer order filters

The design of integer order filters is a well established theme. For now, there is no corresponding development for fractional filters. We are going to describe the most important versions of integer order filters [100]. These can be used as starting points for obtaining fractional versions. In this section we will consider lowpass filters with cut-off frequency ω_L; other types of filters will be addressed in the next section.

1. Butterworth filters

    Butterworth filters are all-pole LS, such that the corresponding frequency response has a maximally flat magnitude. Assuming that the cut-off frequency is 1 rad/s, a Butterworth filter of order N has as poles the roots of the equation:

$$(-s^2)^N + 1 = 0 \tag{4.93}$$

These roots are all in the left complex half plane, are uniformly spaced, and assume the form:

$$s_k = ie^{i\pi\frac{2k-1}{2N}}, \quad \begin{cases} k = 1, 2, \ldots, (N+1)/2 \text{ if } N \text{ is odd} \\ k = 1, 2, \ldots, N/2 \text{ if } N \text{ is even} \end{cases} \tag{4.94}$$

The amplitude spectrum is given by $\frac{1}{\sqrt{1+|\omega/\omega_C|^N}}$. The general formula for the TF is:

$$H(s) = \frac{1}{\left(1 + \frac{s}{\omega_C}\right)^P \prod_{k=1}^{\lfloor\frac{N}{2}\rfloor} \left\{1 + 2\sin\left[(2k-1)\frac{\pi}{2}N\right]\frac{s}{\omega_C} + \left(\frac{s}{\omega_C}\right)^2\right\}} \tag{4.95}$$

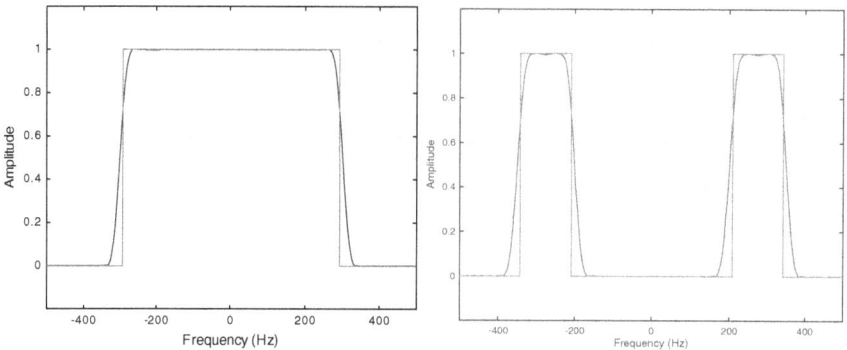

Fig. 4.10: Ideal and designed lowpass and bandpass filters.

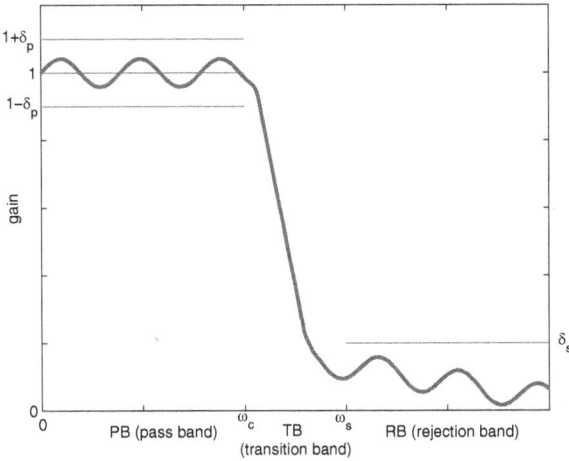

Fig. 4.11: Filter specifications, presented for the case of a lowpass filter.

where $\lfloor \frac{N}{2} \rfloor$ is the integer part of $\frac{N}{2}$ and $P = 0$ if N is even or $P = 1$ if N is odd. The denominators for orders $N = 1, 2, 3, 4$, e 5 are given by:

$$s + 1 \tag{4.96}$$

$$s^2 + \sqrt{2}s + 1 \tag{4.97}$$

$$(s^2 + s + 1)(s + 1) \tag{4.98}$$

$$(s^2 + 0.76537s + 1)(s^2 + 1.84776s + 1) \tag{4.99}$$

$$(s^2 + 0.61803s + 1)(s^2 + 1.61803s + 1)(s + 1) \tag{4.100}$$

2. Chebyshev filters

The good behaviour of Butterworth filters at low frequencies may not be compensated by the degradation of their characteristics in the rejection band. This means that, by changing the characteristics of the approach in the passband, we can improve the behaviour in the rejection band. Chebyshev filters minimise the maximum deviation relative to the ideal behaviour, to improve the rejection band. Let A_m be the ripple in the pass band (usually in dB). We define parameter ϵ by

$$\epsilon = \sqrt{10^{0.1A_m} - 1}. \tag{4.101}$$

A Chebyshev filter is also an all-pole TF given by $H(s) = K/P(s)$. The poles of a filter with order N lie on an ellipse and are given by

$$s_k = \sigma_k + i\omega_k \quad k = 0, 1, 2, 3, \ldots 2N - 1 \tag{4.102}$$

$$\sigma_k = \pm \sin \left[\frac{\pi}{2} \left(\frac{2k + 1}{n} \right) \right] \sinh \left(\frac{y}{N} \right) \tag{4.103}$$

$$\omega_k = \pm \cos \left[\frac{\pi}{2} \left(\frac{2k + 1}{n} \right) \right] \cosh \left(\frac{y}{N} \right) \tag{4.104}$$

$$\epsilon = \sinh(y) \tag{4.105}$$

To exemplify, consider the case $A_m = 1$ dB. For orders $N = 1, 2, 3, 4$ and 5, we obtain for K and $P(s)$:

K	Denominator
1.96523	$s + 1.96523$
0.98261	$s^2 + 1.09773s + 1.10251$
0.49130	$(s^2 + 0.49417s + 0.9942)(s + 0.49417)$
0.24565	$(s^2 + 0.27907s + 0.9865)(s^2 + 0.67374s + 0.2794)$
0.12283	$(s^2 + 0.17892s + 0.98831)(s^2 + 0.46841s + 0.4293)(s + 2.8949)$

3. Elliptic filters

   In the foregoing approximations, the attenuation in the rejection band grows much more than needed. This is a source of inefficiency. The elliptical or Cauer approach seeks to remedy this situation. It is the most common form of filter design. A distinctive feature of such filters is that they have attenuation poles in the rejection band. The elliptic approach leads to a rational function with finite poles and zeros. Let Ω be the quotient between the cut-off frequencies of the rejection and pass bands. For $\Omega = 3$ and $N = 2, 3, 4$ and 5, we obtain

Numerator	Denominator
$s^2 + 17.48528$	$s^2 + 1.35715s + 1.55532$
$s^2 + 11.82718$	$(s^2 + 0.58942s + 1.14559)(s + 0.65263)$
$(s^2 + 10.4554)(s^2 + 58.471)$	$(s^2 + 0.32979s + 1.063281)(s^2 + 0.86258s + 0.37787)$
$(s^2 + 10.4554)(s^2 + 58.471)$	$(s^2 + 0.21066s + 1.0351)(s^2 + 0.58441s + 0.496388)(s + 0.37452)$

4. Bessel filters

   Contrary to previous approaches, where attention has been paid to the amplitude only, this approach also pays attention to the phase and the delay. A Bessel filter is an all-pole TF, $H(s) = K/P(s)$, where the denominator is a Bessel polynomial:

$$P_N(s) = \sum_{k=0}^{N} a_k s^k \tag{4.106}$$

$$a_k = \frac{(2N - k)!}{2^{N-k} k! (N - k)!} \tag{4.107}$$

Next we present some of the polynomials and the corresponding constant K.

K	Denominator
1	$s + 1$
3	$s^2 + 3s + 3$
15	$(s^2 + 3.67782s + 6.45944)(s + 2.32219)$
105	$(s^2 + 5.79242s + 9.14013)(s^2 + 4.20758s + 11.4878)$
945	$(s^2 + 6.7039s + 14.2725)(s^2 + 4.64934s + 18.15631)(s + 3.64674)$

4.6.3 Transformations of frequencies

Till now, nothing has been said about other than lowpass filters. Let us plug the gap using frequency transformations. In this way, we can convert filters from a type to a different one [100].

Assume we have a lowpass filter with cut-off frequency W_c and want to obtain another one with cut-off frequency Ω_c. It is enough to make the substitution

$$s \Rightarrow \frac{W_c}{\Omega_c} s. \tag{4.108}$$

With this substitution, the new filter remains a lowpass, but with cut-off frequency Ω_c. In particular, the continuous-time lowpass filters of the previous section have a cut-off frequency equal to 1 rad/s; to obtain a filter with cut-off frequency f_C Hz or ω_C rad/s, simply replace s with $2\pi f_C s$ or $\omega_C s$.

Should we want to convert a lowpass into a highpass, the substitution to be made is

$$s \Rightarrow W_c / \Omega_c s^{-1}. \tag{4.109}$$

To pass from a low pass to a bandpass, the substitution is

$$s \Rightarrow W_c \frac{s^2 + \Omega_i \Omega_s}{s(\Omega_s - \Omega_i)}, \tag{4.110}$$

where Ω_i and Ω_s are the inferior and superior cut-off frequencies.

In an analogous way, the passage from a passband filter to a stopband is obtained with the substitution

$$s \Rightarrow W_c \frac{s(\Omega_s - \Omega_i)}{s^2 + \Omega_i \Omega_s}. \tag{4.111}$$

4.6.4 Fractional filters

The design of fractional filters is an open area. New possibilities of designing better filters are available with the introduction of the fractional orders. This allows a further degree of freedom, but the price to pay is a consequently enlarged search domain when performing a parameter optimisation, often carried out numerically. We can have some reduction in the computational burden by starting from one of the above described prototypes.

In general, fractional filters outperform the integer ones. By changing pole locations and finding a suitable order, a fractional filter can perform as a higher or lower order integer filter with some advantages, among which the main ones are:

–   that we can have a smaller quadratic error between the ideal filter taken as goal and the designed filter;
–   that this is frequently achieved with a smaller number of parameters.

Exercises

1. Find numerically the frequency responses of the following transfer functions and represent them in a Bode diagram:

   a) $F_1(s) = \dfrac{1}{s^{1/3} + 10}$

   b) $F_2(s) = \dfrac{1}{s^{1/3} + 100}$

   c) $F_3(s) = \dfrac{1}{(s + 10)^{1/3}}$

   d) $F_4(s) = \dfrac{1}{s^{4/3} + 10}$

   e) $F_5(s) = \dfrac{1}{s^{7/3} + 10}$

   f) $F_6(s) = \dfrac{1}{s^{10/3} + 10}$

2. Do the same for the transfer functions in exercise 3.1. Use a convolution to find again each transfer function's frequency response from its impulse response; see (2.1).

3. Find CRONE, Carlson's and Matsuda's approximations for the transfer functions in exercise 3.1. Approximations should be valid at least one decade on each side of the cut-off frequency.

4. Find numerically the frequency responses of Butterworth, Chebyshev, elliptic and Bessel filters with 2 poles and a 1 rad/s cut-off frequency. Plot them in a Bode diagram for comparison.

5 State-space representation

5.1 General considerations

We have studied three equivalent representations of linear continuous-time invariant systems: the differential equation, the impulse response, and the transfer function. If the system is variable in time, the differential equation will have time variant coefficients and we can still easily generalise the notion of impulse response, but this is no longer the case when a TF is used. On the other hand, these representations "forget" what is happening on the "inside" of a given system, since the system is considered a "black box" that relates an input with an output.

There is, however, a formulation of a LS in which this is not the case: the representation in state variables. The popularity of this description — also called internal representation — has increased due to the enormous development of the mathematical methods and associated computational algorithms. Its matrix formulation, in addition to being clear, allows the use of relatively simple computational techniques. On the other hand, it allows to formulate, in an identical way, a great number of different systems, such as non-linear, time-varying, or multivariate [10, 11, 35, 36, 119]

Describing systems in state space allows the disclosure of properties that remain invisible when the system is described by an input-output representation or by a frequency analysis. It should be noticed that this description has the drawback of eventually using state variables without any connection to physical reality. On the other hand, is more suitable for dealing with transient responses. However, in general, the disadvantages are insignificant when compared to the advantages.

To put it in other words, the dynamic of a system is described by its mathematical model. In a classical representation, this model describes, as said, an input-output relationship. In the present context, an intermediate variable appears: the state. A state represents the "memory" of the system, that is, the influence of the past in the future. Before we define a state more rigorously, let us give a mathematically more correct idea of what a system is.

A dynamic system is a mathematical concept defined by the following axioms:

1. Consider the following sets: set of instants $\mathcal{T}$, set of states $\mathcal{X}$, set of input signals $\mathcal{U}$, and set of output signals $\mathcal{Y}$.
2. $\mathcal{T}$ is an ordered subset of real numbers.
3. The elements of $\mathcal{U}$ and $\mathcal{Y}$ have bounded variations and are almost everywhere continuous.
4. Assume that the system is allowed to evolve from a state $x(\tau) \in X$ at an initial instant $\tau \in \mathcal{T}$, under the action of an input $v(t) \in \mathcal{U}$. Let $x(t)$ be the state of the system at instant $t \in \mathcal{T}$. Then, there exists a function ϕ that represents the state $x(t) = \phi(t, \tau, x(\tau), v) \in \mathcal{X}$ that has the following properties:

DOI 10.1515/9783110624588-005

(a) It is defined for all the values of $t > \tau$, but not necessarily for all values $t < \tau$ (time orientation).
(b) $\phi(t, t, x, u) = x(t)$, $t \in \mathcal{T}$, $x \in \mathcal{X}$, $u \in \mathcal{U}$ (consistency).
(c) For any three instants t_1, t_2, t_3 we have (composition):

$$\phi(t_3, t_1, x(t_1), u(t_1)) = \phi(t_3, t_2, \phi(t_2, t_1, x(t_1), u(t_1)), u(t_2)), \quad x \in X, \; u \in \mathcal{U} \tag{5.1}$$

Here, $u(t_1)$ is an input beginning at time t_1, and $u(t_2)$ is the input beginning at time t_2.
(d) If $u = u' \in \mathcal{U}$ in the interval $[\tau, t[$, then $\phi(t, \tau, x, u) = \phi(t, \tau, x, u')$ (causality).
5. The output representation $\psi : \mathcal{T} \times \mathcal{X} \times \mathcal{V} \in \mathcal{Y}$ defines the system output:

$$y(t) = \psi(t, x(t), u(t)) \tag{5.2}$$

Definition 5.1.1. *If the output $y(t)$ and state $x(t)$ at any instant t can be determined exactly from the state at $t_0 < t$ (initial state) and from the knowledge of the input in interval $[t_0, t[$, a system is said to be* deterministic.

If, on the other hand, the state and output at $t > t_0$ can only be determined with a given probability or by other statistical methods, the system is said to be stochastic.

In what follows we will consider deterministic systems. For them, if we know the state $x(t_1)$ in $t_1 > t_0$, this suffices to know the behaviour of the system for $t > t_1$. We can then say that the state summarises essential information about the past, necessary to determine the future behaviour of the system.

Remark 5.1.1. *It should be noted that different inputs can lead to the same state.*

5.2 Standard form of the equations

The set of equations describing the relations among input, state and output is constituted by two matrix equations:

$$T[\mathbf{x}(t)] = \phi[t, \mathbf{x}(t), \mathbf{v}(t)] \tag{5.3}$$

$$y(t) = \psi[t, \mathbf{x}(t), \mathbf{v}(t)] \tag{5.4}$$

where

$$T[\mathbf{x}(t)] = \frac{d^\alpha \mathbf{x}(t)}{dt^\alpha} \tag{5.5}$$

Equation (5.3) is called *state equation* or *dynamic equation*, and equation (5.4) is the *output equation* or *observation equation*. In the following, we will assume that the system at hand is linear, whereby ϕ and ψ must be linear functions. This leads to

$$T[\mathbf{x}(t)] = \mathbf{A}(t)\mathbf{x}(t) + \mathbf{B}(t)\mathbf{v}(t) \tag{5.6}$$

$$\mathbf{y}(t) = \mathbf{C}(t)\mathbf{x}(t) + \mathbf{D}(t)\mathbf{v}(t) \tag{5.7}$$

where $\mathbf{x}(t)$ is a $N \times 1$ vector, whereby the matrix $\mathbf{A}(t)$ in an $N \times N$ matrix. The dimensions of the other matrices are chosen in agreement with the type of system:

- SISO system:

$$T[\mathbf{x}(t)] = \underset{N \times N \text{ matrix}}{\mathbf{A}(t)} \; \underset{N \times 1 \text{ vector}}{\mathbf{x}(t)} + \underset{N \times 1 \text{ vector}}{\mathbf{B}(t)} \; \underset{\text{scalar}}{\mathbf{v}(t)} \qquad (5.8)$$

$$\mathbf{y}(t) = \underset{1 \times N \text{ vector}}{\mathbf{C}(t)} \; \underset{N \times 1 \text{ vector}}{\mathbf{x}(t)} + \underset{\text{scalar}}{\mathbf{D}(t)} \underset{\text{scalar}}{\mathbf{v}(t)} \qquad (5.9)$$

- MIMO system with n_v inputs and n_y outputs:

$$T[\mathbf{x}(t)] = \underset{N \times N \text{ matrix}}{\mathbf{A}(t)} \; \underset{N \times 1 \text{ vector}}{\mathbf{x}(t)} + \underset{N \times n_v \text{ matrix}}{\mathbf{B}(t)} \; \underset{n_v \times 1 \text{ vector}}{\mathbf{v}(t)} \qquad (5.10)$$

$$\mathbf{y}(t) = \underset{n_y \times N \text{ matrix}}{\mathbf{C}(t)} \; \underset{N \times 1 \text{ vector}}{\mathbf{x}(t)} + \underset{n_y \times n_v \text{ matrix}}{\mathbf{D}(t)} \; \underset{n_v \times 1 \text{ vector}}{\mathbf{v}(t)} \qquad (5.11)$$

The above equations represent the standard form of expressing a linear system in *state variables*, which are the elements of the state $\mathbf{x}(t)$. Because the state is in fact a vector, it is also known as *state vector*.

Definition 5.2.1. *A system is* time invariant *if matrices* **A**, **B**, **C**, *and* **D** *are constant. These systems are also called* autonomous.

Other interesting notions:
- The space generated by all linear combinations of the state components is called *state space*. The temporal evolution of state in the state space can be of great use in the study of systems, even nonlinear ones.
- When matrix **A** assumes a Jordan form, the system is said to be expressed in the *normal form*.
- State variables do not form a unique set. If $\mathbf{x}(t)$ is a $N \times 1$ vector with the system state variables, and **P** is a $N \times N$ invertible matrix, then the variables in vector $\mathbf{w}(t) = \mathbf{P}\mathbf{x}(t)$ are also state variables.

The use of the usual symbology of the block diagrams allows us to obtain a general representation of (5.6)–(5.7). Each matrix is represented by a block. A new symbol is introduced to represent the fractional integration. The graphic representation of the linear system can be seen in figure 5.1 where the parallel lines indicate that the quantities flowing therein are vectorial.

In the following, we will assume that we are dealing with a linear time invariant system, formally introduced below.

Definition 5.2.2 (State-space representation of a LTI system). *Given a MIMO LTI system, its state-space representation is*

$$D^{\alpha}\mathbf{x}(t) = \mathbf{A}\mathbf{x}(t) + \mathbf{B}\mathbf{v}(t) \qquad (5.12)$$

$$\mathbf{y}(t) = \mathbf{C}\mathbf{x}(t) + \mathbf{D}\mathbf{v}(t) \qquad (5.13)$$

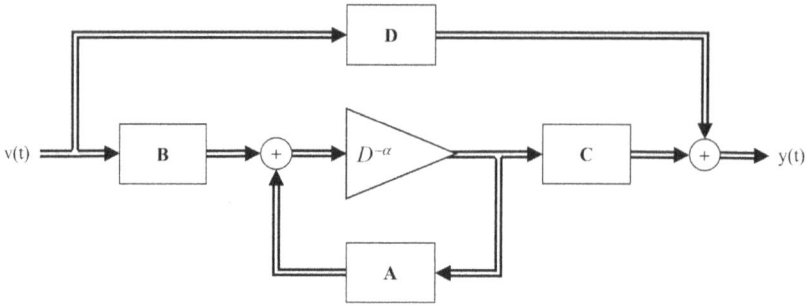

Fig. 5.1: Block representation of a state variables system

where $\boldsymbol{\alpha} = [\alpha_1 \; \alpha_2 \; \ldots \alpha_N]^T$ is a vector with (positive) differentiation orders, and the derivative $D^{\boldsymbol{\alpha}}$ of vectorial order $\boldsymbol{\alpha}$ is defined as

$$D^{\boldsymbol{\alpha}}\mathbf{x}(t) = \left[D^{\alpha_1}x_1(t) \; D^{\alpha_2}x_2(t) \; \ldots \; D^{\alpha_n}x_n(t)\right]^T \tag{5.14}$$

Define the diagonal matrix $\mathrm{diag}\,(s^{\boldsymbol{\alpha}})$ with diagonal elements s^{α_k}, $k = 1, 2, \cdots, N$. The Laplace transformation of (5.12)-(5.13) is

$$\mathrm{diag}\,(s^{\boldsymbol{\alpha}})\,\mathbf{X}(s) = \mathbf{A}\mathbf{X}(s) + \mathbf{B}\mathbf{U}(s) \tag{5.15}$$

$$\mathbf{Y}(s) = \mathbf{C}\mathbf{X}(s) + \mathbf{D}\mathbf{U}(s) \tag{5.16}$$

From (5.15),

$$\mathbf{X}(s) = (\mathrm{diag}\,(s^{\boldsymbol{\alpha}}) - \mathbf{A})^{-1}\mathbf{B}\mathbf{U}(s) \tag{5.17}$$

and replacing this in (5.16) we get

$$\frac{\mathbf{Y}(s)}{\mathbf{U}(s)} = \mathbf{C}\left[\mathrm{diag}\,(s^{\boldsymbol{\alpha}}) - \mathbf{A}\right]^{-1}\mathbf{B} + \mathbf{D}, \tag{5.18}$$

which is a MIMO fractional transfer function matrix.

Definition 5.2.3 (Commensurable state-space representation). *The state space representation is said to be commensurable when all the orders* α_k, $k = 1, 2 \ldots n$ *are equal, in which case the diagonal matrix in (5.18) can be written thus:*

$$\mathrm{diag}\,(s^{\boldsymbol{\alpha}}) = s^{\alpha}\mathbf{I}. \tag{5.19}$$

5.3 State transition operator

Let us go back to the dynamic equation

$$D^{\boldsymbol{\alpha}}\mathbf{x}(t) = \mathbf{A}\mathbf{x}(t) + \mathbf{B}\mathbf{u}(t) \tag{5.20}$$

and assume that the input is null for $t > 0$. Using the results introduced in 3.4, we can write

$$\mathbf{diag}\,(s^{\alpha})\,\mathbf{X}(s) - \mathrm{diag}\,(s^{\alpha-1})\,\mathbf{x}(0) = \mathbf{A}\mathbf{X}(s). \tag{5.21}$$

Therefore, the transition operator verifies

$$\mathbf{x}(t) = \Phi(0,t) \cdot \mathbf{x}(0) \tag{5.22}$$

and is given by the inverse LT

$$\Phi(0,t) = \mathcal{L}^{-1}\left\{[\mathrm{diag}\,(s^{\alpha}) - \mathbf{A}]^{-1} \cdot \mathrm{diag}\,(s^{\alpha-1})\right\}. \tag{5.23}$$

Although there is a closed form for such inverse, we will not present it, since it is a bit involved [34, 35]. In the commensurate case, such inverse is nothing else than the multidimensional MLF

$$\Phi(0,t) = \sum_{0}^{\infty} \mathbf{A}^{n} \frac{t^{n\alpha}}{\Gamma(n\alpha+1)} \varepsilon(t) \tag{5.24}$$

and leads easily to the general state transition operator.

Definition 5.3.1 (State transition operator). *The states at two instants t and τ are related by the **state transition operator** $\Phi(\tau,t)$:*

$$\mathbf{x}(t) = \Phi(\tau,t) \cdot \mathbf{x}(\tau), \tag{5.25}$$

The state transition operator is defined by

$$\Phi(\tau,t) = \Phi(0,t) \cdot \Phi^{-1}(0,\tau). \tag{5.26}$$

with $\Phi(0,0) = \mathbf{I}$.

It can be shown that this operator verifies the usual properties, namely the semi-group property. To finish this study we only need to point out how we can compute $\Phi^{-1}(0,\tau)$. If we write

$$\Phi^{-1}(0,\tau) = \sum_{0}^{\infty} \mathbf{B}_{n} \frac{\tau^{n\alpha}}{\Gamma(n\alpha+1)} \varepsilon(\tau) \tag{5.27}$$

we can show that $\mathbf{B}_{0} = \mathbf{I}$ and

$$\mathbf{B}_{n} = \sum_{k=0}^{n} \binom{n\alpha}{k\alpha} \mathbf{A}^{k} \mathbf{B}_{n-k}, \quad n = 1, 2, \cdots \tag{5.28}$$

allowing us to compute $\Phi^{-1}(0,\tau)$ in a recursive way.

5.4 Canonical representations of SISO systems

Because the state of a system is not unique, as seen in section 5.2, a SISO commensurable transfer function given by

$$\frac{Y(s)}{U(s)} = \frac{b_0 + b_1 s^\alpha + b_2 s^{2\alpha} + \ldots + b_N s^{N\alpha}}{a_0 + a_1 s^\alpha + a_2 s^{2\alpha} + \ldots + s^{N\alpha}} \tag{5.29}$$

has infinite state-space representations, but some are particularly useful and have special names.

Definition 5.4.1. *The* controllable canonical form *of (5.29) is*

$$
s^\alpha
\left\{
\begin{bmatrix}
X_1 \\ X_2 \\ \vdots \\ X_{n-1} \\ X_n
\end{bmatrix}
=
\overbrace{
\begin{bmatrix}
0 & 1 & 0 & \cdots & 0 \\
0 & 0 & 1 & \cdots & 0 \\
\vdots & \vdots & \vdots & \ddots & \vdots \\
0 & 0 & 0 & \cdots & 1 \\
-a_0 & -a_1 & -a_2 & \cdots & -a_{n-1}
\end{bmatrix}
}^{A}
\begin{bmatrix}
X_1 \\ X_2 \\ \vdots \\ X_{n-1} \\ X_n
\end{bmatrix}
+
\overbrace{
\begin{bmatrix}
0 \\ 0 \\ \vdots \\ 0 \\ 1
\end{bmatrix}
}^{B}
U
\right.
$$

$$
Y = \underbrace{\begin{bmatrix} b_0 - a_0 b_n & b_1 - a_1 b_n & \cdots & b_{n-1} - a_{n-1} b_n \end{bmatrix}}_{C}
\begin{bmatrix}
X_1 \\ X_2 \\ \vdots \\ X_n
\end{bmatrix}
+ \underbrace{b_n}_{D} U
\tag{5.30}
$$

Definition 5.4.2. *The* observable canonical form *of (5.29) is*

$$
s^\alpha
\left\{
\begin{bmatrix}
X_1 \\ X_2 \\ \vdots \\ X_{n-1} \\ X_n
\end{bmatrix}
=
\overbrace{
\begin{bmatrix}
0 & 0 & \cdots & 0 & -a_0 \\
1 & 0 & \cdots & 0 & -a_1 \\
\vdots & \vdots & \vdots & \ddots & \vdots \\
0 & 0 & \cdots & 1 & -a_{n-1}
\end{bmatrix}
}^{A}
\begin{bmatrix}
X_1 \\ X_2 \\ \vdots \\ X_n
\end{bmatrix}
+
\overbrace{
\begin{bmatrix}
b_0 - a_0 b_n \\ b_1 - a_1 b_n \\ \cdots \\ b_{n-1} - a_{n-1} b_n
\end{bmatrix}
}^{B}
U
\right.
$$

$$
Y = \underbrace{\begin{bmatrix} 0 & 0 & \cdots & 0 & 1 \end{bmatrix}}_{C}
\begin{bmatrix}
X_1 \\ X_2 \\ \cdots \\ X_{n-1} \\ X_n
\end{bmatrix}
+ \underbrace{b_n}_{D} U
\tag{5.31}
$$

Remark 5.4.1. *There are other interesting representations, namely the* diagonal canonical form *and the* Jordan canonical form.

All the state-space representations of a given system have state matrices **A** with the same eigenvalues.

Remark 5.4.2. *The roots of the polynomial* $A(p) = p^n + \sum_{k=0}^{n-1} a_k p^k$, *built with the denominator coefficients of transfer function $G(s)$, given by (5.18), are the eigenvalues of matrix* **A** *of any of the state-space representations of a commensurate $G(s)$.*

Exercises

1. Find the canonical controllable and observable representations of the transfer functions in exercise 3.1.
2. Prove the statement in Remark 5.4.2. *Hint:* start by showing that matrixes **A** of the canonical and controllable canonical forms have the same eigenvalues.

6 Feedback representation

6.1 Why feedback?

In chapter 1, we introduced the notion of system, using the most general definition, and presented also three different ways of associating systems. One of them was feedback, which is such a very important way of combining systems. We present in this chapter an additional study of this interconnection structure [36, 66, 95, 119].

Consider a tank fed through a tap with a given flow rate c_i, e.g. in L/s. Suppose that the outlet valve allows a flow c_o L/s. If $c_i \leq c_o$, all the water that enters the tank flows out and nothing happens: the tank is only a place of passage. No accumulation takes place. However, if $c_o < c_i$, the water accumulates in the tank till it is full. After that, the water overflows and is lost. To avoid this, we need either someone manipulating the faucet to reduce the flow, or an automatic mechanism producing the same action as shown in figure 6.1.

This kind of device can be seen in our homes. In the bathroom, there is a cistern which has a float that drives the inlet valve. As the water level rises in the tank, the valve is progressively closed, reducing the flow of water entering the tank. When the level reaches a pre-specified height, the flow is cut off.

This is a very simple example of a feedback system. Let us present a more suitable description. Assume that our input is the desired level — the reference input $r(t)$. The float measures the actual level — which is the system output $y(t)$. As long as the difference between the desired and actual output is not null, the faucet continues to pour water into the cistern, stoping when $e(t) = r(t) - y(t) = 0$. In figure 6.2 we describe schematically the system.

A more general situation is depicted in figure 6.3, where another block is shown, normally called *controller* [11, 95, 119]. This block is a system designed to lead system

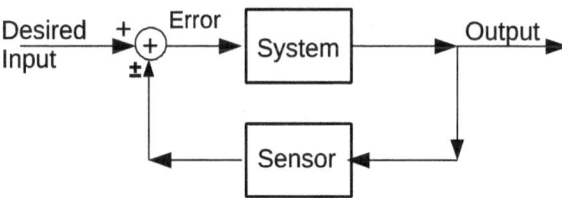

Fig. 6.1: Block representation of a feedback system.

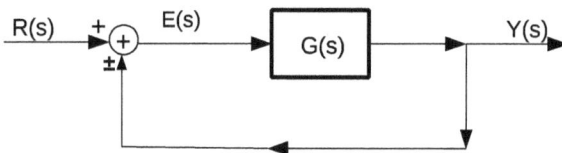

Fig. 6.2: Block representation of a simple feedback system.

DOI 10.1515/9783110624588-006

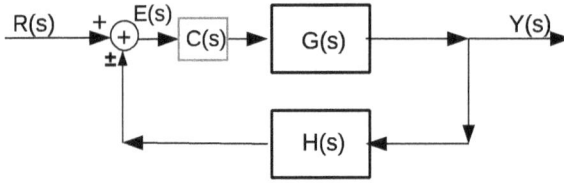

Fig. 6.3: Block representation of a feedback control system.

$G(s)$ (frequently called *plant*) to produce the output $y(t)$ we want. In other words, $C(s)$ is designed so that the output $y(t)$ will follow the reference $r(t)$ and the error $e(t)$ is zero.

The smaller system with input $e(t)$ and output $y(t)$ is called *open loop*, to distinguish it from the feedback system that is in *closed loop*. The open loop transfer function is

$$\frac{Y(s)}{E(s)} = C(s)G(s) \tag{6.1}$$

From

$$Y(s) = C(s)G(s)E(s) \tag{6.2}$$

$$E(s) = R(s) \pm H(s)Y(s), \tag{6.3}$$

the closed loop system has transfer function

$$\frac{Y(s)}{R(s)} = \frac{C(s)G(s)}{1 \mp C(s)G(s)H(s)} \tag{6.4}$$

Notice that figure 6.3 allows for a *positive feedback*, corresponding to the sign + in the sum point on the left. In this case the input of the controller is no longer an error $e(t) = r(t) - y(t)$, but rather the sum $y(t) + r(t)$. When the − sign is employed in the sum point, we then have *negative feedback*. We will, in all that follows, consider negative feedback only. It is intuitive from the discussion above that negative feedback is far more useful for control purposes. A more detailed discussion is outside the scope of this book. In any case, it is not very difficult to redo the following developments for the positive feedback case, since results are rather similar.

6.2 On controllers

A controller is a tool that physically monitors and alters the operations of a dynamic system. This apparatus can be implemented using different technological means: mechanical, hydraulic, pneumatic, or electronic. However, nowadays, the vast majority of controllers are implemented using microprocessors and computers. Some simple examples of controllers:

– The Watt centrifugal regulator;
– The thermostat of a heater;

- The autopilot in a plane;
- The pH control at a water treatment plant.

There are many possible mathematical laws that can be used in controllers. For example, the thermostat belongs to the class of so-called "on-off" controllers, with an output that can only assume two values (and is consequently related to the controller input, the error $e(t)$, through a non-linear relation). However, the proportional-integral (PI) and the proportional-integral-derivative (PID) controllers are the most widely used, since they are good enough for a vast number of applications, and enjoy some degree of simplicity. Indeed, in many applications of modern industry, because of the complexity of the control systems, computers are needed to implement advanced control algorithms, while PID controllers can be implemented in a variety of ways including simple electronic components.

PID controllers have an output which is continuous and depends linearly on the controller input, given by the transfer function

$$C(s) = K_p \left(1 + \frac{1}{T_i s} + T_d s\right), \quad K_p, T_i, T_d, \lambda, \mu > 0. \tag{6.5}$$

The transfer function comprises three parts. The first part provides an output proportional to the closed loop error:

$$u(t) = K_p e(t) \tag{6.6}$$

This makes sense because larger errors are expected to require larger control actions to be corrected. However, small errors will lead to small control actions and this may cause practical problems, such as a too slow decrease of the error as it approaches zero. So, to make the output $y(t)$ converge better to the reference $r(t)$ in such cases, the PID provides a second part of the control action:

$$u(t) = \frac{K_p}{T_i} \int e(t) \, dt \tag{6.7}$$

This integral control action will also be able to eliminate steady-state errors in some cases (a detailed analysis of this effect falls outside the scope of this book). The third part of the control action attempts to make the controller respond faster when there is a large increase of the error (measured by its derivative):

$$u(t) = K_p T_d \frac{de(t)}{dt} \tag{6.8}$$

This derivative action, however, in the presence of noise (especially high frequency noise), risks resulting in an erratic control action. Because of this, the derivative action is often omitted, resulting in a PI controller. When the derivative control action exists, it is implemented in practice together with a lowpass filter.

Fractional PID controllers are generalisations of PID controllers and their two possible transfer functions have already been introduced in section 4.3.2, together with their frequency response:

$$G_5(s) = K_p \left(1 + \frac{1}{(T_i s)^\lambda} + (T_d s)^\mu \right), \quad K_p, T_i, T_d, \lambda, \mu > 0 \tag{6.9}$$

$$G_6(s) = K_p \left(1 + \frac{1}{T_i s} \right)^\lambda (1 + T_d s)^\mu, \quad K_p, T_i, T_d, \lambda, \mu > 0 \tag{6.10}$$

Both reduce to (6.5) when $\lambda = \mu = 0$, and both implement a three-part control action: there is a proportional part, a part that grows with accumulated error, and a part that grows with error fluctuations.

6.3 The Nyquist stability criterion

Let us consider the problem of knowing whether or not a closed loop is stable. For this purpose, we need to find the locations of the zeroes of the denominator $D(s)$ of the closed loop transfer function (6.4) function, which is $D(s) = 1 + C(s)G(s)H(s)$.

In chapter 3 we saw the relation between the location of the pseudo-poles of a system and its stability. The Matignon theorem is often used as a criterion of stability for commensurate systems. As referred in section 3.2, a system is unstable if it has at least a pseudo-pole in the region delimited by the contour in figure 6.4, defined as $\mathfrak{R}_m = \{s \in \mathbb{C} : |\arg(s)| < \frac{\pi}{2}\alpha\}$, where α is the commensurate order. The contour consists in two half-straight lines defined by angles $\pm\frac{\pi}{2}\alpha$ in relation to the positive half real axis, and is closed by an arc of infinite radius.

Remember that any pole of the system must be given by $p^{\frac{1}{\alpha}}$, where p is a pseudo-pole. Therefore, by showing that there are no pseudo-poles of the closed loop transfer function (6.4) in $\mathfrak{R}_m$, we are actually showing that its denominator $D(s)$ has no roots in the region $\mathfrak{R} = \{s \in \mathbb{C} : |\arg(s)| < \frac{\pi}{2}\}$, independently of order α. This region $\mathfrak{R}$ is delimited by the contour γ depicted in figure 6.5. The idea is to formulate a criterion to test the stability of the feedback system, without calculating explicitly the closed loop transfer function (6.4). The result is the Nyquist stability criterion, which is based on *Cauchy's argument principle* [11, 95, 119]:

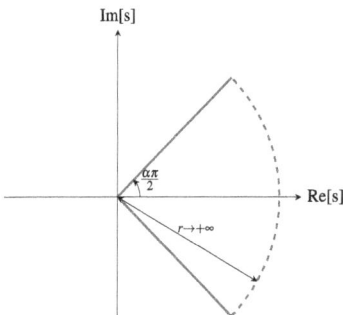

Fig. 6.4: Complex contour used to build the Nyquist diagram referring to the Matignon theorem.

Definition 6.3.1 (Cauchy's argument principle). *Consider a function $F(s)$, $s \in \mathbb{C}$, which is analytic on and inside a complex contour γ, except for a finite number of points. Let Z and P be respectively the number of zeros and poles of $F(s)$ that lie inside contour γ. As variable s travels the contour γ, in a given direction, its image $F(s)$ will travel a contour $F(\gamma)$, that travels around the origin a number of turns N equal to*

$$N = Z - P \tag{6.11}$$

To verify the stability of transfer function

$$F(s) = \frac{\sum\limits_{k=0}^{M} b_k s^{\alpha k}}{\sum\limits_{k=0}^{N} a_k s^{\alpha k}} \tag{6.12}$$

we have to map the complex contour given by

$$\rho = \lim_{r \to +\infty} \{s \in \mathbb{C} : s = i\omega, \ \omega \in [-r, r]\} \cup \{s \in \mathbb{C} : s = r e^{i\theta}, \ \theta \in [-\tfrac{\pi}{2}, +\tfrac{\pi}{2}]\} \tag{6.13}$$

or, in alternative, we use the related integer transfer function

$$F_{int}(z) = \frac{\sum\limits_{k=0}^{M} b_k z^k}{\sum\limits_{k=0}^{N} a_k z^k} \tag{6.14}$$

to map the contour given by

$$\rho_{frac} = \lim_{r \to +\infty} \{z \subset \mathbb{C} : z = w e^{\pm i a \pi / 2}, \ w \in [0, r]\} \cup \{z \in \mathbb{C} : z = r e^{i\theta}, \ \theta \in [-\tfrac{a\pi}{2}, +\tfrac{a\pi}{2}]\} \tag{6.15}$$

which results in exactly the same. Both contours are shown in Figure 6.5. If there are poles of (6.12) on contour ρ (or pseudo-poles of (6.14), which are poles of (6.12), on contour ρ_{frac}), circular indentations around them must be added to the contour, as seen in Figure 6.5 as well. The image of the contour is the *Nyquist diagram*, that we mentioned in chapter 4.

Let us now study the stability of a closed loop transfer function given by (6.4), using its poles and contour γ. We know P: it is the number of poles of $C(s)G(s)H(s)$ on $\mathbb{R}$ (the unstable poles). Therefore, we only need to compute N. To do it, we remark that, instead of computing the number of turns of $D(\gamma)$ around the origin, we can compute the number of turns of $C(s)G(s)H(s)$ around -1: clockwise encirclements count as positive, counterclockwise encirclements as negative. Consequently, the closed-loop TF $\frac{G(s)}{1+C(s)G(s)H(s)}$ will be stable if and only if $N = -P$. This is a very interesting result, since it shows that the feedback can be used to obtain stable systems from unstable ones.

Should we rather use contour γ_{frac}, the pseudo-poles of (6.4) are used instead.

Suppose now that the gain of the open loop changes. This may be because a proportional controller $C(s) = k$ is used, or because the controller $C(s)$ has fixed

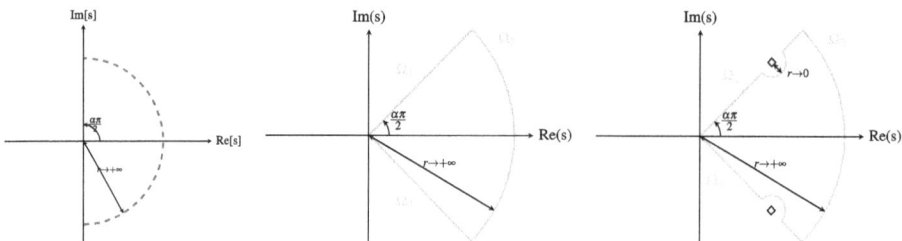

Fig. 6.5: Complex contours used to build the Nyquist diagram. Left: contour ρ (6.13); centre: contour ρ_{frac} (6.15); right: indentations around poles on a contour.

zeros and poles but its gain k may vary, or because the gain of the plant $G(s)$ or the sensor $H(s)$ may vary. In any of these cases, all points in the Nyquist diagram will be multiplied by this constant k. We can study the stability of the closed loop by counting the turns no longer around -1, but rather around any point in the real axis, given by $-1/k$, as seen in the following example.

Example 6.3.1. *The pseudo-zeros and pseudo-poles of plant*

$$G(s) = \frac{s - s^{\frac{1}{2}} - 6}{s^{\frac{1}{2}}(s+4)}, \tag{6.16}$$

are the roots of polynomials $\lambda^2 - \lambda - 6$ (i.e. the zeros are $\lambda = 3$ and $\lambda = -2$) and $\lambda^3 + 4\lambda$ (i.e. the poles are $\lambda = 0$ and $\lambda = \pm 2i$). Figure 6.6 shows on the top the corresponding map of poles and zeros, and a contour to build the Nyquist diagram. The Matignon theorem shows that the system is marginally stable in open-loop, since two of the roots of $\lambda^3 + 4\lambda$, $\lambda = \pm 2i$, fall outside the contour, and $\lambda = 0$ falls on the border. Even though this contour has arcs with finite radii, it suffices to plot the Nyquist diagram on the bottom of Figure 6.6, clearly showing that the diagram, that diverges to infinity, must be closed on the left. From here, the following table can be drawn, to see if a closed loop consisting of $G(s)$ and a gain k will be stable:

	$]-\infty, 0[$	$]0, 0.23[$	$]0.23, +\infty[$
$-\frac{1}{k}$			
k	$]0, +\infty[$	$]-\infty, -4.35[$	$]-4.35, 0[$
N	1	2	0
P	0	0	0
Stable in closed-loop	**No**	**No**	**Yes**

In fact, the closed loop will be stable only if $-4.35 < k < 0$, as can be easily verified computing its TF and using Matignon theorem.

6.3.1 Drawing the Nyquist diagram

To draw the Nyquist diagram of the open loop $C(s)G(s)H(s)$, which is the easiest way of applying the Nyquist criterion to study the stability of the corresponding closed loop, the following considerations should be taken into account:

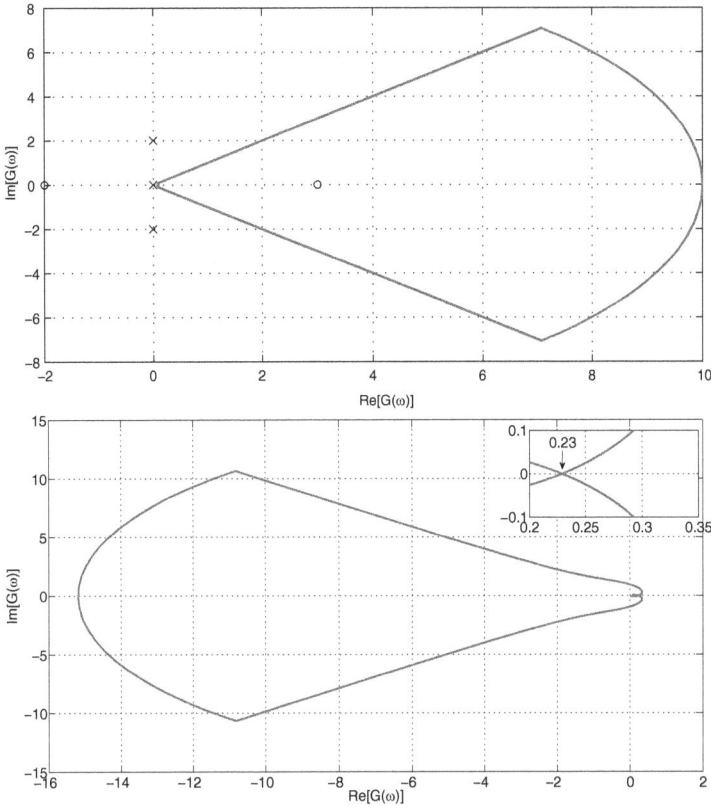

Fig. 6.6: Top: pole-zero map of example 6.3.1, and complex contour for the Nyquist diagram; bottom: the Nyquist diagram obtained with the contour on the top.

1. The image of the upper half straight line of the Nyquist contour, $s = \omega e^{ia\pi/2}$, $\omega > 0$, corresponds to the frequency response of $G(s)H(s)$. The determination of the frequency response can be subdivided into the following tasks:
   (a) Calculate the static gain, $\lim_{\omega \to 0} C(i\omega)G(i\omega)H(i\omega)$;
   (b) Calculate the high frequency response, $\lim_{\omega \to \infty} C(i\omega)G(i\omega)H(i\omega)$;
   (c) Obtain the real axis crossing points, $Im[C(i\omega)G(i\omega)H(i\omega)] = 0$;
   (d) Obtain the imaginary axis crossing points, $Re[C(i\omega)G(i\omega)H(i\omega)] = 0$;
   (e) For all frequencies ω:
   - The absolute value (amplitude spectrum), $|C(i\omega)G(i\omega)H(i\omega)|$, corresponds to the distance from the origin of the complex plane to the Nyquist diagram.
   - The phase spectrum, $\arg\{C(i\omega)G(i\omega)H(i\omega)\}$, corresponds to the angular position of the Nyquist diagram in relation to the real axis.
2. The diagram with the graphical representation of the frequency response in the complex plane is known as a polar diagram;

3. The image of the lower half straight line of the Nyquist contour, $s = \omega e^{-ia\pi/2}$, $\omega > 0$, is the complex conjugate of the polar diagram. Its graphical representation can be obtained by the reflection of the polar diagram that uses the real axis as axis of symmetry;

4. The image of the curve with radius $r \to \infty$ consists of only one point, $\lim_{s\to\infty} C(s)G(s)H(s)$. In the most usual case, where the open loop transfer function has more poles than zeros, the resulting point is the origin of the complex plane;

5. If $C(s)G(s)H(s)$ has poles on the imaginary axis, it is necessary to modify the Nyquist contour with indentations, as we referred above. At these points, the Nyquist plot increases to infinity, and the corresponding asymptotes must be calculated.

6.3.2 Gain and phase margins

The Nyquist criterion has another interesting feature: *it allows us to discover how far from stability (or from unstability) we are*. This can be done through two new quantities: the gain and phase margins, which give the degree of relative stability. As seen above, point $-1 = e^{\pm i\pi}$ is fundamental for stability analysis. It gives us two references: gain = 1 (or 0 dB) and phase π (or $-\pi$). Therefore, at any point of the Nyquist diagram we can measure how far we are from those quantities. We proceed as follows [11, 95, 119]:

1. The *phase crossover frequency*, ω_{cp}, is the frequency corresponding to a phase of $\pm\pi$. Some systems have more than one phase crossover frequency. Others have none.

2. The *gain margin* (GM) is found from the amplitude spectrum at the phase crossover frequency as

$$GM = -|C(\omega_{cp})G(\omega_{cp})H(\omega_{cp})|_{dB}. \tag{6.17}$$

3. The *gain crossover frequency*, ω_{cg}, is the frequency corresponding to an amplitude equal to 1, or, which is the same, to an amplitude of 0 dB. Some systems have more than one gain crossover frequency. Others have none.

4. The *phase margin* (PM) is found from the phase spectrum at the gain crossover frequency as

$$PM = \pi + \arg\left[C(\omega_{cg})G(\omega_{cg})H(\omega_{cg})\right]. \tag{6.18}$$

The gain margin and the phase margin are known as *stability margins*. From the above, it is clear that they can be read in the Nyquist plot as follows:

1. Let g be a point in the complex plane where the Nyquist plot crosses the negative real axis. Then $GM = -20\log_{10}|g|$.

2. Let θ be the argument of a point in the complex plane where the Nyquist plot crosses a unit radius circle centred at the origin. Then $PM = \pi - \theta$.

See figure 6.7.

From figure 6.7 it is also clear that, because there should be no encirclements of point -1, the closed loop system will be stable if both margins are positive, and unstable when both are negative. No conclusions can be taken when one of the margins is positive and the other negative, or when there are several crossover frequencies. Example 6.3.2 and Figure 6.8 show one example of each possible case.

Example 6.3.2. *There are four possible cases for the values GM and PM may assume. The Bode diagrams of the examples below for these four cases are given in Figure 6.8.*

1. *When both GM> 0 and PM> 0, the closed loop is stable. Consider for instance*

$$G_1(s) = \frac{10}{s+1}, \tag{6.19}$$

for which PM> 0. As its phase never goes down to −180°, the gain margin is measured at frequency $\omega_{cp} = +\infty$, and since the phase decreases to −∞ dB then GM=+∞ dB. So the corresponding closed loop should be stable, and in fact $\frac{10}{s+11}$ is stable.

2. *When both GM< 0 and PM< 0, the closed loop is unstable. That is the case, for instance, of*

$$G_2(s) = \frac{3}{(s+0.1)^3}. \tag{6.20}$$

So the corresponding closed loop should be unstable, and in fact $\frac{3}{s^3+0.3s^2+0.03s+3.001}$ is unstable, as can be seen using the Routh-Hurwitz criterion.

3. *When GM> 0 and PM< 0, or GM< 0 and PM> 0, nothing can be said about stability. That is the case, for instance, of*

$$G_3(s) = \frac{0.05(s+10)^3}{(s+0.1)^3}, \quad G_4(s) = \frac{0.5(s+10)^3}{(s+0.1)^3}. \tag{6.21}$$

In both cases the two margins have opposite signs. The closed loop for $G_3(s)$, which is $\dfrac{0.05s^3+1.5s^2+15s+50}{1.05s^3+1.8s^2+15.03s+50}$, happens to be unstable, but it might just as well be stable. The closed loop for $G_4(s)$, which is $\dfrac{0.5s^3+15s^2+150s+500}{1.5s^3+15.3s^2+150s+500}$, happens to be stable, but it might just as well be unstable.

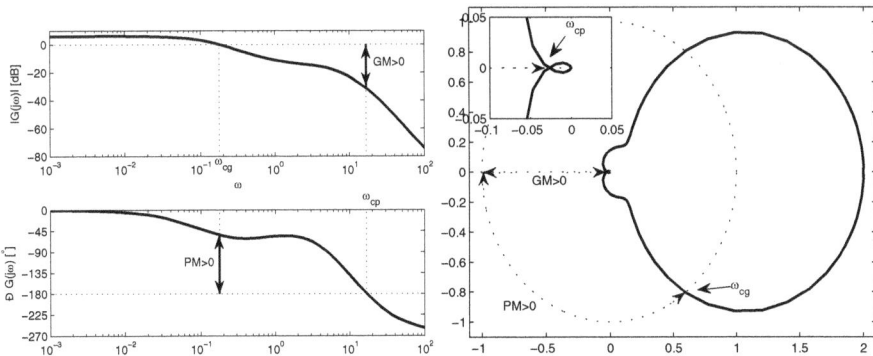

Fig. 6.7: Stability margins in the Bode diagram and in the Nyquist diagram.

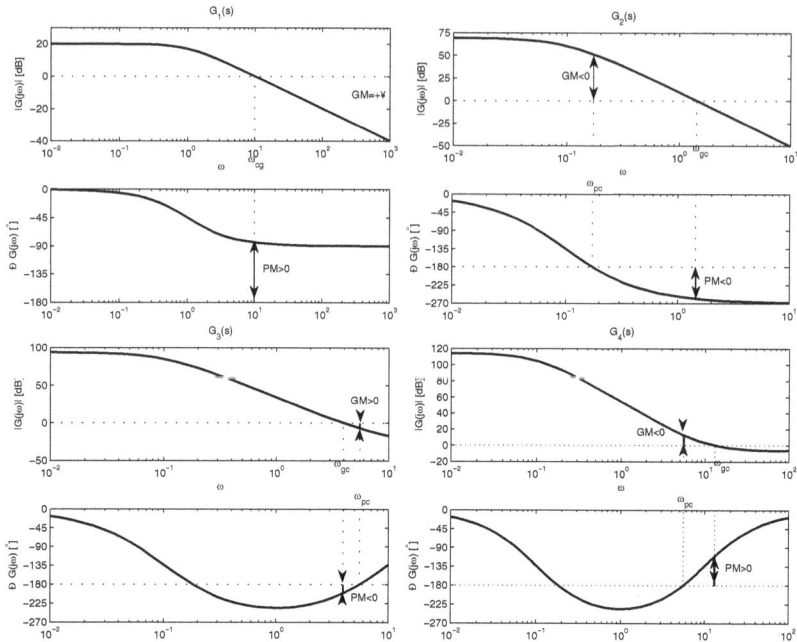

Fig. 6.8: The four open loop plants from example 6.3.2.

Exercises

1. Draw the Nyquist diagrams of all the systems studied in chapter 4.
2. Draw the Nyquist diagrams of the following plants:

    a) $F_1(s) = \dfrac{1}{s^{1/3} + 10}$

    b) $F_2(s) = \dfrac{1}{s(s^{1/3} + 10)}$

    c) $F_3(s) = \dfrac{1}{s^2 + s + 1}$

    d) $F_4(s) = \dfrac{1}{s(s^2 + s + 1)}$

    e) $F_5(s) = \dfrac{s - 2}{s^2 + s + 1}$

    f) $F_6(s) = \dfrac{s - 2}{s^2 - s + 1}$

3. Use the Nyquist stability criterion to find out if the closed loops with the open loops given by the transfer functions of the previous exercise are stable.
4. Confirm the results, whenever this is possible, from the stability margins of the each plant.
5. Suppose that the closed loops of the previous exercises comprise a variable gain k. Use again the Nyquist stability criterion to find for each case the values of k that stabilise the closed loop.

7 On fractional derivatives

7.1 Introduction

Fractional derivative (FD) is the name assigned to several mathematical operators, that generalise derivatives and integrals to orders that are not integer. In spite of the name, orders need in fact not be fractional: they can be irrational or complex. For this reason, the name *non–integer order derivative* is also used. While the formulation adopted in chapter 2 was enough for our objectives, some additional remarks are now in order.

The operators collectively named as FD are the Grünwald-Letnikov (GL), Liouville (L), Riemann-Liouville (RL), Caputo (C), Marchaud, Hadamard, and Riesz operators, plus some other less frequent definitions [4, 28, 37–39, 54, 60, 64, 87, 98, 99, 107, 114]. Among these formulations, we consider here those that can be useful for the introduction of fractional linear systems and their characterisation, in the perspective of compatibility with integer orders. This was motivated by problems appearing in applied sciences where proposed formulations are frequently not compatible with classic integer order counterparts. This leads to a frequently posed question: "Are mathematical models with FD consistent with the laws of physics?" [52]. The answer is generally *no*, because in the literature we find fractional-order models that are proposed without considering a backward compatibility with classic laws. Moreover, this lack of compatibility often leads to uncommon results, motivating their derogatory characterisation as "metaphysical derivatives" suggested by Stéphane Dugowson [12].

The aim of this chapter is to describe a coherent basis for establishing fractional operators compatible with the corresponding classic integer order. In particular, the formulation developed below fully justifies the procedures introduced in previous chapters, where we proposed definitions of Impulse Response, Transfer Function and Frequency Response in a very general framework that include the classic operators (obtained when the orders become integers).

7.2 Some historical considerations

The idea of differentiation of non–integer order seems to have origin in some reflections expressed by Leibniz in a letter addressed to Bernoulli and in his answer [12]. The subject was first discussed between the two and later with other mathematicians, but no formula or procedure for the computation was proposed.

Although Euler and Fourier made attempts in the field, the first really important contribution in fractional calculus (FC) was accomplished by Liouville, that proposed several formulas for computing the derivatives of any order [46–50]. However, Liouville based his theory on the expansion of the functions as a sum of exponentials, and this created several difficulties because, at that time, the Bromwich integral for

The original version of this chapter was revised. An Erratum is available at DOI 10.1515/9783110624588-021
DOI 10.1515/9783110624588-007

the inverse LT was not yet formulated. Among the proposed formulas he included generalisations of the incremental ratio. However, the importance of this result was overlooked, and only later did Grünwald [24] and Letnikov [41] return to the subject, but considering only the particular case of functions defined on $\mathbb{R}^+$. The name "Grünwald-Letnikov derivatives" is currently adopted for derivatives based on the incremental ratio. Later, Riemann in an unpublished work proposed an integral formulation that led to the usually called Riemann-Liouville (RL) FD.

Until recently, the RL definition for fractional order derivatives was favoured in relation to the others, mainly because it is the more natural definition of fractional derivative after the Grünwald-Letnikov definition, because explicit expressions for some fractional derivatives can be more easily found than with the Grünwald-Letnikov definition, and because of the rigourous study presented in the excellent references listed in [107]. However, in recent years the Caputo derivative attracted an increasing attention and received the preference of many users, that assumed it solved the initial conditions problem.

However, neither RL, nor C, FD are suitable for being the base for the theory we presented in the previous chapters. In the following we will describe a more correct framework.

7.2.1 Returning to the differintegrator

In chapter 2, we introduced the *differintegrators* as linear systems characterised by the TF

$$H(s) = s^\alpha \tag{7.1}$$

with $Re(s) > 0$, causal, or $Re(s) < 0$, anti-causal, as ROC. We wrote also that if $\alpha > 0$ we speak of "differentiator", while with $\alpha < 0$ we use the term "integrator". In (2.12), we showed that the impulse responses corresponding to integrators are fractional powers:

$$\mathcal{L}^{-1}\left[\frac{1}{s^\alpha}\right] = \left[\pm\frac{t^{\alpha-1}}{\Gamma(\alpha)}\varepsilon(\pm t)\right] \tag{7.2}$$

with $\alpha > 0$.

Now, we are in conditions of introducing the fractional integration as the output of an integrator. For the causal case, we have

$$D_f^{-\alpha}f(t) = \frac{1}{\Gamma(\alpha)}\int_0^\infty f(t-\tau)\tau^{\alpha-1}\,d\tau. \tag{7.3}$$

We will call it "forward integrator". If $\alpha = n \in \mathbb{N}$, we obtain a repeated integration according to the Riemann definition of integral. This formula, although motivated by Liouville's results, was first introduced by Serret in 1844. Liouville had really proposed in 1832 the anti-causal version, that we will call "backward integrator":

$$D_b^{-\alpha}f(t) = \frac{1}{(-1)^\alpha\Gamma(\alpha)}\int_0^\infty f(t+\tau)\tau^{\alpha-1}\,d\tau. \tag{7.4}$$

We will study the problem of obtaining similar formulas for derivatives. Consider the causal case. Assume that we write:

$$D_f^\alpha f(t) = \frac{1}{\Gamma(\alpha)} \int_0^\infty f(t - \tau) \tau^{-\alpha-1} \, d\tau. \tag{7.5}$$

But in this expression the integrand is hypersingular at the origin. There are three ways of avoiding the difficulty:
1. Write $s^\alpha = s^N s^{\alpha-N}$, where N is any integer greater than α (normally, we use a value such that $N - 1 < \alpha \le N$), to get

$$D_f^\alpha f(t) = \frac{1}{\Gamma(N - \alpha)} \frac{d^N}{dt^N} \int_0^\infty f(t - \tau) \tau^{N-\alpha-1} \, d\tau \tag{7.6}$$

   that is currently called "Liouville derivative" [107]. The particular case valid for right signals is called "Riemann-Liouville derivative".
2. Similarly, write $s^\alpha = s^{\alpha-N} s^N$, where N is again any integer greater than α (and again the usual choice of N is $N - 1 < \alpha \le N$). We have

$$D_f^\alpha f(t) = \frac{1}{\Gamma(N - \alpha)} \int_0^\infty \frac{d^N}{dt^N} f(t - \tau) \tau^{N-\alpha-1} \, d\tau \tag{7.7}$$

   that is currently called "Liouville-Caputo derivative" [28], but was also introduced by Liouville. The so-called "Caputo derivative" is a particular case valid only for right signals, and was introduced by Gerasymov in 1948 and by Caputo in 1967 [105].
3. Another alternative consists of prolonging the validity of (7.3) for negative values of α. This leads to a "regularised Liouville derivative" that we can formulate in a way that includes (7.3). It reads

$$D_f^\alpha f(t) = \frac{1}{\Gamma(-\alpha)} \int_0^\infty \tau^{-\alpha-1} \left[f(t - \tau) - \sum_{m=0}^{N-1} \frac{(-1)^m f^{(m)}(t)}{m!} \tau^m \right] d\tau. \tag{7.8}$$

From a practical point of view, we can say that the Liouville-Caputo is the worst way of doing derivative computation. In fact, consider the folowing operations $D_f^N [f(t) * g(t)]$ and $f(t) * D_f^N [g(t)]$. If the involved functions are enough well behaved, the two operations will give the same result that we state as a property of the convolution:

$$D_f^N [f(t) * g(t)] = f(t) * D_f^N [g(t)]. \tag{7.9}$$

This would mean that the Liouville and Liouville-Caputo derivatives would be equivalent, which is known to be incorrect. To understand why we must reallize that
- The convolution has a smoothing effect. Therefore, in the left side of (7.9) we are computing the derivative of a function with "better behaviour" than each of the convolution factors.

-   $D_f^N[g(t)]$ has a worst analytic behaviour than $g(t)$; eventually it can be discontinuous.

Therefore, it seems clear that *if the right hand side exists, then it also occurs for the left hand side*. The reverse is not true. This allows us to do a fair comparison of the three derivatives leading us to conclude that:

-   If $f(t)$ has a Laplace transform with a nondegenerate region of convergence, the 3 derivatives give the same result.
-   The Liouville-Caputo derivative is too demanding from an analytical point of view, since it needs the unnecessary existence of the Nth order derivative.
-   If $f(t) = 1$, $t \in \mathbb{R}$, the Riemann-Liouville derivative does not exist, since the integral is divergent.

The corresponding backward derivatives are easily obtained.

7.3 Fractional incremental ratia approach

7.3.1 The derivative operators and their inverses

So that forward FD formulations should be consistent with the laws of physics, we recall the most important results from classic calculus. The standard definition of derivative is

$$Df(t) = f'(t) = \lim_{h \to 0} \frac{f(t) - f(t-h)}{h}, \tag{7.10}$$

or

$$Df(t) = f'(t) = \lim_{h \to 0} \frac{f(t+h) - f(t)}{h}. \tag{7.11}$$

Substituting $-h$ for h interchanges the definitions, meaning that we only have to consider $h > 0$. In this situation (7.10) uses the present and past values, while (7.11) uses present and future values. In the following we will distinguish the two cases by using the subscripts f (forward – in the sense that we go from past into future, a direct time flow) and b (backward – meaning a reverse time flow).

It is straightforward to invert the above equations to obtain

$$D_f^{-1}f(t) = \lim_{h \to 0} \sum_{n=0}^{\infty} f(t - nh) \cdot h, \tag{7.12}$$

and

$$D_b^{-1}f(t) = \lim_{h \to 0} \sum_{n=0}^{\infty} f(t + nh) \cdot h. \tag{7.13}$$

In these relations the different time flow shows its influence, since causality or anti-causality are clearly stated. We have

$$D_f^{-1}D_f f(t) = D_f D_f^{-1}f(t) = f(t) \tag{7.14}$$

$$D_b^{-1}D_b f(t) = D_b D_b^{-1}f(t) = f(t). \tag{7.15}$$

We will call D^{-1} "anti-derivative" [87].

Higher order derivatives and corresponding inverses are usually obtained recursively, but closed formulas can also be established. Considering N a positive integer, such expressions assume the forms

$$D_f^{\pm N} f(t) = \lim_{h \to 0^+} \frac{\sum_{n=0}^{\infty} \frac{(\mp N)_n}{n!} f(t - nh)}{h^{\pm N}} \qquad (7.16)$$

$$D_b^{\pm N} f(t) = (-1)^N \lim_{h \to 0^+} \frac{\sum_{n=0}^{\infty} \frac{(\mp N)_n}{n!} f(t + nh)}{h^{\pm N}}, \qquad (7.17)$$

where $(a)_k = a(a + 1)(a + 2) \ldots (a + k - 1)$ denotes the Pochammer symbol.

These relations motivate the comments:

-   The one-step computation of derivatives using (7.16) and (7.17) may not give the result that we would obtain in a recursive procedure. However, in the conditions stated in section 7.1, they give the same result.
-   In the derivative case, the summation goes only to N, since $(-N)_n$ becomes null for $n > N$.
-   Let n_1 and n_2 be two integer values. With (7.16) and under the assumed functional space, we can write

$$D_f^{n_1} D_f^{n_2} f(t) = D_f^{n_2} D_f^{n_1} f(t) = D_f^{n_1 + n_2} f(t). \qquad (7.18)$$

For backward derivatives the situation is similar. This result is straightforward using the properties of the binomial coefficients or the $\mathcal{Z}$ transform [100].

-   It is possible to combine the forward and backward derivatives to obtain so-called two-sided (centred) derivatives [69, 87], as we will see at chapter 11.

7.3.2 Fractional incremental ratia

Return back to $H(s) = s^\alpha$ and rewrite it as

$$H(s) = s^\alpha = \begin{cases} \lim_{h \to 0^+} \left(\frac{1 - e^{-sh}}{h} \right)^\alpha & Re(s) > 0 \\ \lim_{h \to 0^+} \left(\frac{e^{sh} - 1}{h} \right)^\alpha & Re(s) < 0 \end{cases} \qquad (7.19)$$

Consider the well-known binomial series

$$(1 - w)^\alpha = \sum_{n=0}^{\infty} \binom{\alpha}{n} (-1)^n w^n = \sum_{n=0}^{\infty} \frac{(-\alpha)_n}{n!} w^n, \quad |w| < 1, \qquad (7.20)$$

together with any function $f(t)$ defined on $\mathbb{R}$, and let α be any real number. We can define the forward and backward derivatives given by:

$$D_f^\alpha f(t) = \lim_{h \to 0^+} \frac{\sum_{n=0}^{\infty} \frac{(-\alpha)_n}{n!} f(t - nh)}{h^\alpha} \qquad (7.21)$$

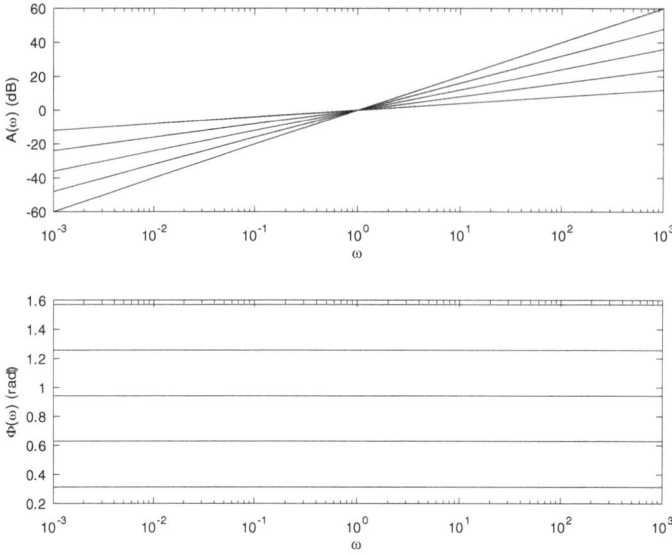

Fig. 7.1: Bode diagrams of the fractional derivators of orders $\alpha = 0.2, 0.4, \cdots, 1$.

$$D_b^\alpha f(t) = e^{-i\pi\alpha} \lim_{h \to 0^+} \frac{\sum_{n=0}^{\infty} \frac{(-\alpha)_n}{n!} f(t + nh)}{h^\alpha}. \tag{7.22}$$

Although introduced by Liouville in 1832, these are called Grünwald-Letnikov derivatives (GL), bearing the names of the mathematicians that produced more results about these derivatives. From the point of view of applied sciences, the Grünwald-Letnikov FD seem to be a natural way for generalising the notion of derivative. They have a solid meaning for any real (or complex) order, and let us recover the classic results for integer orders.

From the procedure followed to arrive at the GL FD, we can say that they exist for functions with LT. This statement can be enlarged to the Fourier transform for functions with LT convergent on a region that includes the imaginary axis. It can be shown that this already happens with sinusoids or periodic signals. In fact, it can be shown that

$$\begin{cases} D_f^\alpha \cos(\omega t) = \omega^\alpha \cos(\omega t + \alpha \frac{\pi}{2}) \\[2mm] D_f^\alpha \sin(\omega t) = \omega^\alpha \sin(\omega t + \alpha \frac{\pi}{2}) \end{cases} \quad , \ t \in \mathbb{R}, \omega \in \mathbb{R}^+, \tag{7.23}$$

showing that the "differintegrators" produce a phase shift of $\alpha \frac{\pi}{2}$. In general, it is not a simple task to formulate the weakest conditions that ensure the existence of GL derivatives, although we can give some necessary conditions for their existence. We must care about the behaviour of the function along the half straight-line $t \pm nh$ with $n \in \mathbb{Z}^+$. If the function is zero for $t < a$ (respectively, when $t > a$) the forward (backward) derivative exists at every finite point of $f(t)$. In the general case, we must have in mind the behavior of the binomial coefficients. They verify

$$\left| \begin{pmatrix} \alpha \\ n \end{pmatrix} \right| \leq \frac{A}{n^{\alpha+1}}, \tag{7.24}$$

meaning that $f(t)\frac{A}{n^{\alpha+1}}$ must decrease, at least as $f(t)\frac{A}{n^{|\alpha|+1}}$ when n goes to infinite. For the forward case, if $\alpha > 0$, it is enough to ensure that $f(t)$ is bounded for $t < 0$, but, if $\alpha < 0$, $f(t)$ must decrease to zero so that the series is convergent. In particular, they should be used for the causal (anti-causal) functions. This is very interesting, since we conclude that the existence of the fractional derivative depends only on what happens in one half of the real line, left or right. Consider for instance $f(t) = t^N$, with $N \in \mathbb{N}$. If $N > \alpha$, we conclude immediately that $D^\alpha (t^N)$ defined for every $t \in \mathbb{R}$ does not exist, unless α is a positive integer, because the summation in (7.21) is divergent.

There are several properties exhibited by (7.21)-(7.22) [76]:
-   Linearity.
-   Additivity and Commutativity of the orders. If we apply (7.21) twice for any two orders, we have [87]

$$D_f^\alpha D_f^\beta f(t) = D_f^\beta D_f^\alpha f(t) = D_f^{\alpha+\beta} f(t). \tag{7.25}$$

-   Neutral element.

$$D_f^\alpha D_f^{-\alpha} f(t) = D_f^0 f(t) = f(t). \tag{7.26}$$

From (7.26) we conclude that there is always an inverse element, that is, for every α there is always the $-\alpha$ order derivative.
-   Backward compatibility ($n \in \mathbb{N}$). If $\alpha = n$, then:

$$D_f^n f(t) = \lim_{h \to 0} \frac{\sum_{k=0}^n (-1)^k \binom{n}{k} f(t - kh)}{h^n} \tag{7.27}$$

We obtain this expression repeating the first order derivative. If $\alpha = -n$, then:

$$D_f^{-n} f(t) = \lim_{h \to 0} \sum_{k=0}^n \frac{(n)_k}{k!} f(t - kh) \cdot h^n,$$

that corresponds to a n-th repeated integration [87].
-   We can apply the two-sided LT to (7.21) and (7.22) to obtain

$$\mathcal{L} [D^\alpha f(t)] = s^\alpha \mathcal{L} [f(t)], \tag{7.28}$$

with $Re(s) > 0$ in the first and $Re(s) < 0$ in the second. This means that there are two systems (one causal and one anti-causal differintegrator) with TF given by $H(s) = s^\alpha$. This result agrees with the discussion in section 7.3.
-   The generalised Leibniz rule for the product. The generalised Leibniz rule gives the FD of the product of two functions. It is one of the most important character-

istics of the FD and assumes the form [107]

$$D_f^\alpha \left[f(t)g(t) \right] = \sum_{k=0}^{\infty} \binom{\alpha}{k} D_f^k f(t) D_f^{\alpha-k} g(t). \tag{7.29}$$

The formula is identical for the backward case.

7.3.3 Some examples

We now consider the particular cases mentioned in section 7.3.
1. Constant function.
    If $f(t) = 1$, for every $t \in \mathbb{R}$ and $\alpha \in \mathbb{R}$, then we have

$$D^\alpha f(t) = \lim_{h \to 0^+} \frac{\sum_{k=0}^{\infty} \binom{\alpha}{k}(-1)^k}{h^\alpha} = \begin{cases} 0, & \text{if } \alpha > 0 \\ \\ \infty, & \text{if } \alpha < 0 \end{cases}. \tag{7.30}$$

2. Causal power function.
    We calculate the FD of the Heaviside function. Starting from [107]

$$\sum_{k=0}^{n} \binom{\alpha}{k}(-1)^k = \binom{\alpha-1}{n}(-1)^n = \frac{1}{\Gamma(1-\alpha)} \frac{\Gamma(-\alpha+n+1)}{\Gamma(n+1)}, \tag{7.31}$$

we can show that

$$D^\alpha \varepsilon(t) = \frac{t^{-\alpha}}{\Gamma(1-\alpha)} \varepsilon(t). \tag{7.32}$$

As $\delta(t) = D\varepsilon(t)$, we deduce using (7.25) that

$$D^\alpha \delta(t) = \frac{t^{-\alpha-1}}{\Gamma(\alpha+1)} \varepsilon(t). \tag{7.33}$$

With (7.32) it is possible to obtain the derivative of any order of the continuous function $p(t) = t^\beta \varepsilon(t)$, with $\beta > 0$. The LT of $p(t)$ yields $P(s) = \frac{\Gamma(\beta+1)}{s^{\beta+1}}$ for $Re(s) > 0$, and the FD of order α is given by $s^\alpha \frac{\Gamma(\beta+1)}{s^{\beta+1}}$. Therefore, the expression

$$D^\alpha t^\beta \varepsilon(t) = \frac{\Gamma(\beta+1)}{\Gamma(\beta-\alpha+1)} t^{\beta-\alpha} \varepsilon(t) \tag{7.34}$$

generalises the integer order formula (see section 7.3) for any $\alpha \in \mathbb{R}$ [38]. In particular, with $\beta = \alpha - 1$, the result is:

$$D^\alpha \frac{t^{\alpha-1}}{\Gamma(\alpha+1)} \varepsilon(t) = \frac{t^{-1}}{\Gamma(0)} \varepsilon(t) = \delta(t). \tag{7.35}$$

7.3.4 Classic Riemann-Liouville and Caputo derivatives

The commutative property of the convolution allows us to write, from (7.6) and (7.7):

$$D_f^\alpha f(t) = D_f^N \left[\frac{1}{\Gamma(-\alpha+N)} \int_{-\infty}^{t} (t-\tau)^{N-\alpha-1} f(\tau)\, d\tau \right] \tag{7.36}$$

and

$$D_f^\alpha f(t) = \frac{1}{\Gamma(-\alpha+N)} \int_{-\infty}^{t} (t-\tau)^{N-\alpha-1} f^{(N)}(\tau)\, d\tau. \tag{7.37}$$

These expressions are the general formulations of Liouville and Liouville-Caputo derivatives. The usual formulations of Riemann-Liouville (RL) and Caputo (C) derivatives are obtained from the above relations assuming that the function is defined in a given interval $[a,b]$. Therefore, for $t \in [a,b]$ the RL and C forward derivatives are given by

$$^{RL}D_f^\alpha f(t) = D_f^N \left[\frac{1}{\Gamma(-\alpha+N)} \int_a^t (t-\tau)^{N-\alpha-1} f(\tau)\, d\tau \right] \tag{7.38}$$

$$^{C}D_f^\alpha f(t) = \frac{1}{\Gamma(-\alpha+N)} \int_a^t (t-\tau)^{N-\alpha-1} f^{(N)}(\tau)\, d\tau. \tag{7.39}$$

for $t > a$. These derivatives do not enjoy most of the important properties of the classic derivative previously presented, namely the results referring the derivatives of sinusoids, that following (7.38) and (7.39) are no longer sinusoids [91]. Another example is given by the following differential equation:

$$^{C}D_f^\alpha f(t) + af(t) = {}^{C}D_f^\beta \varepsilon(t). \tag{7.40}$$

If the initial conditions are null the output is also null. In fact, the C derivative of the unit step is identically null. Therefore, we have $^{C}D_f^\alpha f(t) + af(t) = 0$ implying that $f(t) = 0$, if the initial condition is zero.

7.4 On complex order derivatives

In previous chapters, we considered only real orders of differentiation. Here we will consider complex orders. We will study them in the frequency domain, due to its simplicity and facility in getting results. Therefore, consider the frequency response:

$$H(\omega) = (i\omega)^{u+iv} = (i\omega)^u (i\omega)^{iv} \tag{7.41}$$

The first term is well known and we have studied it before. It is an hermitian operator. So, we are going to study the second term, $\Phi(\omega) = (i\omega)^{iv}$. To start, we remark that

$i\omega = |\omega|e^{i\frac{\pi}{2}\mathrm{sgn}(\omega)}$. Then

$$\Phi(\omega) = (i\omega)^{iv} = e^{-v\frac{\pi}{2}\mathrm{sgn}(\omega)}e^{iv\log|\omega|}. \tag{7.42}$$

This frequency response has an amplitude

$$A(\omega) = |\omega|^{u}e^{-v\frac{\pi}{2}\mathrm{sgn}(\omega)} \neq A(-\omega) \tag{7.43}$$

and a phase

$$\phi(\omega) = v\log|\omega| \neq -\phi(-\omega). \tag{7.44}$$

We therefore conclude that the derivative of complex order is not hermitian. This may lead to a strange result as the next example shows.

Example 7.4.1. *Compute the fractional complex order derivative of the sinusoid $x(t) = \cos(t)$. As known, the Fourier transform of $x(t)$ is $X(\omega) = \pi\delta(\omega - 1) + \pi\delta(\omega + 1)$. Thus the Fourier transform of the derivative is*

$$Y(\omega) = (i\omega)^{u+iv}\left[\pi\delta(\omega - 1) + \pi\delta(\omega + 1)\right]. \tag{7.45}$$

Using the properties of the Dirac impulse, we can write

$$\begin{aligned} Y(\omega) &= \pi(i)^{u+iv}\delta(\omega - 1) + \pi(-i)^{u+iv}\delta(\omega + 1) \\ &= \pi e^{i\frac{u\pi}{2}}e^{-v\frac{\pi}{2}}\delta(\omega - 1) + \pi e^{-i\frac{u\pi}{2}}e^{v\frac{\pi}{2}}\delta(\omega + 1), \end{aligned} \tag{7.46}$$

that, after the inverse Fourier transform, leads to

$$y(t) = e^{-v\frac{\pi}{2}}e^{i\left(t - u\frac{\pi}{2}\right)} + e^{v\frac{\pi}{2}}e^{i\left(t + u\frac{\pi}{2}\right)}, \tag{7.47}$$

that is a complex signal.

In attempting to avoid the non-hermitian character of the complex order derivative, hermitian operators can be created using the conjugate derivative. Consider thus the operators defined by:

$$(i\omega)^{u}(i\omega)^{iv} \pm (i\omega)^{u}(i\omega)^{-iv}. \tag{7.48}$$

As $(i\omega)^{u}$ is equal in both terms and is hermitian, we will study only

$$\Psi(\omega) = (i\omega)^{iv} \pm (i\omega)^{-iv}. \tag{7.49}$$

Some manipulations lead to

$$\Psi(\omega) = e^{-v\frac{\pi}{2}\mathrm{sgn}(\omega)}e^{iv\log|\omega|} \pm e^{v\frac{\pi}{2}\mathrm{sgn}(\omega)}e^{-iv\log|\omega|}. \tag{7.50}$$

Setting $\Psi(\omega) = A(\omega)e^{i\Phi(\omega)}$, we obtain

$$A(\omega) = [2\cosh(v\pi) \pm 2\cos(2v\log|\omega|)]^{\frac{1}{2}} \tag{7.51}$$

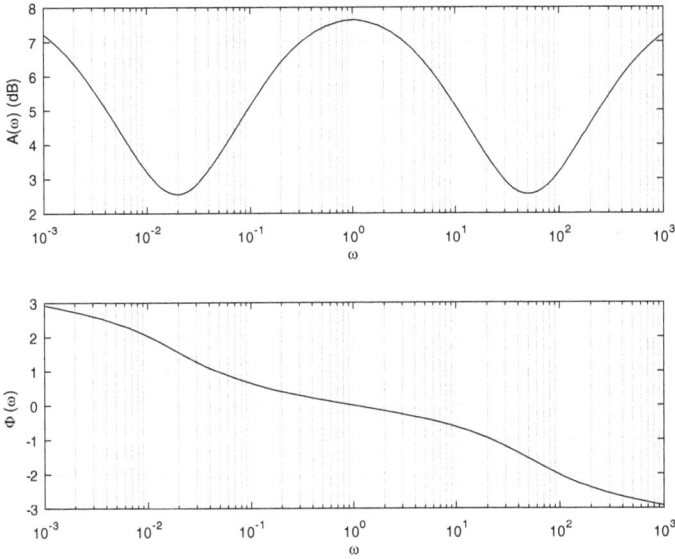

Fig. 7.2: Bode diagrams of (7.50), for $v = 0.4$.

and

$$\theta(\omega) = -\arctan\left\{\left[\tanh\left(\frac{v\pi}{2}\right)\right]^{\pm 1}\tan(v\log|\omega|)\right\}\mathrm{sgn}(\omega). \tag{7.52}$$

This leads us to conclude that

1. The operator with frequency response given by (7.50) is hermitian;
2. Its amplitude spectrum is oscilating, but bounded; it behaves like an allpass filter;
3. The corresponding phase spectrum decreases almost linearly (see Figure 7.2);
4. It cannot be considered as a fractional derivative.

Part II: **Discrete-time**

8 Discrete-time linear systems. Difference equations

8.1 On time scales

In this and the two following chapters, we will study the discrete-time linear systems. However, we will present a framework proposed by Aulbach and Hilger as an attempt to unify the mathematical treatment of the continuous/discrete-time cases, and generalise the results obtained in one of the situations to the other. Aulbach and Hilger introduced the calculus on *measure chains* [29] and the (more popular) *calculus on time scales* [7]. These are nonempty, closed subsets $\mathbb{T} \subset \mathbb{R}$. This requires some care, since the term *scale* is used in Signal Processing with a somehow different meaning. We will first describe the language of time scale calculus.

Let t be the current instant on a time scale $\mathbb{T}$. The previous next instant is denoted by $\rho(t)$. Similarly, the next following instant is denoted by $\sigma(t)$. We define two graininess functions, $v(t)$ and $\mu(t)$,

$$v(t) = t - \rho(t), \quad \mu(t) = \sigma(t) - t. \tag{8.1}$$

Frequently, the terms forward and backward graininess are used. We will not adopt this terminology, since these words have meanings that are different from the ones used in Signal Processing applications.

Let us define

$$v^0(t) = 0, \quad v^n(t) = v^{n-1}(t) + v\left(t - v^{n-1}(t)\right), \quad n \in \mathbb{N}. \tag{8.2}$$

We will have $\rho^0(t) := t$ and $\rho^n(t) := \rho\left(\rho^{n-1}(t)\right)$, $n \in \mathbb{N}$. Note that $v^1(t) = v(t)$ and $\rho^1(t) = \rho(t)$. When moving into the past, we have

$$\rho^0(t) = t = t - v^0(t), \tag{8.3}$$

$$\rho^1(t) = \rho(t) = t - v(t) = t - v^1(t), \tag{8.4}$$

$$\rho^2(t) = \rho(\rho(t)) = \rho(t) - v(\rho(t)) = t - v(t) - v(t - v(t)) = t - v^2(t), \tag{8.5}$$

$$\vdots \tag{8.6}$$

$$\rho^n(t) = t - v^n(t). \tag{8.7}$$

Moving into the future, the definitions and results are similar:

$$\mu^0(t) := 0, \quad \mu^n(t) := \mu^{n-1}(t) + \mu\left(t + \mu^{n-1}(t)\right), \quad n \in \mathbb{N}. \tag{8.8}$$

We will have $\sigma^0(t) := t$ and $\sigma^n(t) := \sigma\left(\sigma^{n-1}(t)\right)$, as well as $\sigma^n(t) = t + \mu^n(t)$.

This formulation states the most general case that embraces continuous-time, discrete-time and mixed-time. In what follows, we will only consider the discrete-time case and the particular situation of *constant graininess*; that is to say, there is constant *inter-sample interval*. In this case, which is the most usual in practice, we let $\mu^n(t) =$

DOI 10.1515/9783110624588-008

$v^n(t) = nT$. Then $\sigma^n(t) = t + nT$, and $\rho^n(t) = t - nT$, $n = 0, 1, 2, \ldots$ Thus, we conclude that, if there is constant graininess, $\mathbb{T} = T\mathbb{Z}$, $T > 0$. This is called a *uniform time scale*, and T is the inter-sample interval.

8.2 Systems on uniform time scales: difference equations

When $\mathbb{T} = T\mathbb{Z}$, $T > 0$, and the time instants are given by $t_n = nT$, $n \in \mathbb{Z}$, we frequently omit the graininess T and write $f(n)$ instead of $f(nT)$. With this notation we are able to introduce the traditional discrete-time linear systems that are characterised by linear difference equations with the general form [31, 61, 65, 100]

$$\sum_{k=0}^{N_0} a_k y(n-k) = \sum_{k=0}^{M_0} b_k x(n-k) \tag{8.9}$$

where $n, M_0, N_0 \in \mathbb{Z}^+$ and the coefficients a_k, $k = 0, 1, \ldots, N_0$ and b_k, $k = 0, 1, \ldots, M_0$ are real constants.

Remark 8.2.1. *Of course, we could use advances $y(n + k)$ instead of delays $y(n - k)$. However, in many situations this is not correct, since we will be using past values, not future values, to know the present. This will always be the case when data is treated in real time. Advances can be used only when data is measured first, and then, later on, treated in batch. There are also predictor problems, in which we are trying to predict a future value from past ones. In those, we use $y(n + 1)$ for one step-ahead prediction; or even $y(n + 2)$, $y(n + 3)$, $\ldots$, $y(n + k)$ for k-step ahead prediction.*

Remark 8.2.2. *As mentioned at the beginning of chapter 1, there are signals that depend on variables other than time. For those signals, the difference between advances and delays may be irrelevant. Think for instance about image processing, in which these two operations correspond to moving in opposite directions around a given point of the image.*

The constant coefficient ordinary difference equations [13] have a long tradition in applied sciences and have a large amount of engineering applications, mainly in Signal Processing [31, 61, 65, 100], where they are referred as ARMA (Autoregresive–Moving Average) models.

8.3 Exponentials as eigenfunctions

The discrete convolution was introduced in chapter 1 and was defined by:

$$x(n) * y(n) = \sum_{k=-\infty}^{\infty} x(k)y(n-k), \; n \in \mathbb{Z} \tag{8.10}$$

We will use it to compute the output of the system defined by (8.9). Consider a particular input $x(n) = \delta(n)$. The corresponding output $h(n)$ is the *impulse response* of the system. So, it verifies the equation

$$\sum_{k=0}^{N_0} a_k h(n-k) = \sum_{k=0}^{M_0} b_k \delta(n-k) \tag{8.11}$$

Now convolve both sides in (8.11) with some other arbitrary input $x(n)$.

$$\sum_{k=0}^{N_0} a_k h(n-k) * x(n) = \sum_{k=0}^{M_0} b_k \delta(n-k) * x(n) \tag{8.12}$$

Using the properties of the convolution we can write

$$\sum_{k=0}^{N_0} a_k [h(n-k) * x(n)] = \sum_{k=0}^{M_0} b_k x(n-k) \tag{8.13}$$

A comparison of this equation with (8.9) allows us to conclude that its solution is given by

$$y(n) = h(n) * x(n) \tag{8.14}$$

This means that the solution of (8.9) is the convolution of $x(n)$ with the impulse response.

Theorem 8.3.1. *Let $x(n) = z^n$, $z \in \mathbb{C}$, $n \in \mathbb{Z}$. The corresponding output is given by*

$$y(n) = H(z)z^n \tag{8.15}$$

provided that $H(z)$ exists.

This theorem shows that the exponentials are the eigenfunctions of constant coefficient ordinary difference equations.

Proof. Insert $x(n) = z^n$ into (8.14) to get

$$y(n) = \sum_{k=-\infty}^{\infty} h(k)z^{n-k} = \sum_{k=-\infty}^{\infty} h(k)z^{-k}z^n. \tag{8.16}$$

Making

$$H(z) = \sum_{k=-\infty}^{\infty} h(k)z^{-k} \tag{8.17}$$

we obtain (8.15). $H(z)$ is the *transfer function* of the system defined by the difference equation (8.9) and is the $\mathcal{Z}$-transform of the impulse response. $\square$

Inserting (8.15) and $x(n) = z^n$ into (8.9) we conclude immediately that

$$H(z) = \frac{\sum_{k=0}^{M_0} b_k z^{-k}}{\sum_{k=0}^{N_0} a_k z^{-k}} \tag{8.18}$$

In the following, we will consider that the *characteristic polynomial* in the denominator is not zero for the particular value of z at hand. Later we will consider the cases where the characteristic polynomial is zero (i.e. z is a pole).

Example 8.3.1. *Let $x(n) = 2^n$ and consider the equation*

$$y(n) = x(n) + x(n-1) \tag{8.19}$$

We have $H(z) = 1 + z^{-1}$. So the output is given by $y(n) = H(2)2^n = \frac{3}{2}2^n$. Let now $x(n) = (-1)^n$. We have $y(n) = H(-1)(-1)^n = 0$.

Example 8.3.2. *Consider the difference equation*

$$y(n) + y(n-1) - 4y(n-2) + 2y(n-3) = x(n) + 2x(n-1). \tag{8.20}$$

Let $x(n) = (1/2)^n$. The solution is given by

$$y(n) = \frac{1 + 2(1/2)^{-1}}{1 + (1/2)^{-1} - 4(1/2)^{-2} + 2(1/2)^{-3}}(1/2)^n = \frac{5}{3}(1/2)^n. \tag{8.21}$$

Example 8.3.3. *Consider the difference equation*

$$y(n) + y(n-1) - 4y(n-2) + 2y(n-3) = x(n) - x(n-1) \tag{8.22}$$

Let $x(n) = 3^n$. The solution is given by:

$$y(n) = \frac{1 - 3^{-1}}{1 + 3^{-1} - 4 \times 3^{-2} + 2 \times 3^{-3}}3^n = \frac{9}{13}3^n \tag{8.23}$$

8.3.1 Determination of the impulse response from the difference equation

Consider the simple equation $y(n) + ay(n-1) = x(n)$, let $y(-1) = 0$ (it is the initial condition), and make $x(n) = \delta_n$. If the system is causal, we have $y(0) = 1$ and $y(n) = -ay(n-1)$ for $n > 0$. Then,

$$y(1) = -a, y(2) = a^2, y(3) = -a^3, \ldots \tag{8.24}$$

If the system is anti-causal, we have $y(0) = 1$ and $y(n-1) = -\frac{1}{a}y(n)$ for $n < 0$. Then,

$$y(-1) = -1/a, y(-2) = 1/a^2, y(-3) = -1/a^3 \ldots \tag{8.25}$$

This suggests how we proceed in practice. We make $x(n) = \delta_n$ and calculate the output recursively. Using the general form (8.9)

$$\sum_{k=0}^{N_0} a_k y(n-k) = \sum_{k=0}^{M_0} b_k x(n-k) \tag{8.26}$$

with $x(n) = \delta_n$ and attending to $a_0 = 1$, we have

$$y(n) = -\sum_{k=1}^{N_0} a_k y(n-k) + b_n \qquad \text{if } n \le M_0 \tag{8.27}$$

$$y(n) = -\sum_{k=1}^{N_0} a_k y(n-k) \qquad \text{if } n > M_0 \tag{8.28}$$

As can be seen, the right side of the difference equation only affects the first M points of the impulse response; so we conclude that the left part, corresponding to the denominator of the transfer function, is determinant in the characteristics of the system.

Similar results can be obtained from the step response, which is another way of characterising a system. For this, let us first see the relation between the impulse and the step responses. Let $r_d(n)$ be the step response:

$$r_d(n) = \sum_{-\infty}^{\infty} \varepsilon_n h(n-k) = \sum_{0}^{\infty} h(n-k) = \sum_{-\infty}^{n} h(k). \tag{8.29}$$

Consequently,

$$r_d(n) = \sum_{-\infty}^{n} h(k). \tag{8.30}$$

As $\delta_n = \varepsilon_n - \varepsilon_{n-1}$,

$$h(n) = r_d(n) - r_d(n-1) \tag{8.31}$$

These relationships are useful in practice. To obtain $r_d(n)$, we now make $x(n) = \epsilon_n$ and proceed as in the previous case. We get

$$r_d(n) = -\sum_{k=1}^{N_0} a_k y(n-k) + \sum_{k=0}^{min(M_0,n)} b_k \tag{8.32}$$

8.3.2 From the impulse response to the difference equation

Rewrite (8.9) for the particular case of the impulse response

$$\sum_{k=0}^{N_0} a_k h(n-k) = \sum_{k=0}^{M_0} b_k \delta_{n-k}. \tag{8.33}$$

From here we conclude immediately that

$$h(n) = -\sum_{k=1}^{N_0} a_k h(n-k) + b_n \qquad \text{if } n \le M_0 \tag{8.34}$$

or

$$b_n = \sum_{k=0}^{N_0} a_k h(n-k) \qquad \text{if } n \le M_0, \tag{8.35}$$

and

$$h(n) = -\sum_{k=1}^{N_0} a_k h(n-k) \qquad \text{if} \quad n > M_0. \tag{8.36}$$

This is in fact a linear system of equations, that is easy to solve because the matrix is triangular and is, in addition, a Toeplitz matrix that allows a recursive solution. In practice, the existence of a recursive solution alleviates the problem of ignorance of M_0 and N_0. Once we get the solution $a_i, i = 1, \ldots, N_0$ of (8.36), we substitute it into (8.35) to compute the b_n, $n = 0, 1, \ldots$.

8.4 The frequency response

As stated in a previous chapter, the steady-state response is obtained assuming that the excitation of the system has started a long time ago (theoretically an infinite number of time instants away). The most appropriate way to carry out this study is to use periodic signals, such as sinusoids, as input. As we saw earlier, the response of a transfer function system $H(z)$ to a sinusoid $e^{\omega_0 n}$ is given by

$$y(n) = H(e^{i\omega_0}) e^{i\omega_0 n}. \tag{8.37}$$

We will now study the importance of the frequency response.

Example 8.4.1. *Consider the difference equation*

$$y(n) + y(n-1) - 4y(n-2) + y(n-3) = x(n). \tag{8.38}$$

Let $x(n) = e^{i\frac{\pi}{2}n}$. The solution is given by

$$y(n) = \frac{1}{1 + i^{-1} - 4i^{-2} + i^{-3}} e^{i\frac{\pi}{2}n} = \frac{1}{5} e^{i\frac{\pi}{2}n}. \tag{8.39}$$

This is very interesting since it allows us to compute easily the solution when $x(n) = \cos(\omega_0 t)$ or $x(n) = \sin(\omega_0 t)$. For the first case, we have

$$x(n) = \cos(\omega_0 t) = \frac{1}{2} e^{i\omega_0 n} + \frac{1}{2} e^{-i\omega_0 n}, \tag{8.40}$$

that leads to

$$y(n) = H(e^{i\omega_0}) \frac{1}{2} e^{i\omega_0 n} + H(e^{-i\omega_0}) \frac{1}{2} e^{-i\omega_0 n}. \tag{8.41}$$

$H(e^{i\omega}) = |H(e^{i\omega})| e^{i\varphi(e^{i\omega})}$ is called *frequency response*. If the coefficients in (8.9) are real, function $|H(e^{i\omega})|$ is the *amplitude spectrum* and is an even function, while $\varphi(e^{i\omega})$ is the *phase spectrum* and is an odd function.

Theorem 8.4.1. *The output of (8.9) when $x(n) = \cos(\omega_0 n)$ is given by*

$$y(n) = |H(e^{i\omega_0})| \cos[\omega_0 n + \varphi(e^{i\omega_0})] \tag{8.42}$$

Proof. According to what we said above, $\left|H(e^{-i\omega})\right| = \left|H(e^{i\omega})\right|$ and $\varphi(e^{-i\omega}) = -\varphi(e^{i\omega})$, which leads to

$$y(n) = \left|H(e^{i\omega_0})\right| \frac{1}{2}\left[e^{i\omega_0 n}e^{i\varphi(e^{i\omega})} + e^{-i\omega_0 n}e^{-i\varphi(e^{i\omega})}\right] \qquad (8.43)$$

that is the desired result.   □

It can be seen that the action of a filter with transfer function $H(z)$ over a sinusoidal input can be summarised as:
- Changing the amplitude of the input sinusoid, that can be amplified or attenuated, by a factor $A(e^{i\omega}) = \left|H(e^{i\omega})\right|$, called *amplitude spectrum* or *gain*;
- Offsetting the phase of the input sinusoid by $\phi(e^{i\omega})$, called *argument*, or *phase spectrum*, or simply *phase*.

Example 8.4.2. *Consider again the above equation, but change the second member:*

$$y(n) + y(n-1) - 4y(n-2) + y(n-3) = 3x(n) - 4x(n-1). \qquad (8.44)$$

Also assume that $x(n) = \sin\left(\frac{\pi}{2}n\right)$. Then

$$H(z) = \frac{3 - 4z^{-1}}{1 + z^{-1} - 4e^{-2} + z^{-3}} \qquad (8.45)$$

and

$$y(n) = \frac{1}{2i}\frac{3 - 4e^{-i\pi/2}}{1 + e^{-i\pi/2} - 4e^{-i\pi} + e^{-i3\pi/2}}e^{i\frac{\pi}{2}n} - \frac{1}{2i}\frac{3 - 4e^{i\pi/2}}{1 + e^{i\pi/2} - 4e^{i\pi} + e^{3\pi/2}}e^{-i\frac{\pi}{2}n}, \qquad (8.46)$$

leading to

$$y(n) = \sin\left(\frac{\pi}{2}n + \varphi\right) \qquad (8.47)$$

with $\varphi = \arctan(4/3)$.

The calculation of the frequency response of a given system from the difference equation is quite easy. Calculate the discrete-time Fourier transform of the numerator and denominator coefficients and divide them:

$$H(e^{i\omega}) = \frac{\sum\limits_{k=0}^{M_0} b_k e^{-i\omega k}}{\sum\limits_{k=0}^{N_0} a_k e^{-i\omega k}} \qquad (8.48)$$

This is very simple using the FFT (fast Fourier transform) algorithm [100]. However, we often have a special interest in being able to quickly sketch at least the absolute value of the frequency response. Now, gain and phase can be represented by:

$$A(e^{i\omega}) = b_0 \frac{\prod\limits_{k=1}^{M_0} \left|e^{i\omega} - z_k\right|}{\prod\limits_{k=1}^{N_0} \left|e^{i\omega} - p_k\right|} \qquad (8.49)$$

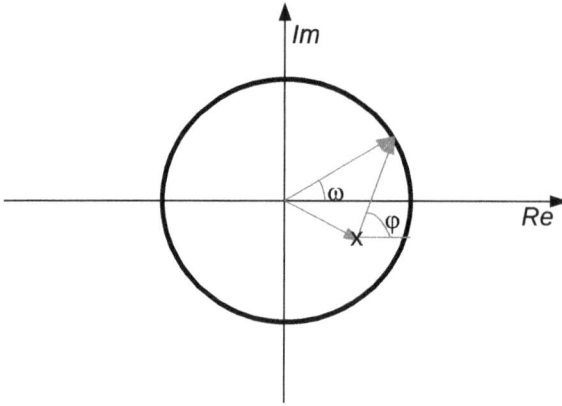

Fig. 8.1: Phasor diagram – Vector difference between the exponential, $e^{i\omega}$, and the complex representing the pole (or zero)

$$\phi(e^{i\omega}) = \omega(N_0 - M_0) + \sum_{k=1}^{M_0} \arg\left(e^{i\omega} - z_k\right) - \sum_{k=1}^{N_0} \arg\left(e^{i\omega} - p_k\right), \qquad (8.50)$$

where the parameters z_k and p_k are, respectively, the zeroes and poles of the system.

On the complex plane, $\left|e^{i\omega} - p_k\right|$ represents the distance from the pole p_k to $e^{i\omega}$, while $\arg\left(e^{i\omega} - p_k\right)$ is the angle between the vector that connects the point p_k with $e^{i\omega}$ and an horizontal straight line (see figure 8.1). So we can define the gain as

$$A(e^{i\omega}) = K \frac{\text{Product of the distances from the zeroes to } e^{i\omega} \text{ on the unit circle}}{\text{Product of the distances from the poles to } e^{i\omega} \text{ on the unit circle}} \qquad (8.51)$$

or, in dB,

$$A(e^{i\omega})_{dB} = K_{dB} + \text{Sum of distances from zeroes (dB)} - \text{Sum of distances from the poles (dB)}. \qquad (8.52)$$

The phase is given by

$$\phi(e^{i\omega}) = \omega(N_0 - M_0) + \text{Sum of angles from the zeroes} - \text{Sum of angles from the poles}. \qquad (8.53)$$

These angles are defined by the vector from the pole or zero to the unit circumference, and a horizontal line passing through that pole or zero.

8.5 Classification of systems

Let us return to equation (8.9). The integers N_0 and M_0 are called system orders. If N_0 and M_0 are non-null, the system is called *autoregressive moving-average* or ARMA. When it is necessary to specify the orders, we will write ARMA(N_0, M_0). If $M_0 = 0$, the system only has memory of the outputs, so it is called *autoregressive* or AR. If $N_0 = 0$,

the system only has memory of the past entries and it is said to be a *moving average* or MA. In terms of the transfer function, and knowing that a zero or pole located at the origin only introduces delay or advance, we can state that:
- an ARMA(N_0, M_0) system has N_0 poles and M_0 zeroes;
- an AR(N_0) system has only N_0 poles;
- an MA(M_0) system has only M_0 zeroes.

Remark 8.5.1. *Modernly, the AR and ARMA designations are also used for continuous-time systems (CAR, continuous autoregressive, and CARMA, continuous autoregressive moving-average).*

Considering now that the inverse system has an FT that is the inverse of the FT of the system, we can state that:
- any ARMA system has an equivalent AR(∞) or MA(∞);
- any AR(N_0) system has an equivalent MA(∞);
- any MA(M_0) system has an equivalent AR(∞);
- the inverse of an AR(∞) system may not be an ARMA or a finite-order MA;
- the inverse of an MA(∞) system may not be an ARMA or a finite-order AR.

It should be noted that these statements do not take account of stability issues. In general, the inverse of a stable system is not stable. It is not difficult to verify that the AR and ARMA systems have an impulse response with infinite duration, whereas the MA systems have an impulse response with finite duration. These reasons lead to call them respectively *infinite impulse response* (IIR) and *finite impulse response* (FIR) systems. FIR systems are extremely important and responsible for many day-to-day practical achievements, such as mobile communications. Indeed, the fact that they have a finite impulse response, and therefore a short memory, causes them to have short transients. In addition, there are very efficient designing methods that allow to control the shape of the frequency response, namely the phase.

We will consider only the case of causal and stable linear systems. The knowledge of the poles and zeros defines the amplitude (aside from a constant factor) and the phase (aside from a constant). On the other hand, given an amplitude, there may be several phases for which we have a causal and stable linear system. The different phases depend on the positions of the zeros. If the system has all its zeros within the unit disk, the system is said to be *minimum phase* and there is a relationship between the amplitude and phase: one is the Hilbert transform of the other. If the system has all the zeros outside the unit circle, the system is said to be *maximum phase*. In other cases, it is said that they are *mixed phase*. Maximum and mixed phase systems are also termed *non-minimum phase*.

To understand why stable causal systems with all zeros inside (outside) the unit disk are called minimum (maximum) phase systems; see again what was said in section 8.4 about the influence of poles and zeros in the phase. Using simple systems with one zero that we can move from inside to outside the unit circle, it is easy to discover the effect on the phase. With some suitable experiments, it is straightforward to state the following rules, where it is admitted that the frequency increases from zero to π:

– zeros outside unit circle cause negative phase transitions;
– poles inside the unit circle cause negative phase transitions;
– zeros inside unit circle cause positive phase transitions;
– poles outside the unit circle cause positive phase transitions.

There is a class of FIR systems that deserves special consideration: that of linear phase filters. The impulse response of a linear phase FIR with order N_0 verifies

$$h(n) = \pm h(N_0 - n). \tag{8.54}$$

Therefore, we can show that:

– non-real zeroes of a linear-phase exist in sets of four, given by $(z_0, 1/z_0^*), (z_0^*, 1/z_0)$;
– zeroes on the unit circle, $z_0 = e^{i\omega_0}$, exist in complex conjugate pairs, as long as $z_0 \neq \pm 1$;
– zeroes on the real line, $z_0 = a$, exist in sets of two, given by $z_0 \neq \pm 1$;
– zeroes at $z_0 = \pm 1$ form single sets.

8.6 Singular systems

In previous sections we dealt with the regular case when the exponentials z^n, $n \in \mathbb{Z}$, $z \in \mathbb{C}$ are the eigenfunctions of the systems. The regularity is expressed on the finiteness of $H(z_0)$. The situation is not so simple in the singular case, $H(z) = \infty$. When $z = 1$, the system is called an *auto-regressive integrated moving average* (ARIMA) system. We will state without proof the result that allows us to deal with the problem [71].

Theorem 8.6.1. *The output of the system described by equation (8.9) when* $x(n) = (-1)^k(-n)_k z_0^n$ *with* $A(z_0) = 0$ *is given by*

$$y(n) = z_0^n \left[\sum_{j=0}^{K} \binom{K}{j} \bar{H}^{(j)}(1)(-1)^{K-j}(-n)_{K-j} \right] \tag{8.55}$$

with

$$\bar{H}(z) = \frac{\bar{B}(z)}{\bar{A}(z)} = \frac{(1 - z^{-1})^m B(z_0 z)}{A(z_0 z)} \tag{8.56}$$

It is not difficult to show that (8.55) can be written as

$$y(n) = z_0^n \left[\sum_{j=0}^{K} \binom{K}{j} \bar{H}^{(j)}(1) \frac{(K-j)!}{(K+m-j)!}(-1)^{K+m-j}(-n)_{K+m-j} \right]. \tag{8.57}$$

If $K = 0$ (pure exponential input), we obtain

$$y(n) = z_0^n \bar{H}(1)\frac{1}{m!}(-1)^m(-n)_m.$$
(8.58)

If we make $z_0 = e^{i\omega_0 n}$ we are led to conclude that the response of the ARIMA model to a pure sinusoid is never a pure sinusoid: the amplitude increases with time. This is the reason why this model is used for modelling non-stationary situations.

Example 8.6.1. *Consider the following equation with $x(n) = n(-1)^n$:*

$$y(n) - y(n-1) - 4y(n-2) - 2y(n-3) = x(n).$$
(8.59)

The point $z = -1$ is a pole of the transfer function, $A(-1) = 0$, of order $m = 1$. On the other hand, $\bar{H}(z) = \frac{1+z^{-1}}{1-z^{-1}-4z^{-2}-2z^{-3}} = \frac{1}{1-2z^{-1}-2z^{-2}}$ and $\bar{H}'(z) = -\frac{-2z^{-2}-4z^{-3}}{(1-2z^{-1}-2z^{-2})^2}$, leading to $\bar{H}(-1) = 1$ and $\bar{H}'(-1) = 2$. The solution is $y(n) = [1/2(n)_2 + 2n](-1)^n$.

Example 8.6.2. *Consider the following ARIMA equation with $x(n) = 1$:*

$$y(n) - 2y(n-1) + 3y(n-2) - 2y(n-3) = x(n)$$
(8.60)

The point $z = 1$ is a pole of the transfer function, $A(1) = 0$, of order $m = 1$. On the other hand,

$$\bar{H}(z) = \frac{1}{1-z^{-1}+2z^{-2}},$$
(8.61)

leading to $\bar{H}(1) = 1/2$. The solution is $y(n) = n/2$.

Example 8.6.3. *The oscillator is a very interesting system that can be defined by the equation*

$$y(n) - 2\cos(\omega_0)y(n-1) + y(n-2) = x(n) - \cos(\omega_0)x(n-1).$$
(8.62)

Now, let $x(n) = e^{i\omega_0 n}$. The system has two simple ($m = 1$) poles at $e^{\pm i\omega_0 n}$. So, $\bar{H}(e^{i\omega_0}) = 1/2$ and the output is easily obtained:

$$y(n) = \frac{1}{2}ne^{i\omega_0 n}.$$
(8.63)

As we said above, this is a non-stationary model.

Exercises

1. Find the impulse responses of the following transfer functions:

   (a) $G_1(z) = \dfrac{4z^{-2} + 3z^{-1} + 2}{7z^{-2} + 6z^{-1} + 5}$

   (b) $G_2(z) = \dfrac{2z^{-2} + 3z^{-1} + 4}{5z^{-2} + 6z^{-1} + 7}$

(c) $G_3(z) = \dfrac{1}{7z^{-2} + 6z^{-1} + 5}$

(d) $G_4(z) = 4z^{-2} + 3z^{-1} + 2$

2. Find the responses of each of the transfer functions above to the following inputs:

(a) $x(n) = \epsilon(n)$

(b) $x(n) = n$

(c) $x(n) = \epsilon(n) - \epsilon(n - 5)$

(d) $x(n) = \sin\left(\dfrac{n\pi}{8}\right)$

3. Find numerically the frequency responses of each of the transfer functions above. Plot them in a Bode diagram.

4. Show that the response of a discrete transfer function to $x(n) = \cos(\omega_o t)$ is the same as its response to $x(n) = \cos((\omega_o + 2k\pi)t)$, $k in \mathbb{Z}$.

5. Explain from the result of exercise 4 why Bode diagrams of discrete transfer functions are represented only up to $\omega_o = \pi$, or rather $\omega_o = \dfrac{\pi}{T}$, where T is the sample time.

6. Prove (8.57) from (8.55).

9 Z transform. Transient responses

9.1 Z transform

9.1.1 Introduction

When acting on a system, either by introducing or by withdrawing an excitation, the system normally produces a response having a transient component, in the sense that it tends to zero more or less rapidly, depending on the stability characteristics of the system. If it is asymptotically stable, this component of the response tends effectively to zero. If the system is stable in the broad sense, the transient response may not tend to zero. The system response when there is no input is called a *free regime* response. The study of the transient response is very important, for example, in telecommunications, where it originates the so-called intersymbol interference. Normally, the behaviour of transient systems is studied using, as input, either a step or a rectangular pulse. These tasks are easily performed with the use of the Z transform that we will study next [31, 65, 100].

9.1.2 Definition

The Z transform, or ZT, is defined by

$$H(z) = \mathcal{Z}[h(n)] = \sum_{n=-\infty}^{\infty} h(n)z^{-n}, \quad z \in \mathbb{C}. \tag{9.1}$$

In some scientific domains, z instead of z^{-1} is used. In other domains, it is often called "generating function" or "characteristic function". Definition (9.1) is the bilateral ZT to distinguish from the unilateral ZT, defined by

$$H_u(z) = \sum_{n=0}^{\infty} h(n)z^{-n}, \tag{9.2}$$

and frequently used in the study of systems. The existence conditions of the ZT are similar to the ones of the LT. Therefore:

Definition 9.1.1. *If function $h(n)$ is such that there are finite positive real numbers, r_- and r_+, for which*

$$\sum_{n=0}^{\infty} |h(n)|r_-^n < \infty \text{ and } \sum_{n=-\infty}^{-1} |h(n)|r_+^n < \infty, \tag{9.3}$$

then the ZT exists, and the range of values for which those series converge defines a region of convergence that is an annulus.

It should be borne in mind that this condition is sufficient, but not necessary. The signals that verify it are the *exponential order* signals:

DOI 10.1515/9783110624588-009

Definition 9.1.2. *A discrete-time signal $x(n)$ is called* exponential order signal *if there exist integers n_1 and n_2, and positive real numbers a, b, A, B, such that $A\,a^{n_1} < |x(n)| < B\,b^{n_2}$ for $n_1 < n < n_2$.*

For these signals the ZT exists and the ROC is an annulus centred at the origin, generally delimited by two circles of radius r_- and r_+ such that $r_-|z| < r_+$. However, there are cases where the annulus can become infinite:

- If the signal is right, the ROC is the exterior of a circle centered at the origin ($r_+ = \infty$): $|z| > r_-$.
- Similarly, if the signal is left, the ROC is the interior of a circle centered at the origin ($r_- = 0$): $|z| < r_+$.
- If the signal is a pulse, the ROC is the complex plane, possibly with the exception of the origin. In the ROC, the ZT defines an analytical function.

It should be noted that the ROC is part of the definition of a given ZT. This means that there may be different signals with the same function as ZT, but different ROCs.

If the convergence region contains the unit circle, by making $z = e^{i\omega}$ we obtain the Fourier Transform of discrete-time signals, which we will simply call Fourier Transform (FT), unless there is danger of confusion. This means that not all signals with ZT have FT. The signals with ZT that do have also FT are those for which the ROC is non-degenerate and contains the unit circle. For some signals, the ROC degenerates in the unit circumference, so there is no ZT; this is the case of the sinusoids or the constant function.

Example 9.1.1. *The ZT of the rectangular pulse*

$$x(n) = \varepsilon(n) - \varepsilon(n - N), \ N \in \mathbb{N} \tag{9.4}$$

is given by

$$X(z) = 1 + z^{-1} + z^{-2} + z^{-3} + \cdots z^{-N-1} = \frac{1 - z^{-N}}{1 - z^{-1}} \tag{9.5}$$

and the ROC is the whole complex plane, excluding the point $z = 0$.

Example 9.1.2. *The right exponential*

$$x(n) = a^n \varepsilon(n) \tag{9.6}$$

is an exponential order signal, and so for its ZT we have

$$X(z) = \sum_{n=0}^{\infty} h(n).z^{-n} = \frac{1}{1 - a.z^{-1}} \qquad if\,|a.z^{-1}| < 1 \tag{9.7}$$

since the summation is a geometric series, if $|az^{-1}| < 1$. Therefore, the ROC is the exterior of the circle defined by $|z| > |a|$. According to what has been said above, if $|a| < 1$, by making $z = e^{i\omega}$, we obtain the FT of $x(n) = a^n \varepsilon(n)$. In particular, we can get the ZT of the *unit step $\varepsilon(n)$. It is enough to make $a = 1$ to get*

$$\mathcal{Z}[\varepsilon(n)] = \frac{1}{1 - z^{-1}} \qquad if \ |z| > 1 \tag{9.8}$$

Example 9.1.3. *We will now calculate the ZT of the two-sided signal*

$$x(n) = -\varepsilon(-n-1) + \left(\frac{1}{2}\right)^n \varepsilon(n). \tag{9.9}$$

As in the previous examples,

$$X(z) = -\sum_{n=\infty}^{-1} z^{-n} + \sum_{n=0}^{\infty} \left(\frac{1}{2}\right)^n z^{-n} = \frac{z}{z-1} + \frac{1}{1 - \frac{1}{2} \cdot z^{-1}} \tag{9.10}$$

where the convergence of the first series imposes that $|z| < 1$ and the second that $|z| > 1/2$. Therefore, the ROC is an annulus centred on the origin and delimited by the circumferences of radius $1/2$ and 1. Finally, we can write

$$X(z) = \frac{z(2z - 3/2)}{(z-1)(z-1/2)} \qquad \frac{1}{2} < |z| < 1. \tag{9.11}$$

Example 9.1.4. *If we consider the signal $x(n) = -\left(\frac{1}{2}\right)^n \varepsilon(-n-1) + \left(\frac{1}{2}\right)^n \varepsilon(n)$, the ROC is the empty set. So, $x(n)$ does not have ZT.*

In general, and remembering what was said above:

- If the signal is right side, the respective ROC is the exterior of the circle, centred at 0, which passes through the farthest pole from the origin.
- If the signal is left side, its ROC is the interior of the circle, centred at 0, passing through the pole closest to the origin.
- If the signal is bilateral and has ZT, its ROC will always be a circular ring that can degenerate into the empty set.
- If the signal is of finite duration, its ROC is the entire complex plane, with the possible exception of $z = 0$.
- The ROC of the Z transform of a signal does not contain any pole.

9.2 Main Properties of the ZT

The properties of the ZT are very useful both in the calculation of transformations and in their inversion. We will present them below, without proofs [65, 100].
1. Linearity and homogeneity:

$$\mathcal{Z}[ax(n) + by(n)] = a\mathcal{Z}[x(n)] + b\mathcal{Z}[y(n)] \tag{9.12}$$

Example 9.2.1. *As an application, we will calculate the ZT of $\cos(\omega)\varepsilon(n)$ and $\sin(\omega)\varepsilon(n)$. Because $e^{i\omega n}\varepsilon(n) = \cos(\omega)\varepsilon(n) + i\sin(\omega)\varepsilon(n)$, by linearity,*

$$\mathcal{Z}[e^{i\omega n}.\varepsilon(n)] = \mathcal{Z}[\cos(\omega).\varepsilon(n)] + i.\mathcal{Z}[\sin(\omega).\varepsilon(n)] \tag{9.13}$$

Computing the ZT

$$\mathcal{Z}[e^{i\omega n}\,\varepsilon(n)] = \frac{1}{1 - e^{i\omega}\,z^{-1}}, \qquad |z| > 1 \tag{9.14}$$

we obtain

$$\frac{1}{1 - e^{i\omega}.z^{-1}} = \frac{1 - \cos\omega z^{-1}}{1 - 2\cos\omega.z^{-1} + z^{-2}} + i\frac{\sin\omega.z^{-1}}{1 - 2\cos\omega.z^{-1} + z^{-2}} \tag{9.15}$$

from where we conclude that

$$\mathcal{Z}[\cos(\omega n).\varepsilon(n)] = \frac{1 - \cos\omega z^{-1}}{1 - 2\cos\omega.z^{-1} + z^{-2}} \tag{9.16}$$

and

$$\mathcal{Z}[\sin(\omega n).\varepsilon(n)] = \frac{\sin\omega.z^{-1}}{1 - 2\cos\omega.z^{-1} + z^{-2}} \tag{9.17}$$

As can be seen, we can consider the sinusoids as impulse responses of systems with two poles on the unit circumference, having arguments equal to ±ω.
When using this property, we have to take into account the intersection of the ROC.

2. Modulation:

$$\mathcal{Z}\left[z_0^n x(n)\right] = X(z_0^{-1} z) \tag{9.18}$$

If $|z_0| = 1$, there is only a rotation.

Example 9.2.2. *If $z_0 = e^{i\omega n}$, we obtain easily the ZT of $e^{i\omega n}\varepsilon(n)$ from (9.8).*

Example 9.2.3. *From (9.16) and (9.17) we can show that*

$$\mathcal{Z}[\rho^n \cos(\omega n)\varepsilon(n)] = \frac{1 - \rho\cos\omega z^{-1}}{1 - 2\rho\cos\omega z^{-1} + \rho^2 z^{-2}} \tag{9.19}$$

and

$$\mathcal{Z}[\rho^n \sin(\omega n)\varepsilon(n)] = \frac{\rho\sin\omega z^{-1}}{1 - 2\rho\cos\omega z^{-1} + \rho^2 z^{-2}} \tag{9.20}$$

3. Time shift:

$$TZ[x(n - k)] = z^{-k}X(z), \qquad k \in \mathbf{Z} \tag{9.21}$$

4. Differentiation in time: let $Dx(n) = x(n) - x(n - 1)$; then

$$\mathcal{Z}\left[D^k x(n)\right] = [1 - z^{-1}]^k X(z) \tag{9.22}$$

Example 9.2.4. *As application, we have*

$$\mathcal{Z}\delta_n = \mathcal{Z}\varepsilon(n) - \mathcal{Z}\varepsilon(n - 1) = [1 - z^{-1}]\frac{1}{[1 - z^{-1}]} = 1 \tag{9.23}$$

and so we conclude that

$$\mathcal{Z}[\delta_n] = 1 \tag{9.24}$$

5. Differentiation in the complex plane: let $\mathcal{Z}[x(n)] = X(z)$; then

$$\mathcal{Z}\left[(-1)^k \frac{(n-1+k)!}{(n-1)!} x(n)\right] = z^k \frac{d^k}{dz^k} X(z) \tag{9.25}$$

6. Time convolution: let $x(n)$ and $y(n)$ be two functions with ZT $X(z)$ and $Y(z)$; we can show that

$$\mathcal{Z}[z(n)] = TZ[x(n) * y(n)] = X(z).Y(z) \tag{9.26}$$

Example 9.2.5. *Consider the triangular signal*

$$x(n) = \begin{cases} 3 & n = 0 \\ 2 & |n| = 1 \\ 1 & |n| = 2 \\ 0 & |n| > 2 \end{cases} \tag{9.27}$$

Its ZT is

$$X(z) = \frac{z^4 + 2z^3 + 3z^2 + 2z + 1}{z^2} \tag{9.28}$$

The ROC is $\mathbb{C}\backslash\{0\}$. However, the triangle is the auto-convolution of the rectangle

$$r(n) = \begin{cases} 1 & |n| \le 1 \\ 0 & |n| > 1 \end{cases} \tag{9.29}$$

having the ZT

$$R(z) = \frac{z^2 + z + 1}{z}, \tag{9.30}$$

from where

$$X(z) = R(z)^2 \tag{9.31}$$

The dual property of the time convolution is the following:

$$\mathcal{Z}[x(n)y(n)] = \frac{1}{2\pi i} \oint_\gamma X(z/w)Y(w)w^{-1}dw \tag{9.32}$$

The convolution in the transform domain is done along a circle of radius 1 centred at the origin. In order to demonstrate this property we have to rely on the inversion integral that will be presented later.

7. Accumulation:

$$\mathcal{Z}\left[\sum_{n=-\infty}^{n} x(k)\right] = \frac{X(z)}{1-z^{-1}} \tag{9.33}$$

It is a particular case of the convolution:

$$\sum_{n=-\infty}^{n} x(k) = x * \varepsilon \tag{9.34}$$

8. Correlation: let $x(n)$ and $y(n)$ be two real functions with ZT $X(z)$ and $Y(z)$. Its correlation, $R_{xy}(n) = \sum_{n=-\infty}^{\infty} x(k)y(k+n)$, has ZT given by

$$\mathcal{Z}[R_{xy}(n)] = X(z).Y(z^{-1}) \tag{9.35}$$

Example 9.2.6. *If* $x(n) = a^n \varepsilon(n)$ *and* $y(n) = b^n \varepsilon(n)$,

$$\mathcal{Z}[R_{xy}(n)] = \frac{1}{1 - a.z^{-1}} \frac{1}{1 - b.z} \tag{9.36}$$

9. Time reversal: if $\mathcal{Z}[x(n)] = X(z)$ for $b < |z| < a$, with $a, b \in \mathbb{R}_0^+$, then

$$\mathcal{Z}[x(-n)] = X(z^{-1}) \tag{9.37}$$

for $1/a < |z| < 1/b$. Eventually, we can have $a = +\infty$ and/or $b = 0$. In the causal case, $a = +\infty$ and $b > 0$; in the anti-causal case, it will be $a > 0$ and $b = 0$.

9.3 Signals whose transforms are simple fractions

Let us consider two exponentials, one causal $(a^n \varepsilon(n))$ and the other anti-causal $(-a^n \varepsilon(-n))$. Their transforms are

$$\sum_{n=0}^{\infty} a^n.z^{-n} = \frac{1}{1 - a.z^{-1}} \qquad |z| > |a| \tag{9.38}$$

$$-\sum_{n=-\infty}^{-1} a^n.z^{-n} = \frac{1}{1 - a.z^{-1}} \qquad |z| < |a| \tag{9.39}$$

Those signals have the same analytical expression for TZ, but different ROCs.

The fractions of the second members of (9.38) and (9.39) are particular cases of simple or partial fractions that are of the type

$$F(z) = \frac{1}{(1 - a.z^{-1})^n} \tag{9.40}$$

or

$$F(z) = \frac{1}{(z - a)^n} \tag{9.41}$$

Let us continue with the fractions of the type (9.40) and try to find the signals of which they are transforms. We will start with (9.38). Proceeding step by step, by computing the derivatives of both members in order to z (we could obtain the same result differentiating in order to a), we get

$$\mathcal{Z}[(n+1)_k a^n \varepsilon(n)] = \frac{k!}{(1 - a.z^{-1})^{k+1}}, \qquad |z| > |a| \tag{9.42}$$

The relation (9.42) is general and, if used reversally, shows how we can obtain the inverse ZT of a simple fraction of order k. With a similar procedure, we obtain

$$\mathcal{Z}[-(n+1)_k a^n \varepsilon(-n-k-1)] = \frac{1}{(1 - a.z^{-1})^{k+1}}, \qquad |z| < |a|. \tag{9.43}$$

For fractions of type (9.39), we get

$$\mathcal{Z}\left[(n-k)_k a^{n-k-1}\varepsilon(n-k-1)\right] = \frac{k!}{(z-a)^{k+1}} \qquad |z| > |a|. \qquad (9.44)$$

From (9.43), it comes

$$\mathcal{Z}\left[-(n-k)_k a^{n-k-1}\varepsilon(-n)\right] = \frac{k!}{(z-a)^{k+1}} \qquad |z| < |a|. \qquad (9.45)$$

With (9.42) and (9.44) or (9.43) and (9.45), we can carry out the inversion of rational transforms, namely, transfer functions. For this particular case, the above results allow us to draw some conclusions regarding the influence of the poles on the time response of causal systems:

- The closer to the unit circle the poles are, the slower the decay of the impulse response.
- In the limiting case of the poles being on the circumference, the impulse response does not tend to zero:
  - If the order of the pole is greater than 1, the impulse response grows indefinitely.
  - If the order of the pole is equal to 1:
    * the impulse response remains constant, if the pole is equal to 1;
    * the impulse response oscillates, if the pole is complex.

9.4 Inversion of the ZT

9.4.1 Inversion Integral

Definition 9.4.1. *The inverse ZT can be obtained by the inversion integral defined by*

$$x(n) = \frac{1}{2\pi i} \oint_\gamma X(z)z^{n-1}dz \qquad (9.46)$$

where γ is a circle centred at the origin located in the ROC of the transform and described in the direct sense.

The calculation uses Cauchy's theorem of complex variable functions. However, most cases found in practice do not require such a procedure. We can use a partial fraction decomposition or a McLaurin series expansion.

Example 9.4.1. *Let $X(z) = \frac{z}{(z-0.75)(z+0.5)}$. The integrand has 2 or 3 poles, depending on the variable n and the chosen ROC. We have three situations. Let r be the radius of the integration path γ.*
1. *Causal solution:*
   The ROC is defined by $|z| > 0.75$. Therefore, we need to choose $r > 0.75$. As the poles

are simple (i.e. order 1),

$$x(n) = (4/5)\left[0.75^n - (-0.5)^n\right]\varepsilon(n) \tag{9.47}$$

2. *Anti-causal solution:*
   The ROC is defined by $|z| < 0.5$. Therefore, we need to choose $r < 0.5$. In this case, we have for $n \leq 0$ a pole at the origin with multiplicity equal to $|n|$. For $n = 0$, the residue assumes the value $-8/3$. In the other cases with $n < 0$, the computation of the residue involves the (potentially tedious) computation of a derivative. We can avoid this work by making a change of variable in the integral, $z = 1/w$, and so obtain a situation similar to the previous one.
3. *Bilateral solution:*
   If we choose a circle with radius $0.5 < r < 0.75$, the solution is, as expected, the sum of a causal term (corresponding to the pole at -0.5) and of an anti-causal term (corresponding to the pole at 0.75).

9.4.2 Inversion by decomposition into a polynomial and partial fractions

As said before, we are interested in transfer functions of the type

$$H_d(z) = \frac{\sum_{m=0}^{M} b_m z^{-m}}{\sum_{m=0}^{N} a_m z^{-m}} \tag{9.48}$$

where we assume $a_0 = 1$, without loosing generality. In a perfectly general way, we can decompose the previous TF into the sum of a polynomial and a fraction, both functions of z^{-1}:

$$H_d(z) = \sum_{m=0}^{M-N+1} \gamma_m z^{-m} + \frac{\sum_{m=0}^{N-1} \beta_m z^{-m}}{\sum_{m=0}^{N} a_m z^{-m}} \tag{9.49}$$

The polynomial case is extremely simple and is solved by mere visual inspection.

$$X(z) = \sum_{m=0}^{M-N+1} \gamma_m z^{-m} \Rightarrow x(n) = \begin{cases} \gamma_n & \text{for } 0 \leq n \leq M - N + 1 \\ 0 & \text{for } 0 < n \text{ and } n > M - N + 1 \end{cases} \tag{9.50}$$

The fraction can always be decomposed into a linear combination of simple fractions (see Appendix E):

$$Q(z) = \sum_{i=1}^{N_p} \sum_{j=1}^{m_i} \frac{a_{ij}}{(1 - p_i z^{-1})^j}. \tag{9.51}$$

Here N_p i is the number of disctict poles and m_i are the corresponding orders. Making $K = \sum_{1}^{N_p} m_n$, we get

$$Q(z) = z^K \sum_{i=1}^{N_p} \sum_{j=1}^{m_i} \frac{a_{ij}}{(z - p_i)^j}. \tag{9.52}$$

Example 9.4.2. *Suppose that a given ZT has two zeros at the points $z = \pm i$ and three poles at the points $z = -0,9$ and $2.e^{\pm i\theta}, \theta \in]-\pi, \pi]$:*

$$H(z) = \frac{(z-j).(z+j)}{(z - 2e^{-i\theta}).(z - 2e^{i\theta}).(z+0.9)} \tag{9.53}$$

Noting that the residues corresponding to pairs of conjugate complex poles are also conjugate, the partial fraction decomposition is

$$H(z) = \frac{A}{z - 2.e^{i\theta}} + \frac{A^*}{z - 2.e^{-i\theta}} + \frac{1}{z+0.9} \tag{9.54}$$

The two first fractions have the associate ROC defined by $|z| > 2$ and $|z| < 2$. The last one has ROCs $|z| > 0.9$ and $|z| < 0.9$. This means that the function $H(z)$ represents a given ZT if and only if we assign it a given ROC. Since we have three ROCs, there are three different signals having $H(z)$ as their ZT:

1. *Region A: $ROC = \{z \in \mathbb{C} : |z| < 0.9\} \Rightarrow x_A(n)$, left signal,*
2. *Region B: $ROC = \{z \in \mathbb{C} : 0.9 < |z| < 2\} \Rightarrow x_B(n)$, two-sided signal,*
3. *Region C: $ROC = \{z \in \mathbb{C} : |z| > 2\} \Rightarrow x_C(n)$, right signal.*

The coefficients A and C are given by

$$A = \frac{4e^{-i2\theta} + 1}{4i\sin(\theta)(2e^{-i\theta} + 0.9)} \tag{9.55}$$

$$C = \frac{0.9^2 + 1}{(0.9 - 2e^{-i\theta})(0.9 - 2e^{i\theta})} = \frac{0.81 + 1}{0.81 - 4\cos(\theta) + 4} \tag{9.56}$$

In particular, for $\theta = \pi/4$, we get $A = 0.5375.e^{-i0.3056}$ and $C = 0.9134$. With these values and using (9.44) and (9.45), the causal and anti-causal systems will be unstable, while the corresponding response to the stable acausal system will be given by:

$$h_n = -A\,2^{n-1}e^{i(n-1)\pi/4}\varepsilon(-n) - A^*\,2^{n-1}e^{-i(n-1)\pi/4}\varepsilon(-n) + C\,0.9^{n-1}\varepsilon(n) \tag{9.57}$$

The first two terms can be combined resulting in a real sinusoid.

Example 9.4.3. *Consider the signal with ZT*

$$X(z) = \frac{3 - \frac{5}{6}z^{-1}}{(1 - \frac{1}{4}z^{-1})(1 - \frac{1}{3}z^{-1})} \tag{9.58}$$

with ROC $|z| > \frac{1}{3}$. To make the terms in z^{-1} disappear, first of all multiply the numerator and denominator of $X(z)$ by z^2, resulting in

$$X(z) = z\frac{3z - \frac{5}{6}}{(z - \frac{1}{4})(z - \frac{1}{3})} \tag{9.59}$$

The term z that has been left out since it is useful in the following step. Using any method of partial fraction decomposition, the above expression can be written successively as

$$X(z) = z \left[\frac{1}{z - \frac{1}{4}} + \frac{2}{z - \frac{1}{3}} \right] = \frac{z}{z - \frac{1}{4}} + \frac{2z}{z - \frac{1}{3}} = \frac{1}{1 - \frac{1}{4}z^{-1}} + \frac{2}{1 - \frac{1}{3}z^{-1}} \tag{9.60}$$

The inversion is now immediate. It is enough to use (9.38), to get

$$x(n) = \left[\left(\frac{1}{4} \right)^n + 2 \left(\frac{1}{3} \right)^n \right] \varepsilon(n) \tag{9.61}$$

We leave as an exercise to obtain the inverse transforms with the other ROCs.

9.4.3 Inversion by series expansion

The method of inversion by series expansion consists in obtaining the McLaurin or Laurent series in z^{-1} or z.

Example 9.4.4. *Consider the ZT*

$$X(z) = \frac{z^5 + 2z^4 - 3z^2 - 2z + 1}{z^3} \tag{9.62}$$

with ROC = $\{z \in \mathbb{C} : |z| > 0\}$. X(z) can be written as

$$X(z) = z^2 + 2z - 3z^{-1} - 2z^{-2} + z^{-3} \tag{9.63}$$

Inverting term by term, we obtain

$$x(n) = \delta(n+2) + 2\delta(n+1) - 3\delta(n-1) - 2\delta(n-2) + \delta(n-3) \tag{9.64}$$

Example 9.4.5. *Let H(z) defined by*

$$H(z) = \frac{1}{1 - z^2} \tag{9.65}$$

If $|z| < 1$, we have

$$H(z) = \sum_{n=0}^{\infty} z^{2n} = \sum_{n=-\infty}^{0} z^{-2n} \tag{9.66}$$

from where we conclude that

$$h(n) = \begin{cases} 1 & \text{if } n \le 0 \text{ and is even} \\ 0 & \text{if } n \text{ is odd or if } n \ge 2 \text{ and is even} \end{cases} \tag{9.67}$$

Similarly, if $|z| > 1$, we have

$$H(z) = -z^{-2} \sum_{n=0}^{\infty} z^{-2n} = \sum_{n=1}^{\infty} z^{-2n} \tag{9.68}$$

from where we obtain

$$h(n) = \begin{cases} -1 & \text{if } n \geq 2 \text{ and is even} \\ 0 & \text{if } n \text{ is odd or if } n \leq 0 \text{ and is even} \end{cases} \tag{9.69}$$

Example 9.4.6. *Let $X(z) = e^{1/z}$ for $|z| > 0$. Attending to*

$$X(z) = \sum_{n=0}^{\infty} \frac{1}{n!} z^{-n} \tag{9.70}$$

we conclude immediately that

$$x(n) = \frac{1}{n!} \varepsilon(n) \tag{9.71}$$

9.4.4 Step response

We saw earlier how to get the impulse response. The step response, $r_\varepsilon(n)$, can be easily obtained from it. Just note that

$$\delta(n) = \varepsilon(n) - \varepsilon(n-1) \Rightarrow \varepsilon(n) = \sum_{k=0}^{n} \delta(k) \tag{9.72}$$

Therefore

$$r_\varepsilon(n) = \sum_{k=0}^{n} h(k) \tag{9.73}$$

Obviously we can start from the transfer function. Let $H(z) = \frac{B(z)}{A(z)}$ be a TF. The ZT of the step response is

$$R_\varepsilon(z) = \frac{1}{1 - z^{-1}} \frac{B(z)}{A(z)} \tag{9.74}$$

Suppose that $H(z)$ is a proper fraction. If not, proceed as usual: decompose into the sum of a polynomial with a proper fraction. Given what has been done above, we can write

$$R_\varepsilon(z) = \frac{H(1)}{1 - z^{-1}} + \frac{C(z)}{A(z)} \tag{9.75}$$

where $C(z) = \sum_{k=0}^{N-1} c_k z^{-k}$. From previous relations we obtain

$$\frac{H(1)}{1 - z^{-1}} + \frac{C(z)}{A(z)} = \frac{1}{1 - z^{-1}} \frac{B(z)}{A(z)} \tag{9.76}$$

Now we reduce the fractions of the first member to the same denominator and we get

$$B(z) = H(1)A(z) + (1 - z^{-1})C(z) \tag{9.77}$$

By recursive identification of the coefficients corresponding to equal powers of z^{-1}, we obtain

$$b_0 = H(1)a_0 + c_0 \tag{9.78}$$

$$b_1 = H(1)a_1 + c_1 - c_0 \tag{9.79}$$

$$b_2 = H(1)a_2 + c_2 - c_1 \tag{9.80}$$

$$\cdots \tag{9.81}$$

$$b_k = H(1)a_k + c_k - c_{k-1} \tag{9.82}$$

$$\cdots \tag{9.83}$$

from which the following relation is immediately deduced:

$$c_k = \sum_{j=0}^{k} [b_j - H(1)a_j] \qquad k = 0, 1, \ldots, N-1 \tag{9.84}$$

This allows completely defining $R_d(z)$ in (9.75). From here we conclude the following:

Remark 9.4.1. *The step response has two components. One of them is proportional to the step (the proportionality constant is the static gain). The other is the impulse response of a modified ARMA model that will tend to zero if the system is stable in the strict sense.*

Example 9.4.7. *Let $H(z) = \frac{1}{1-pz^{-1}}$ for $|z| > |p|$ and $p \neq 1$. The static gain of the system is $H(1) = \frac{1}{1-p}$ and $C(z) = c_0 = \frac{-p}{1-p} = -pH(1)$, and so*

$$R_\varepsilon(z) = \frac{1}{1-z^{-1}} \frac{1}{1-pz^{-1}} = \frac{H(1)}{1-z^{-1}} - pH(1)\frac{1}{1-pz^{-1}} \tag{9.85}$$

where it is concluded that

$$r_\varepsilon(t) = H(1)\left[1 - p^{n+1}\right]\epsilon(n) \tag{9.86}$$

9.4.5 Response to a causal sinusoid

Let us repeat the development done in the previous section but now for the case where the input is a causal sinusoid. This question is not very different from the previous one in that a cosine or sine can always be expressed as a sum of cisoids, according to Euler's formula. So, let us start by solving the problem corresponding to an input type $s(n) = e^{i\omega_0 n}\epsilon(n)$. As we saw earlier, its ZT is $S(z) = \frac{1}{1-e^{i\omega_0}z^{-1}}$, $|z| > 1$. We will have

$$R_s(z) = \frac{1}{1-e^{i\omega_0}z^{-1}}\frac{B(z)}{A(z)} = \frac{H(e^{i\omega_0})}{1-e^{i\omega_0}z^{-1}} + \frac{C(z)}{A(z)} \tag{9.87}$$

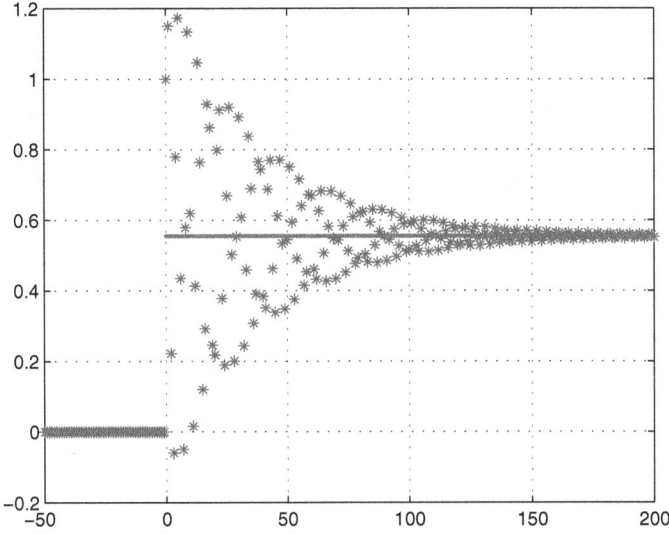

Fig. 9.1: Response of a first order system with $p = 0.9$ to a step.

with $C(z) = \sum\limits_{k=0}^{N-1} c_k z^{-k}$ and where

$$B(z) = H(e^{i\omega_0})A(z) + (1 - e^{i\omega_0}z^{-1})C(z) \tag{9.88}$$

that allows us to write

$$b_0 = H(e^{i\omega_0})a_0 + c_0 \tag{9.89}$$

$$b_1 = H(e^{i\omega_0})a_1 + c_1 - e^{i\omega_0}c_0 \tag{9.90}$$

$$b_2 = H(e^{i\omega_0})a_2 + c_2 - e^{i\omega_0}c_1 \tag{9.91}$$

$$\cdots$$

$$b_k = H(e^{i\omega_0})a_k + c_k - e^{i\omega_0}c_{k-1} \tag{9.92}$$

$$\cdots$$

From here the following relations are deduced:

$$c_0 = b_0 - H(e^{i\omega_0})a_0 \tag{9.93}$$

$$c_k = \sum\limits_{j=0}^{k} \left[b_j - H(e^{i\omega_0})a_j \right] e^{i\omega_0(k-j)}, \qquad k = 1, 2, \ldots, N-1 \tag{9.94}$$

Similarly, we can obtain

$$\frac{1}{1 - e^{-i\omega_0}z^{-1}} \frac{B(z)}{A(z)} = \frac{H(e^{-i\omega_0})}{1 - e^{-i\omega_0}z^{-1}} + \frac{D(z)}{A(z)} \tag{9.95}$$

with $D(z) = \sum\limits_{k=0}^{N-1} d_k z^{-k}$ and also

$$d_0 = b_0 - H(e^{-i\omega_0})a_0 \tag{9.96}$$

$$d_k = \sum_{j=0}^{k} \left[b_j - H(e^{-i\omega_0})a_j \right] e^{-i\omega_0(k-j)}, \qquad k = 1, 2, \ldots, N-1 \tag{9.97}$$

Depending on whether the answer to a cosine or a sine is desired, so is the combination of the previous relationships. Suppose we want the answer to a cosine. We have

$$R_{co}(z) = \frac{H(e^{i\omega_0})/2}{1 - e^{i\omega_0}z^{-1}} + \frac{H(e^{-i\omega_0})/2}{1 - e^{-i\omega_0}z^{-1}} + \frac{C(z) + D(z)}{A(z)} \tag{9.98}$$

Let $A(e^{i\omega_0}) = \left|H(e^{i\omega_0})\right|$ and $\phi(e^{i\omega_0}) = \arg\left[H(e^{i\omega_0})\right]$. The sum of the two first terms gives

$$\frac{H(e^{i\omega_0})/2}{1 - e^{i\omega_0}z^{-1}} + \frac{H(e^{-i\omega_0})/2}{1 - e^{-i\omega_0}z^{-1}} =$$

$$= A\cos(\phi)\frac{1 - \cos\omega_0 z^{-1}}{1 - 2\cos\omega_0 z^{-1} + z^{-2}} - A\sin(\phi)\frac{\sin\omega_0 z^{-1}}{1 - 2\cos\omega_0 z^{-1} + z^{-2}} \tag{9.99}$$

corresponding to the time response

$$r_c(n) = \left[A\cos(\phi)\cos\omega_0 n - A\sin(\phi)\sin\omega_0 n \right]\epsilon(n) = A\cos\left(\omega_0 n + \phi\right)\epsilon(n) \tag{9.100}$$

which is identical to the response that is obtained in a steady state regime. By representing $h_t(n)$ as inverse ZT of $\frac{C(z)+D(z)}{A(z)}$, we can write the response of a discrete-time system to a causal cosine in the form

$$r_{co}(n) = A(e^{i\omega_0})\cos\left(\omega_0 n + \phi(e^{-i\omega_0})\right)\epsilon(n) + h_t(n) \tag{9.101}$$

The term $h_t(n)$ corresponds to the transient part that tends to zero if the system is stable.

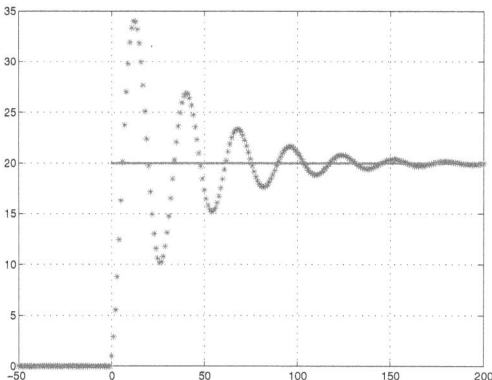

Fig. 9.2: Response of a second order system to a step.

9.5 Stability and Jury criterion

In chapter 1, we introduced the BIBO stability and its relation with the impulse response:

Definition 9.5.1. *A discrete-time linear system is stable, iff its impulse response is absolutely summable:*

$$\sum_{n=-\infty}^{+\infty} |h(n)| < \infty \tag{9.102}$$

According to the ZT theory we can assert an equivalent statement:

Lemma 2. *A discrete-time linear system is stable, iff the ROC of its transfer function contains the unit circle.*

This has the following consequence:

Theorem 9.5.1. *If $h(n)$ is the impulse response of a BIBO stable discrete-time linear system, then $\lim_{n\to\infty} h(\pm n) = 0$.*

This is the reason why we call this kind of systems asymptotically stable.

In way analogous to the continuous case, we define a system stable in the wide sense as one for which the impulse response does not tend to zero, but is bounded: $|h(n)| < A \in \mathbb{R}^+ \; n > n_0$.

Again, there is an intimate relationship between the properties of the impulse response and the location of the poles of the system:

Lemma 3. *A causal (anti-causal) discrete-time linear system is asymptotically stable iff it has all its poles inside (outside) the unit disk.*

In wide sense stability, some poles may lie on the unit circle.

In the following, we shall be concerned with causal systems only. There are several necessary and several sufficient conditions for stability. However, we will not consider them. We will only study the Jury criterion, that is a method for counting the number of zeroes of a polynomial inside the unit disk.

The Jury criterion is based on the reversal of the Levinson recursion [65, 100] and consists in the recursive computation of a decreasing order sequence of polynomials. Let $P_N(z) = A(z) = \sum_{k=0}^{N} a_k^N z^{-k}$; its reverse is $Q_N(z) = z^{-N} A(z^{-1}) = \sum_{k=0}^{N} a_{N-k}^N z^{-k}$. The $(N-1)$–order polynomials are given by

$$\begin{bmatrix} P^{N-1}(z) \\ Q^{N-1}(z) \end{bmatrix} = \frac{1}{1 - C_N^2} \begin{bmatrix} z & -zC_N \\ -C_N & 1 \end{bmatrix} \begin{bmatrix} P^N(z) \\ Q^N(z) \end{bmatrix} \tag{9.103}$$

whereby

$$P^{N-1}(z) = \frac{z}{1 - C_N^2} \left[P^N(z) - C_N Q^N(z) \right] \tag{9.104}$$

or, as the constant $1 - C_N^2$ can be incorporated in $P^{N-1}(z)$:

$$P^{N-1}(z) = z\left[P^N(z) - C_N Q^N(z)\right] \tag{9.105}$$

which can be written in terms of the coefficients in the form:

$$a_i^{N-1} = a_i^N - C_N a_{N-i}^N \qquad i = 0, \ldots, N-1 \tag{9.106}$$

As can be seen, the coefficient C_N is calculated so that the coefficient of order N is null. Therefore, it is given by

$$C_N = \frac{a_N^N}{a_0^N} \tag{9.107}$$

This coefficient is called *reflection coefficient*.

The procedure for obtaining the reflection coefficients and the successively decreasing order polynomials can be systematised in a table, called the Jury table, which is given below. As in the Routh table, we can multiply any polynomial by a positive constant without changing the conclusions of the Jury table.

Jury table

z^N	a_0^N	a_1^N	$\cdots$	$\cdots$	$\cdots$	a_{N-1}^N	a_N^N	
	a_N^N	a_{N-1}^N	$\cdots$	$\cdots$	$\cdots$	a_1^N	a_0^N	$\frac{a_N^N}{a_0^N}$
z^{N-1}	a_0^{N-1}	a_1^{N-1}	$\cdots$	$\cdots$	a_{N-2}^{N-1}	a_{N-1}^{N-1}		
	a_{N-1}^{N-1}	a_{N-2}^{N-1}	$\cdots$	$\cdots$	a_1^{N-1}	a_0^{N-1}		$\frac{a_{N-1}^{N-1}}{a_0^{N-1}}$
z^{N-2}	a_0^{N-2}	a_1^{N-2}	$\cdots$	a_{N-3}^{N-2}	a_{N-2}^{N-2}			
	a_{N-2}^{N-2}	a_{N-3}^{N-2}	$\cdots$	a_1^{N-2}	a_0^{N-2}			$\frac{a_{N-2}^{N-2}}{a_0^{N-2}}$
$\vdots$	$\vdots$	$\vdots$				$\vdots$	$\vdots$	
z^2	a_0^2	a_1^2	a_2^2	$\cdots$				
	a_2^2	a_1^2	a_0^2	$\cdots$				$\frac{a_2^2}{a_0^2}$
z	a_0^1	a_1^1	$\cdots$					
	a_1^1	a_0^1	$\cdots$					$\frac{a_1^1}{a_0^1}$
z^0	a_0^0							

The values of the last column are the reflection coefficients which inform us about the distribution of the zeros of the polynomial.

Suppose there are m reflection coefficients such that $|C_i| > 1$. Let $L_j\, j = 1, \ldots m$ the corresponding subscripts. Then the number of zeroes in the unit disk is given by

$$N_e = \begin{cases} \displaystyle\sum_{j=1}^{m} (-1)^j L_j & \text{if } m \text{ is even} \\[2mm] N + 1 + \displaystyle\sum_{j=1}^{m} (-1)^j L_j & \text{if } m \text{ is odd} \end{cases} \tag{9.108}$$

However, in a similar way to the Routh-Hurwitz criterion, we can state the **Jury criterion**: *if $a_0^N > 0$, the number of positive values in the upper left elements of each line* is equal to the number of zeros inside the unit disk, and the number of negative values is equal to the number of zeroes outside the unit disk.

Example 9.5.1. *Consider the polynomial* $P(z) = z^4 + 2.8z^3 - 4.35z^2 + 1.75z - 0.2.$

z^4	1	2.8	-4.35	1.75	-0.2	CR
	-0.2	1.75	-4.35	2.8	1	-0.2
z^3	0.96	-3.15	-5.22	2.31		
	2.31	-5.22	-3.15	0.96		2.406
z^2	-4.6	15.71	-12.8			
	-12.8	15.71	-4.6			2.406
z	31.03	-28.02				
	-28.02	31.03				-0.905
z^0	5.73					

The upper left elements of the lines are 1, 0.96, -4.6, 31.03 *and* 5.73. *According to the Jury criterion there are 3 roots inside the unit disk and 1 outside the unit disk. Using the reflection coefficients, we find that 2 have absolute values greater than 1, corresponding to subscripts 2 and 3. Then* $N_e = \sum_j 2(-1)^j L_j = -2 + 3 = 1$ *confirming there is only one root in the unit disk.*

9.5.1 Singular cases of Jury table

As in the case of the Routh-Hurwitz table there are two singularity situations in the Jury table:
- Line of zeroes,
- Some null initial values in a line.

This second case corresponds to a situation resulting from the fact that there is no one to one correspondence between the reflection coefficients and polynomial sets. In fact, there are polynomials that can be derived from two different sequences of reflection coefficients. For example, all polynomials of the type $P_3 = z^3 - az^2 + az - 1$ can be obtained using C_1 and C_2 such that $C_1 - C_2 - C_1 C_2 = a$; on the other hand, if $C_1 = 1$, $a = 0$, then, independently of C_2, we get the same result.

Conversely, there are polynomials for which it is not possible to find any sequence of reflection coefficients that generates them. For instance, the polynomials of the family $P_3 = z^3 + az^2 + bz - 1$ with $|a| \neq |b|$ are in this situation. The procedure to continue the table consists of replacing the polynomial with another with identical root distribution. However, these situations only occur when there are zeros outside the unit circle.

On the other hand, the case where a line of zeroes appears is very important, being used, for example, in the modelling of signals known to be sums of sinusoids. This situation arises when the polynomial has two roots on the unit circle, or two reciprocal roots (their product is 1). To solve this problem, as in the case of the Routh table, we proceed to construct the table using the derivative of the polynomial of the previous line.

Example 9.5.2. *Consider polynomial* $P(z) = z^5 - 1.8z^4 - 0.35z^3 - 0.8z^2 + 1.65z - 0.5$.

z^5	1	−1.8	−0.35	−0.8	1.65	−0.5	
	−0.5	1.65	−0.8	−0.35	−1.8	1	−0.5
z^4	0.75	−0.975	−0.75	−0.975	0.75		
	0.75	−0.975	−0.75	−0.975	0.75		1
z^3	0	0	0	0			
	3	−2.925	−1.5	−0.975			
	−0.975	−1.5	−2.925	3			−0.325
z^2	2.68	−3.41	−2.45				
	−2.45	−3.41	2.68				−9.142
z	0.44	−6.527					
	0 − 6.527	10.44					−14.834
z^0	−96.3						

As it can be seen, there is a root inside the unit circle due to polynomial P_4. This polynomial will obviously have one root inside the unit circle and the remaining two on the unit circle. In fact the roots of P5 are: $0.5, 0.5, 2, 0.6 \pm i0.8$. It should be noted that, considering only the part of the table since the appearance of the symmetric or antisymmetric polynomial, we can know the number of roots on the unit circle.

9.6 Initial Conditions

The responses of the linear systems just studied correspond to two situations in which either the system was excited for a long time or the system was at rest at the initial time. In fact, the impulse and step responses are generally obtained with the system in a zero energy state at the time of the excitation. As long as the steady-state response occurs in causal and stable systems, its effect is attenuated over time, the attenuation being greater or lesser, depending on the proximity of the poles to the unit-radius circumference. The effect of non-zero initial conditions can be very important in the instants immediately after the system excitation. Now we will study the response of linear systems when the conditions are nonzero.

What happens is that we intend to solve a given difference equation to values of n greater than (or less than, in the anti-causal case) a given initial moment that, without loss of generality, we can assume to be the origin. To deal with the question, simply multiply the equation by the Heaviside function, which has the role of an observation window, and change the terms in order to temporarily align the signal and the Heaviside function. To see how this happens, let us return to the difference equation (8.9)

$$\sum_{k=0}^{N} a_k y(n-k) = \sum_{k=0}^{M} b_k x(n-k), \tag{9.109}$$

defined for all $n \in \mathbb{Z}$. To find the solution for $n > 0$, we begin by multiplying both members of the equation by $\varepsilon(n)$:

$$\sum_{k=0}^{N} a_k y(n-k)\varepsilon(n) = \sum_{k=0}^{M} b_k x(n-k)\varepsilon(n). \tag{9.110}$$

Now, we need to relate $x(n-k)\varepsilon(n)$ with $x(n-k)\varepsilon(n-k)$ and $y(n-k)\varepsilon(n)$ with $y(n-k)\varepsilon(n-k)$. To do it, we note that if $k > 0$, we have:

$$x(n-k)\varepsilon(n) = x(n-k)\left[\varepsilon(n-k) + \sum_{i=0}^{k-1} \delta_{n-i}\right]$$

$$= x(n-k)\varepsilon(n-k) + \sum_{i=0}^{k-1} x(i-k)\delta_{n-i} \tag{9.111}$$

and then making $X_c(z) = \mathbb{Z}[x(n).\varepsilon(n)]$,

$$\mathbb{Z}[x(n-k)\varepsilon(n)] = z^{-k}X_c(z) + \sum_{i=0}^{k-1} x(i-k)z^{-i} \tag{9.112}$$

Similarly, we have:

$$x(n+k)\varepsilon(n) = x(n+k)\left[\varepsilon(n+k) + \sum_{i=1}^{k} \delta_{n+i}\right]$$

$$= x(n+k)\varepsilon(n+k) + \sum_{i=1}^{k} x(i+k)\delta_{n+i} \tag{9.113}$$

and then

$$\mathbb{Z}[x(n+k)\varepsilon(n)] = z^{k}X_c(z) + \sum_{i=1}^{k} x(i+k)z^{i} \tag{9.114}$$

Therefore, to solve equations with given initial conditions, it is enough to apply the rules deduced from the difference equation. We obtain, successively:

$$A(z)Y_c(z) + \sum_{k=0}^{N} a_k \sum_{i=0}^{k-1} y(i-k)z^{-i} = B(z)X_c(z) + \sum_{k=0}^{M} b_k \sum_{i=0}^{k-1} x(i-k)z^{-i} \tag{9.115}$$

$$A(z)Y_c(z) = B(z)X_c(z) + \sum_{k=0}^{M} b_k \sum_{i=0}^{k-1} x(i-k)z^{-i} - \sum_{k=0}^{N} a_k \sum_{i=0}^{k-1} y(i-k)z^{-i} \tag{9.116}$$

from where one concludes immediately

$$Y_c(z) = H(z)X_c(z) + \frac{\displaystyle\sum_{k=0}^{M} b_k \sum_{i=0}^{k-1} x(i-k)z^{-i} - \sum_{k=0}^{N} a_k \sum_{i=0}^{k-1} y(i-k)z^{-i}}{A(z)} \tag{9.117}$$

which shows that, as mentioned before, the output has two components: one from the forced regime, another other from the free regime. Note that the numerator of the second term has a degree lower than the greater of the orders of poles and zeros. If the system is asymptotically stable, the free regime term tends to zero when $n \to \infty$ as previously mentioned.

9.7 Initial and final value theorems

The theorems of the initial value and final value are very useful in situations where it is not intended to calculate exactly the response of a given system, but it is enough to make a sketch that gives an idea of its evolution.

Let $x(n)$ be any causal signal, with TZ given by $X(z)$. The initial value $x(0)$ can be obtained from

$$x(0) = \lim_{|z| \to \infty} X(z) \tag{9.118}$$

as is immediately deduced from the definition of ZT. In fact,

$$X(z) = x(0) + x(1) z^{-1} + x(2) z^{-2} + x(3) z^{-3} + \dots \tag{9.119}$$

so that all terms, except the first one, tend to zero when $|z|$ tends to infinity. We can also get the values for the 2nd time instant, or the 3rd, the 4th... For example, for $x(1)$, we proceed as follows:

$$x(1) = \lim_{|z| \to \infty} z \left[X(z) - x(0) \right]. \tag{9.120}$$

This expression can be easily generalised for $x(2)$, $x(3)$, and so on.

As to the final value, the relation is somewhat different, but with some degree of similarity to the corresponding relation for continuous systems:

$$x(\infty) = \lim_{z \to 1} \left(1 - z^{-1} \right) X(z) \tag{9.121}$$

To prove this, we use the accumulation property:

$$\mathcal{Z} \left[\sum_{-\infty}^{n} x(k) \right] = \frac{X(z)}{1 - z^{-1}}. \tag{9.122}$$

Let

$$x(n) = \sum_{k=-\infty}^{n} y(k), \tag{9.123}$$

from where

$$x(\infty) = \lim_{z \to 1} Y(z). \tag{9.124}$$

However, $X(z) = Y(z) 1 - z^{-1}$, yielding immediately the result (9.121). As a useful application, we can determine the final value of the response of a given system to a unit step. It is given by

$$r_d(\infty) = \lim_{z \to 1} \left(1 - z^{-1} \right) \frac{H(z)}{1 - z^{-1}} = H(e^{i0}), \tag{9.125}$$

which is nothing else than the static gain.

9.8 Continuous to discrete conversion (s2z)

9.8.1 Some considerations

The importance of continuous-time systems discussed in chapters 2 to 5, and in particular of integer order systems, frequently called continuous-time ARMA systems (CARMA), together with the ease of computation introduced by discrete ARMA models (DARMA), led to the study of ways of inter-relating them. Problems like the identification of continuous-time systems from sampled data, or embedding a DARMA into a CARMA, are very important in practical applications. Several different methods are applied in the discretisation (also known as s to z, or $s2z$) of a continuous-time, linear, time-invariant ARMA system. They consist in one of the following approximations:

- **Approximating the differential equations by difference equations** — This approximation can use one of three schemes: the forward rectangular rule, the backward rectangular rule, or the trapezoidal rule. The first rule is also known as Euler's rule, whereas the last is referred as Tustin's rule. These schemes can be done through s to z transformations in the transfer function.
- **Transfer function matching by input/output sampling** — Both the input and output are sampled, their ZT is calculated, and then the transfer function is obtained. The most important ways of doing this are the invariant impulse and step procedures.
- **Approximating the transfer function by pole-zero mapping techniques** — The approximated equivalent discrete-time system is obtained by matching the poles and zeroes of the transfer functions of the two systems.
- **Approximating the transfer function by the zero-order hold-equivalence technique** — The continuous-time excitation of the system is held constant between consecutive sampling instants by assuming a zero-order hold. The continuous-time system is then subjected to this "rectangular" input, which in fact consists of a sequence of steps. Sampling both input and output and computing the corresponding Z transforms, we obtain the discrete-time transfer function.
- **The technique of system response by a covariance equivalence** — The equivalent discrete-time system is obtained by requiring that the covariance function of the system response for a Gaussian white noise input coincides at all discrete time lags with that of the continuous-time system.

In what follows, we will concentrate on the first two methods. We will assume that the systems are causal and stable. Also let $H_c(s)$ and $H_d(z)$ be respectively the (integer order) continuous transfer function and the discrete transfer function of a system, given respectively by

$$H_c(s) = \frac{\sum_{m=0}^{M} B_m s^m}{\sum_{n=0}^{N} A_n s^n} \qquad (9.126)$$

$$H_d(z) = \frac{\displaystyle\sum_{m=0}^{M} b_m z^{-m}}{\displaystyle\sum_{n=0}^{N} a_n z^{-n}} \tag{9.127}$$

9.8.2 Approximation of derivatives

The first and more intuitive approach to transform a continuous-time system into a discrete-time is through an approximation of the derivative. Essentially, the method substitutes derivatives by the corresponding incremental ratios (therefore eliminating the computation of the limit in the derivative):

$$\dot{f}(t) = \frac{f(t) - f(t-T)}{T} \qquad \dot{f}(t) = \frac{f(t+T) - f(t)}{T} \tag{9.128}$$

These correspond to rectangular approximations of the signals, where we assume that T is "low enough". For reasons we stated before, we must normally choose the first of these two expressions, since we are usually dealing with causal systems. Should our system be anti-causal, we use the second.

This question can be studied from the point of view of Laplace and Z transforms. In fact, letting $F(s) = \mathcal{L}\left[\dot{f}(t)\right]$, we have

$$F(s) = \frac{1 - e^{-sT}}{T} \qquad F(s) = \frac{e^{sT} - 1}{T} \tag{9.129}$$

Making $z = e^{sT}$, we obtain two $s2z$ transformations:

$$s = \frac{1 - z^{-1}}{T} \qquad s = \frac{z - 1}{T} \tag{9.130}$$

corresponding respectively to

$$\dot{f}(nT) = \frac{f(nT) - f((n-1)T)}{T} \qquad \dot{f}(nT) = \frac{f((n+1)T) - f(t)}{T}. \tag{9.131}$$

As can be easily seen, the first transforms the left complex half-plane into a circumference contained in the unit disk, thus ensuring stability. Such is not the case of the second. In the following we will proceed with the causal case.

Suppose then that the transfer function of the continuous-time LTIS has the usual form (9.126), where we are going to perform substitution (9.130) to get

$$G_d(z) = \frac{\displaystyle\sum_{m=0}^{M} B_m T^{-m} (1 - z^{-1})^m}{\displaystyle\sum_{n=0}^{N} A_n T^{-n} (1 - z^{-1})^n}. \tag{9.132}$$

The binomial expansion

$$(1 - z^{-1})^n = \sum_{k=0}^{n} (-1)^k \binom{n}{k} z^{-k} \tag{9.133}$$

suggests that a representation using powers of z^{-1} is sought. After some manipulations, we obtain the following expressions for coefficients of the numerator and denominator of $G_d(z)$ given by (9.127):

$$b_k = \frac{(-1)^k}{k!} \sum_{n=k}^{M} B_n h^{-n} \frac{n!}{(n-k)!} \tag{9.134}$$

$$a_k = \frac{(-1)^k}{k!} \sum_{n=k}^{N} A_n h^{-n} \frac{n!}{(n-k)!} \tag{9.135}$$

Given that, for small values of T, the coefficients determined above can assume very high values, we can reduce the values to be manipulated by dividing the coefficients obtained by the sum of all of them. In this way, a gain K appears, given by

$$K = \frac{\sum_0^M b_k}{\sum_0^N a_k}, \tag{9.136}$$

and so the discrete system can be written in the form

$$G_d(z) = K \frac{\sum_{m=0}^{M} \bar{b}_m z^{-m}}{\sum_{n=0}^{N} \bar{a}_n z^{-n}}, \tag{9.137}$$

where the zero-order coefficients are now normalised to 1, and the coefficients $\bar{a}_m$ and $\bar{b}_m$ are the new normalised coefficients.

Example 9.8.1. *Consider a continuous-time LTIS with transfer function*

$$G(s) = \frac{2s^2 + 6s + 4}{s^4 + 2s^3 + 3s^2 + 4s + 1}. \tag{9.138}$$

Assume a sampling interval $T = 0.1$. With the above procedure, we obtain

$$G_d(z) = 0.021394 \frac{1 - 1.7424z^{-1} + 0.7576z^{-2}}{1 - 3.77958z^{-1} + 5.37277z^{-2} - 3.40357z^{-3} + 0.81045z^{-4}}. \tag{9.139}$$

The poles of $G_d(z)$ are located at $0.97516 \angle \pm 0.044286\pi$, 0.96410 and 0.88401.

It is interesting to study the effect of the variation of T. For this, we considered the case where $T = 1$ and recalculated the denominator polynomial, which is now $1 - 2z^{-1} + 1.5z^{-2} - 0.6z^{-3} + 0.2z^{-4}$. The corresponding poles are at $0.93674 \angle \pm 0.10202\pi$, $0.47741 \angle \pm 0.42534\pi$. As already noted, the decrease of T shifts the poles towards point 1.

This example suggests that this transformation preserves stability. This is indeed so. Let us see how, by returning to the general form of the transfer function of the continuous system and rewriting it in the factorised form:

$$G(s) = K_0 \frac{\prod\limits_{m=1}^{M} (s - Z_m)}{\prod\limits_{n=1}^{N} (s - P_n)}. \tag{9.140}$$

Now make the substitution of s given by (9.130). We get

$$G_d(z) = K_0 h^{N-M} \frac{\prod\limits_{m=1}^{M} (1 - Z_m T - z^{-1})}{\prod\limits_{n=1}^{N} (1 - P_n T - z^{-1})}. \tag{9.141}$$

As can be seen, both poles and zeros have the general form

$$\gamma = \frac{1}{1 - wT} \tag{9.142}$$

where w is a complex number. It is also noted that, aside a possible extra pole at the origin, this is a pole to pole and zero to zero transformation. This means that we are essentially considering a twofold transformation between the s and the z planes:

$$z = \frac{1}{1 - sT} \tag{9.143}$$

As it is not difficult to verify, the negative real half-axis is transformed into the circumference diameter and the left half-plane will "fall" inside the circumference. Faced with these facts, we conclude that:

– The transition from a continuous to a discrete system preserves stability, but it may not preserve the phase characteristic. A minimum phase system is transformed into a minimum phase system, but a mixed or maximum phase system can be transformed into a minimum phase system.
– The reverse situation is different. The continuous system corresponding to any discrete stable system may not be stable. The same goes for the phase.

9.8.3 Tustin rule or bilinear transformation

The transformation described in the previous section has a great inconvenient: the unit circle is not the image of the left complex plane, as seen in figure 9.3. This means that we have stable discrete-time systems that are transformed into an unstable continuous-time system. This is disadvantageous, mainly when trying to embed a DARMA system in a CARMA. Therefore, we need a one to one transformation between

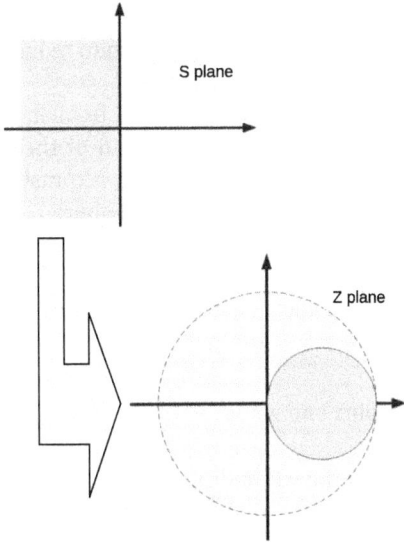

Fig. 9.3: s to z conversion using differences.

the left complex plane and the unit disk. The Tustin rule, based on the bilinear or Möbius transformation, is the required alternative. It is defined by

$$s = \frac{2}{T}\frac{1-z^{-1}}{1+z^{-1}} = \frac{2}{T}\frac{z-1}{z+1} \qquad (9.144)$$

From the mathematical point of view, it corresponds to a trapezoidal integration. This method of conversion is widely used because a stable transfer function always corresponds to a stable, discrete transfer function, the left half plane of the plane of the variable being transformed into the unit circle in z as shown in figure 9.4.

It is not difficult to verify that the negative real half-axis is transformed into the circumference diameter and the left half-plane will fall inside the circle (unit disk). Faced with these facts, we conclude that:

– conversion from a continuous to a discrete system using the Tustin transformation preserves the stability and phase characteristic;
– a system of minimum (mixed, maximum) phase is transformed into a system of the same type;
– the same happens with conversion from a discrete to a continuous system: in the transition from z to s there is preservation of stability and phase characteristic.

According to what we have said above, the imaginary axis $s = i\omega$ is transformed into the unit circle $z = e^{i\Omega}$, $\Omega \in (-\pi, \pi]$. On the other hand, the transformation is one to one. This means that the imaginary axis is compressed and deformed — warped. The relationship between the two frequency axes is given by

$$\omega = \frac{2}{T}\tan\left(\frac{\Omega}{2}\right), \qquad \Omega = 2\arctan\left(\frac{\omega T}{2}\right). \qquad (9.145)$$

In figure 9.5 we can observe the effect of this frequency compression.

When converting from s to z, it is important to choose T small enough to reduce the nonlinear effect in (9.145).

The choice of T is important in the case of filter design with frequency pre-specification. In fact, starting from a pre-specification in the domain of the z variable, transforming to s (pre-warping) and returning to z, we obtain a transfer function that does not depend of T. We can make $2/T = \arctan(\pi f_c / f_a)$, where f_c is the cutoff frequency of the low-pass filter and f_a the sampling frequency. A sampling frequency of $f_a = 4f_c$ is often employed.

The Tustin transformation can be used in a second manner, addressed in the next section.

9.8.4 Direct conversion of poles and zeros into poles and zeros

The fact that the Tustin transformation takes us from the left half-plane to the interior of the circumference of radius 1, guaranteeing, as it was said, the preservation of stability, makes it possible to transform poles into poles and zeros into zeros, while maintaining the characteristics of phase: maximum, minimum or mixed. In this case, we start from the form (9.140), and using the z to s transformation (9.144) we easily see that a binomial of the form $s - p$ becomes

$$\frac{(2 - pT)z - (2 + pT)}{z + 1} = (2 - pT)\frac{z - \frac{(2+pT)}{(2-pT)}}{z + 1}. \tag{9.146}$$

As can be seen, any pole (or zero) p becomes the pole (or zero) $\frac{(2+pT)}{(2-pT)}$. This allows a fast conversion. Each term also contributes a factor of type $(2 - pT)$ for the gain. However,

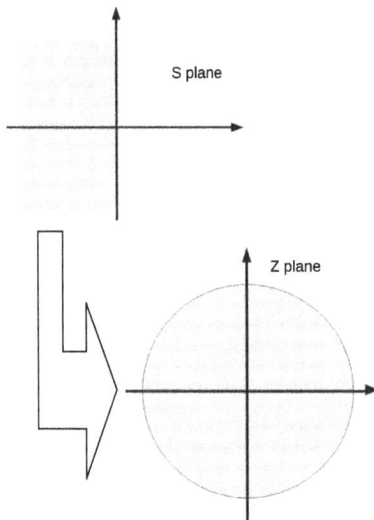

Fig. 9.4: Conversion from s to z using the bilinear transformation.

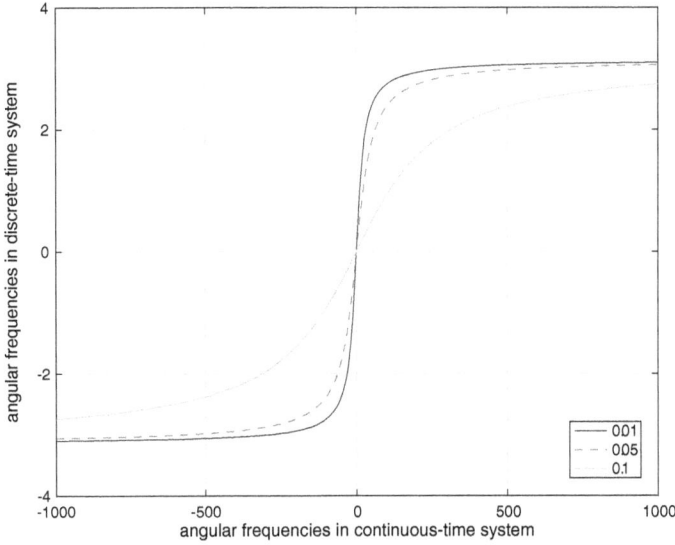

Fig. 9.5: Frequency compression in the bilinear transformation for $T = 0.001, 0.05, 0.1$ in a linear scale.

an extra pole or zero appears at the point $z = -1$, due to the factors $z + 1$. If $N > M$, we have a zero with multiplicity $N - M$; if $M > N$, we have a pole with multiplicity $M - N$. Thus, assuming the case $N > M$, we obtain:

$$G_d(s) = K(z+1)^{N-M} \frac{\displaystyle\prod_{m=1}^{M}\left[z - \frac{2 + z_m T}{2 - z_m T}\right]}{\displaystyle\prod_{n=1}^{N}\left[z - \frac{2 + p_n T}{2 - p_n T}\right]}, \tag{9.147}$$

where

$$K = K_0 \frac{\displaystyle\prod_{m=1}^{M}(2 - z_m T)}{\displaystyle\prod_{n=1}^{N}(2 - p_n T)}. \tag{9.148}$$

Note that this gain does not coincide with the static gain obtained from (9.147) by making $z = 1$.

9.8.5 Invariant response s to z conversion

Transfer function matching by input/output sampling consists in the following steps:
– The input and output are sampled;
– Their ZT is calculated;
– The transfer function is obtained dividing the transforms of output and input.

Well-known examples resulting from the definition of this transformation are the responses to impulse, step, and ramp.

Given an analogue system defined by a transfer function $H(s)$, it is possible to approximate it by a discrete transfer function $H(z)$ using the invariant response transform of the following way:

- The input $x(t)$ is chosen (e.g. impulse, step, ramp, power, exponential).
- The response $y(t)$ is obtained through the inverse LT of $H(s)X(s)$.
- The discrete input $x(n)$ is obtained by sampling $x(t)$, with a sampling interval T; $X(z)$ is the corresponding ZT.
- The discrete response $y(n)$ is the result of sampling $y(t)$ and $Y(z)$ is its ZT.
- The transfer function of the discrete-time system is computed as $H(z) = Y(z)/X(z)$.

The sampling frequency $f_a = 1/T$ must be greater than twice the cut frequency of the filter.

Example 9.8.2 (Invariant impulse). *We are going to exemplify the procedure with transfer function $H(s) = \frac{4}{(s+1)(s+2)}$ $Re(s) > -1$. For $x(t) = \delta(t)$ we have $X(s) = 1$. From*

$$Y(s) = H(s)X(s) = \frac{4}{(s+1)(s+2)} = \frac{4}{s+1} - \frac{4}{s+2}, \quad \mathbb{R}(s) > -1 \tag{9.149}$$

and

$$y(t) = 4[e^{-t} - e^{-2t}]\varepsilon(t), \tag{9.150}$$

that is sampled to give

$$y(n) = 4[e^{-nT} - e^{-2nT}]\varepsilon(n), \tag{9.151}$$

the ZT of which is

$$Y_D(z) = \frac{4z}{z - e^{-T}} - \frac{4z}{z - e^{-2T}}. \tag{9.152}$$

If $x(n) = \delta_n$, then $X_D(z) = 1$, and so

$$H_D(z) = \frac{4z}{z - e^{-T}} - \frac{4z}{z - e^{-2T}}, \quad |z| > e^{-T}. \tag{9.153}$$

Example 9.8.3 (Invariant step). *We consider $H(s) = \frac{4}{(s+1)(s+2)}$ $Re(s) > -1$ again. For $x(t) = \varepsilon(t)$, we have $X(s) = 1/s$, $Re(s) > 0$. Then*

$$Y(s) = H(s)X(s) = \frac{4}{s(s+1)(s+2)} = \frac{2}{s} - \frac{4}{s+1} + \frac{2}{s+2} \quad Re(s) > 0 \tag{9.154}$$

$$y(t) = [2 - 4e^{-t} + 4e^{-2t}]\varepsilon(t). \tag{9.155}$$

By sampling, we obtain

$$y(n) = [2 - 4e^{-nT} + 4e^{-2nT}]\varepsilon(n), \tag{9.156}$$

the ZT of which is

$$Y_D(z) = \frac{2z}{z-1} - \frac{4z}{(z-e^{-T})} - \frac{4z}{(z-e^{-2T})} \quad |z| > e^{-T} \tag{9.157}$$

If $x(n) = \varepsilon(n)$, then $X_D(z) = \frac{z}{z-1}$, $|z| > 1$, and

$$Y_D(z) = H_D(z).X_D(z) = \frac{2z}{z-1} - \frac{4z}{z-e^{-T}} + \frac{2z}{z-e^{-2T}} \quad |z| > 1. \tag{9.158}$$

Therefore, the transfer function is

$$H_D(z) = 2 - \frac{4(z-1)}{z-e^{-T}} + \frac{2(z-1)}{z-e^{-2T}} \quad |z| > e^{-T}. \tag{9.159}$$

Example 9.8.4 (Invariant ramp). *Again for $H(s) = \frac{4}{(s+1)(s+2)}$ $Re(s) > -1$, with $x(t) = t\varepsilon(t)$ we have $X(s) = 1/s^2$, $Re(s) > 0$ and*

$$Y(s) = H(s)X(s) = \frac{4}{s^2(s+1)(s+2)} = -\frac{3}{s} + \frac{2}{s^2} + \frac{4}{s+1} - \frac{1}{s+2}, \quad Re(s) > 0 \tag{9.160}$$

$$y(t) = [-3 + 2t + 4e^{-t} - e^{-2t}]\varepsilon(t). \tag{9.161}$$

By sampling,

$$y(n) = [-3 + 2nT + 4e^{-nT} - e^{-2nT}]\varepsilon(n) \tag{9.162}$$

the ZT of which is

$$Y_D(z) = -\frac{3z}{z-1} + \frac{2zT}{(z-1)^2} + \frac{4z}{(z-e^{-T})} - \frac{z}{(z-e^{-2T})} \quad |z| > 1. \tag{9.163}$$

As $x(n) = nT\varepsilon(n)$ has the ZT $X(z) = \frac{zT}{(z-1)^2}$, $|z| > 1$, then

$$H_D(z) = -\frac{3(z-1)}{T} + 2 - \frac{4(z-1)^2}{z-e^{-T}} + \frac{2(z-1)^2}{z-e^{-2T}} \quad |z| > e^{-T}. \tag{9.164}$$

The invariant response transformation produces a transfer function that is a good approximation only for the response to the chosen signal and not for other inputs. The quality of the approach depends on the sampling interval T, and it is possible to obtain good results if T is reasonably small.

9.8.6 Conversion from partial fraction decomposition

As we saw above, the invariant impulse transformation is a particular case where the input is $x(t) = \delta(t)$. This method allows a direct mapping of $H(s)$ to $H(z)$. Given the transfer function $H(s)$, we begin by decomposing it into simple fractions and transforming each one at a time.

Considering $H(s)$ with N single poles, $p_k, k = 1, \ldots, N$, then $H(s)$ can be described in the form

$$H(s) = \sum_{k=1}^{N} \frac{R_k}{s - p_k}. \tag{9.165}$$

Let $h(t)$ and $h(n)$ be the impulse responses of the continuous and discrete systems, respectively, and assume the system is causal. Then,

$$h(t) = \sum_{k=1}^{N} R_k e^{p_k t} \varepsilon(t) \qquad h(n) = \sum_{k=1}^{N} R_k e^{p_k Tn} \varepsilon(n) \qquad (9.166)$$

where R_k, $k = 1, 2, \cdots$ are the residues.

As the procedure is the same for all the fractions, we will consider one generic case:

$$H(s) = \frac{1}{s - p}, \quad Re(s) > Re(p) \longrightarrow h(t) = e^{pt} \varepsilon(t). \qquad (9.167)$$

Then, $h(n) = e^{pn} \varepsilon(n)$, and its transfer function is

$$H(z) = \frac{1}{1 - e^{pT} z^{-1}} = \frac{z}{z - e^{pT}} \qquad |z| > e^{-pT}. \qquad (9.168)$$

Comparing (9.165) and (9.168) it is easy to see that, for the invariant impulse transformation, the mapping to be performed is $\frac{1}{s-p} = \frac{z}{z-e^{pT}} = \frac{1}{1-e^{pT}z^{-1}}$ for the causal case. For a pole with multiplicity 2, we compute the derivative of $\frac{1}{s-p} = \frac{1}{1-e^{pT}z^{-1}}$ in order to p to obtain

$$\frac{1}{(s-p)^2} = \frac{Te^{pT}z^{-1}}{(1 - e^{pT}z^{-1})^2} \qquad (9.169)$$

For a third order pole, we continue the derivation and get

$$\frac{1}{(s-p)^3} = \frac{1}{2} \frac{T^2 e^{pT} z^{-1}}{(1 + e^{pT}z^{-1})(1 - e^{pT}z^{-1})^2} \qquad (9.170)$$

For higher order poles we can obtain similar formulae by successive derivation. These can be expressed in a general expression:

$$\frac{n!}{(s-p)^{n+1}} = T^n \sum_{0}^{n} (-1)^k \gamma_{(n,k)} \frac{1}{(1 - \theta z^{-1})^{k+1}} - \frac{T}{2} \delta_n \qquad (9.171)$$

where $\gamma_{(n,k)}$, $k = 1, 2, \ldots, n$ numbers [1] are given by

$$\gamma_{(n,k)} = \frac{1}{k} \sum_{l=1}^{k} (-1)^{k-l} \binom{k}{l} l^n \quad k = 1, 2, \ldots, n \qquad (9.172)$$

A better way of expressing the transformation is given by

$$
\begin{bmatrix} 1 \\ \frac{1/T}{s-p} \\ \frac{1/T^2}{(s-p)^2} \\ \frac{2/T^3}{(s-p)^3} \\ \frac{6/T^4}{(s-p)^4} \\ \frac{24/T^5}{(s-p)^5} \\ \vdots \end{bmatrix}
=
\begin{bmatrix}
1 & 0 & 0 & 0 & 0 & 0 & \cdots \\
-1/2 & 1 & 0 & 0 & 0 & 0 & \cdots \\
0 & -1 & 1 & 0 & 0 & 0 & \cdots \\
0 & 1 & -3 & 2 & 0 & 0 & \cdots \\
0 & -1 & 7 & -12 & 6 & 0 & \cdots \\
0 & 1 & -15 & 50 & -60 & 24 & \cdots \\
\vdots & \vdots & \vdots & \vdots & \vdots & \vdots & \ddots
\end{bmatrix}
\begin{bmatrix} 1 \\ \frac{1}{(1-\theta z^{-1})} \\ \frac{1}{(1-\theta z^{-1})^2} \\ \frac{1}{(1-\theta z^{-1})^3} \\ \frac{1}{(1-\theta z^{-1})^4} \\ \frac{1}{(1-\theta z^{-1})^5} \\ \vdots \end{bmatrix}
\qquad (9.173)
$$

where $\theta = e^{pT}$.

Example 9.8.5. *We are going to exemplify the procedure using transfer function $H(s) = \frac{4}{(s+1)(s^2+4s+5)}$, assuming causality. By partial fraction decomposition,*

$$H(s) = \frac{2}{2s+1} + \frac{K}{s+2+i} + \frac{K^*}{s+2-i} \tag{9.174}$$

with $K = -1 - i = \sqrt{2}e^{-i\frac{3\pi}{4}}$. Putting $b = e^{-T}$ and $a = e^{-2T}$, and with help of the above table, we get

$$H(z) = \frac{4\sqrt{2}z}{z-b} \frac{z^2 \cos(\frac{3\pi}{4}) - za \cos(T - \frac{3\pi}{4})}{z^2 - 2az\cos(T) + a^2}. \tag{9.175}$$

Example 9.8.6. *Let now $H(s) = \frac{1}{(s-p)^4}$. We obtain*

$$H(z) = \frac{T^3}{6}\left[\frac{az}{(z-a)^2} + \frac{6a^2z}{(z-a)^3} + \frac{6a^3z}{(z-a)^4}\right] \tag{9.176}$$

that can be written as

$$H(z) = \frac{T^3}{6}\frac{a^3z + 4a^2z^2 + az^3}{(z-a)^4}. \tag{9.177}$$

Remark 9.8.1. *Above, we used the relation*

$$H(s) = \frac{1}{s-p}, \quad Re(s) > Re(p) \longrightarrow h(t) = e^{pt}\varepsilon(t) \tag{9.178}$$

to obtain

$$H(z) = \frac{1}{1 - e^{pT}z^{-1}}. \tag{9.179}$$

However, the procedure is not fully correct. In fact, the value of $h(t)$ for $t = 0$ is not 1, but $1/2$, in agreement with the inversion theorem of the LT. This means that $h(n) = e^{pn}\varepsilon(n) - 1/2\delta_n$. The corresponding ZT is

$$H(z) = \frac{1}{1 - e^{pT}z^{-1}} - \frac{1}{2} = \frac{1}{2}\frac{1 + e^{pT}z^{-1}}{1 - e^{pT}z^{-1}} \tag{9.180}$$

Therefore, we recovered the binomial transformation.

Of course we can obtain the inverse algorithm that allows us to imbed a given discrete-time system into a continuous-time. In matrix formulation we can write

$$\begin{bmatrix} 1 \\ \frac{1}{(1-\theta z^{-1})} \\ \frac{1}{(1-\theta z^{-1})^2} \\ \frac{1}{(1-\theta z^{-1})^3} \\ \frac{1}{(1-\theta z^{-1})^4} \\ \frac{1}{(1-\theta z^{-1})^5} \\ \vdots \end{bmatrix} = \begin{bmatrix} 1 & 0 & 0 & 0 & 0 & 0 & \cdots \\ \frac{1}{2} & 1 & 0 & 0 & 0 & 0 & \cdots \\ \frac{1}{2} & 1 & 1 & 0 & 0 & 0 & \cdots \\ \frac{1}{2} & 1 & \frac{3}{2} & 1 & 0 & 0 & \cdots \\ \frac{1}{2} & 1 & \frac{11}{6} & 2 & 1 & 0 & \cdots \\ \frac{1}{2} & 1 & \frac{25}{12} & \frac{35}{12} & \frac{5}{2} & 1 & \cdots \\ \vdots & \vdots & \vdots & \vdots & \vdots & \vdots & \ddots \end{bmatrix} \begin{bmatrix} 1 \\ 1/T \\ s-p \\ 1/T^2 \\ \frac{2/T^3}{(s-p)^2} \\ \frac{6/T^4}{(s-p)^3} \\ \frac{24/T^5}{(s-p)^4} \\ \frac{(s-p)^5}{} \\ \vdots \end{bmatrix} \tag{9.181}$$

Exercises

1. Repeat example 9.4.7 for the following cases:

   a) $H(z) = \dfrac{1 - az^{-1}}{1 - pz^{-1}}$

   b) $H(z) = \dfrac{1 + pz^{-1}}{1 - pz^{-1}}$

   c) $H(z) = \dfrac{1}{(1 - pz^{-1})(1 - p^*z^{-1})}$

2. Find the ZT of the following signals:

   a) a triangular wave with period 2 and amplitude 2, sampled with $T = 0.1$

   b) a square wave with period 1 and amplitude 1, sampled with $T = \sqrt{\dfrac{2}{100}}$

   c) $\sin t + 0.2\cos 3t - 0.1\sin 7t$, sampled with $T = \dfrac{\pi}{100}$ and then with $T = \dfrac{\pi}{5}$

3. Use the Jury criterion to study the distribution of the zeroes relatively to the unit circle of the following polynomials:

   a) $P_1(z) = z^3 - 0.2z^2 - 0.4z - 0.064$

   b) $P_2(z) = z^3 + z^2 + 0.41z + 0.05$

   c) $P_3(z) = 10z^5 + 24z^4 + 27.4z^3 + 14.2z^2 + 4z + 0.4$

   d) $P_4(z) = z^5 + 3.6z^4 + 10.4z^3 + 2.8z^2 + 17.7z$

   e) $P_5(z) = 1.5z^5 - 13.5z^4 + 28.5z^3 + 3.5z^2 + 4.5z + 0.5$

4. Repeat examples 9.8.2–9.8.4 for the anti-causal and acausal cases.

5. Find discrete equivalents for the following continuous systems using the different methods presented, and verify the stability of both the original and the discretised transfer functions:

   a) $G_1(s) = \dfrac{1}{s + 10}$ with $T = 0.05$ and with $T = 0.5$

   b) $G_2(s) = \dfrac{s + 10}{s^2 + 101s + 100}$ with $T = 0.001$

   c) $G_3(s) = \dfrac{1}{s^3 + 4s^2 + 6s + 4}$ with $T = 0.01$

10 Discrete-time derivatives and transforms

10.1 Introduction: difference vs. differential

In the previous chapter, we introduced several ways of converting a continuous-time system to discrete-time. One of the methods was based on the substitution of derivatives by finite differences. Whenever this was done to integer order differential equations, difference equations appeared naturally. Although the manipulation of difference equations is easier than the manipulation of differential equations, several formulations still use derivatives. It is the case delta systems. These result from substituting the derivative by the incremental ratio $\frac{f(t+h)-f(t)}{h}$ followed by a sampling with interval h. This is essentially what we did in the continuous to discrete conversion, but without getting in the end a difference equation, since finite differences are explicitly kept. These systems are still called differential. While difference equations are not suitable to being fractionalised (they lead to infinite order systems), delta systems have the required features to do so.

Modernly, the approach to differential discrete equations is based on Hilger's works for continuous/discrete unification [29], currently called *calculus on time scales*. His methodology consists in defining a general domain that can be continuous, discrete or mixed (time scales, or more generally measure chains) [5, 7, 21]. He defined two derivatives, delta and nabla, that are either the incremental ratia or their limit to zero when not at an isolated point. These derivative definitions are used to devise corresponding differential equations representing linear systems. In agreement with the used derivatives, we will call them *nabla and delta systems*.

10.2 Derivatives and inverses

10.2.1 Nabla and delta derivatives

As in the previous two chapters, we shall be working in the *time scale* defined by

$$\mathbb{T} = (h\mathbb{Z}) = \{\ldots, -3h, -2h, -h, 0, h, 2h, 3h, \ldots\} \tag{10.1}$$

with $h \in \mathbb{R}^+$ (we will use h instead of T to maintain a closer relation to the continuous-time theory). The results we will obtain can be readily generalised to the shifted time scale $\mathbb{T} = (a + h\mathbb{Z})$, $a < h$ [5, 7]. In the following t will be any generic point in $\mathbb{T} = h\mathbb{Z} = \{kh : k \in \mathbb{Z}\}$.

We begin by introducing causal and anti-causal derivatives.

Definition 10.2.1. *The nabla derivative is given by:*

$$f'_\nabla(t) = \frac{f(t) - f(t-h)}{h} \tag{10.2}$$

DOI 10.1515/9783110624588-010

and the delta derivative by

$$f'_\Delta(t) = \frac{f(t+h) - f(t)}{h} \tag{10.3}$$

Example 10.2.1 (Derivative of unit step). *The nabla derivative of the unit step is*

$$D_\nabla \varepsilon(nh) = \begin{cases} \frac{1}{h} & n = 0 \\ 0 & n \neq 0 \end{cases} \tag{10.4}$$

The delta derivative of the anti-causal unit step $\varepsilon(-nh)$ is

$$D_\Delta \varepsilon(-nh) = \begin{cases} -\frac{1}{h} & n = 0 \\ 0 & n \neq 0 \end{cases} \tag{10.5}$$

Definition 10.2.2. *The results of the example above lead us to introduce the discrete delta (impulse) function by:*

$$\delta(nh) = D_\nabla \varepsilon(nh) \tag{10.6}$$

Example 10.2.2 (Derivative of a causal function). *We are going to obtain the derivative of a causal function. We proceed step by step. For $N = 1$, and remarking that $\varepsilon(t - h) = \varepsilon(t) - h\delta(t)$, we obtain*

$$D_\nabla [f(t)\varepsilon(t)] = f'_\nabla(t)\varepsilon(t) + f(-h)\delta(t) \tag{10.7}$$

$$f'_\nabla(t)\varepsilon(t) = D_\nabla [f(t)\varepsilon(t)] - f(-h)\delta(t) \tag{10.8}$$

that shows how the initial conditions may appear. We are relating the causal part of the derivative with the derivative of the causal part. The first one depends on the past. Repeat the process:

$$\left[f'_\nabla(t)\varepsilon(t)\right]'_\nabla = D^2_\nabla [f(t)\varepsilon(t)] - f(-h)\delta'_\nabla(t) \tag{10.9}$$

Using (10.7), we arrive at

$$f''_\nabla(t)\varepsilon(t) = D^2_\nabla [f(t)\varepsilon(t)] - f(-h)\delta'_\nabla(t) - f'_\nabla(-h)\delta(t), \tag{10.10}$$

that can be generalised for any positive integer order to

$$f^{(N)}_\nabla(t)\varepsilon(t) = D^N_\nabla [f(t)\varepsilon(t)] - \sum_{k=0}^{N-1} f^{(k)}_\nabla(-h)\delta^{(N-1-k)}_\nabla(t) \tag{10.11}$$

that is similar to the well-known jump formula [18].

The theory we developed in chapter 7 suggests us to generalise the above derivatives to fractional orders as follows.

Definition 10.2.3. *The nabla fractional derivative is given by:*

$$f_\nabla^{(\alpha)}(t) = \frac{\sum\limits_{n=0}^{\infty} (-1)^n \binom{\alpha}{n} f(t - nh)}{h^\alpha} \tag{10.12}$$

and the delta fractional derivative by

$$f_\Delta^{(\alpha)}(t) = e^{-i\alpha\pi} \frac{\sum\limits_{n=0}^{\infty} (-1)^n \binom{\alpha}{n} f(t + nh)}{h^\alpha} \tag{10.13}$$

As before, we will call these derivatives respectively *forward* (or direct) and *backward* (or reverse), in agreement with the time flow, from past to future or the reverse. Attending to the fact that $(-1)^n \binom{\alpha}{n} = \frac{(-\alpha)_n}{n!}$, we conclude immediately that these derivatives include as special cases both integer order derivatives and anti-derivatives.

The commutativity of convolution allows us to obtain alternative formulations for derivatives. All we have to do is to put the binomial coefficients under another form and to change the summation variable. We have

$$\frac{(-\alpha)_n}{n!} = \frac{\Gamma(\alpha + 1)}{\Gamma(\alpha - n + 1)n!} = \frac{\Gamma(-\alpha + n)}{\Gamma(-\alpha)n!} \tag{10.14}$$

Therefore, we can write the above derivatives as

$$f_\nabla^{(\alpha)}(t) = \frac{h^{-\alpha}}{\Gamma(-\alpha)} \sum_{k=-\infty}^{n} \frac{\Gamma(-\alpha + n - k)}{(n - k)!} f(kh) \tag{10.15}$$

$$f_\Delta^{(\alpha)}(t) = \frac{(-h)^{-\alpha}}{\Gamma(-\alpha)} \sum_{k=n}^{\infty} \frac{\Gamma(-\alpha + n - k)}{(n - k)!} f(kh) \tag{10.16}$$

with $t = nh$. These formulations for the fractional derivatives must be used with caution for positive integer values of α. They state different forms of expressing the fractional derivatives and should be compared with those we find in current literature, [5, 7, 21].

To compare the above formulae with the continuous-time derivatives, put $k = \tau/h$ and $n = t/h$; we have

$$\frac{\Gamma(-\alpha + n - k)}{(n - k)!} = \frac{\Gamma(-\alpha + (t - \tau)/h)}{\Gamma((t - \tau)/h) + 1)} \tag{10.17}$$

As is known,

$$\frac{\Gamma(-\alpha + (t - \tau)/h)}{\Gamma((t - \tau)/h) + 1)} \approx (t - \tau)/h^{-\alpha-1} \tag{10.18}$$

as $h \to 0$. Treating only the first case, we have

$$\lim_{h \to 0} f_\nabla^{(\alpha)}(t) = \frac{1}{\Gamma(-\alpha)} \int_{-\infty}^{t} (t - \tau)^{-\alpha-1} f(\tau) d\tau \tag{10.19}$$

the forward Liouville derivative. For the delta case, the procedure is equal and leads to the backward Liouville derivative. (Compare these expressions with (7.3).) With these operations, we obtain a "backward compatibility" with continuous-time derivatives.

10.2.2 Existence of fractional derivatives

To study the existence conditions for fractional derivatives we must be aware of the behaviour of the function along the half straight-line $t \pm nh$ with $n \in \mathbb{Z}^+$. If the function has bounded support, meaning that is non zero in a finite domain, both derivatives exist at every finite point of $f(t)$. In the general case, let us remind the asymptotic behaviour of the binomial coefficients

$$\left| \begin{pmatrix} \alpha \\ k \end{pmatrix} \right| \leq \frac{A}{k^{\alpha+1}}, \tag{10.20}$$

and the conclusions we got in 7.3.2, namely that $f(t)\frac{A}{k^{\alpha+1}}$ must decrease, at least as $f(t)\frac{A}{k^{|\alpha|+1}}$ does when $k \to \infty$, and that the existence of the fractional derivative depends only on what happens in one half-line, left or right.

10.2.3 Properties

We are going to present the main properties of the derivatives above presented. The proofs are essentially similar to those of the Grünwald-Letnikov derivatives presented in chapter 7 [87]. So, we will not preset the proofs.

- *Linearity.* The linearity property of the fractional derivative is evident from the above formulae.
- *Causality.* The causality property was already referred to above and can also be obtained easily. We only have to use (10.12). Assume that $f(t) = 0$, for $t < 0$. We conclude immediately from (10.12) that the derivative is also zero for $t < 0$. For the anti-causal case (10.13), the situation is similar.
- *Time reversal.* The substitution $t \Rightarrow -t$ converts the forward (nabla) derivative into the backward (delta) and vice-versa.
- *Time shift.* The derivative operators are shift invariant as it is evident from (10.12) and (10.13).
- *Additivity and commutativity of the orders.*

$$D^\alpha \left[D^\beta f(t) \right] = D^\beta \left[D^\alpha f(t) \right] = D^{\alpha+\beta} f(t) \tag{10.21}$$

- *Neutral element.* This comes from the last property by putting $\beta = -\alpha$:

$$D^\alpha \left[D^{-\alpha} f(t) \right] = D^0 f(t) = f(t). \tag{10.22}$$

This is very important because it states the existence of inverse.
- *Inverse element.* From the last result, we conclude that there is always an inverse element: for every α order derivative, there is always a $-\alpha$ order derivative given by the same formula. Consequently, no primitivation constant has to be added.

- *Associativity of the orders.*

$$D^\gamma \left[D^\alpha D^\beta\right] f(t) = D^{\gamma+\alpha+\beta} f(t) = D^{\alpha+\beta+\gamma} f(t) = D^\alpha \left[D^{\beta+\gamma}\right] f(t) \tag{10.23}$$

This is a consequence of the additivity.
- *Derivative of the product.* The deduction of the formula for derivative of the product is somehow involved:

$$D_\nabla^\alpha [f(t)g(t)] = \sum_{i=0}^\infty \binom{\alpha}{i} f_\nabla^{(i)}(t) g_\nabla^{(\alpha-i)}(t-ih) \tag{10.24}$$

This expression is very similar to the generalised Leibniz rule presented in chapter 7. If α is a positive integer we obtain the well-known classic formula. For the backward case we obtain

$$D_\Delta^\alpha [f(t)g(t)] = \sum_{i=0}^\infty \binom{\alpha}{i} f_\Delta^{(i)}(t) g_\Delta^{(\alpha-i)}(t+ih) \tag{10.25}$$

Example 10.2.3 (Fractional derivatives of the impulses). *It is not difficult do show that the derivative of any order of the impulse is essentially given by the binomial coefficients. In fact from (10.12) and (10.13) we get*

$$D_\nabla^\alpha \delta(n) = h^{-\alpha-1} \frac{(-\alpha)_n}{n!} \varepsilon(nh) \tag{10.26}$$

$$D_\Delta^\alpha \delta(n) = (-h)^{-\alpha-1} \frac{(-\alpha)_{-n}}{(-n)!} \varepsilon(-nh) \tag{10.27}$$

Example 10.2.4 (Fractional derivatives of the unit steps). *According to the above properties, it is easy to obtain the fractional derivative of the step functions. We only have to substitute $\alpha - 1$ for α:*

$$D_\nabla^\alpha \varepsilon(nh) = h^{-\alpha} \frac{(-\alpha+1)_n}{n!} \varepsilon(nh) \tag{10.28}$$

$$D_\Delta^\alpha \varepsilon(nh) = (-h)^{-\alpha} \frac{(-\alpha+1)_{-n}}{(-n)!} \varepsilon(-nh) \tag{10.29}$$

For negative values of α these expressions can be considered the definitions of fractional powers.

Example 10.2.5 (Fractional derivatives of "powers"). *Using the additivity property, we can obtain an important result:*

$$D_\nabla^\beta D_\nabla^\alpha \varepsilon(nh) = D_\nabla^\beta \left[h^{-\alpha+1} \frac{(-\alpha+1)_n}{n!} \varepsilon(nh) \right]$$

$$= h^{-\alpha-\beta+1} \frac{(-\alpha-\beta+1)_n}{n!} \varepsilon(nh), \tag{10.30}$$

leading to the conclusion that

$$D_\nabla^\beta \left[\frac{(a)_n}{n!} \varepsilon(nh) \right] = h^{-\beta+1} \frac{(a-\beta)_n}{n!} \varepsilon(nh). \tag{10.31}$$

This is the analogue of the derivative of the power function. We can obtain a similar result for the delta derivative:

$$D_\Delta^\beta \left[\frac{(a)_{-n}}{(-n)!} \varepsilon(-nh) \right] = (-h)^{-\beta+1} \frac{(a-\beta)_n}{(-n)!} \varepsilon(-nh) \tag{10.32}$$

Example 10.2.6 (The complex sinusoid). *Consider the complex sinusoid $e^{i\omega hn}$ with $\omega \subset \mathbb{R}^+$ and introduce the complex s_0 given by $s_0 = \frac{1-e^{-i\omega h}}{h}$. We obtain*

$$D_\nabla^N e^{i\omega hn} = s_0^N e^{i\omega hn} \tag{10.33}$$

Similarly, with $\sigma_0 = \frac{e^{i\omega h}-1}{h}$,

$$D_\Delta^\alpha e^{i\omega hn} = \sigma_0^\alpha e^{i\omega hn} \tag{10.34}$$

These relations show that complex sinusoids are eigenfunctions of the derivatives.

10.2.4 The nabla and delta exponentials

As in the other kinds of linear systems, the nabla and delta exponentials are the eigenfunctions of the corresponding systems.

Definition 10.2.4. *The **nabla generalised exponential** is defined by*

$$e_\nabla(t, s) = [1 - sh]^{-t/h} \tag{10.35}$$

*and the **delta generalised exponential** by*

$$e_\Delta(t, s) = [1 + sh]^{t/h} \tag{10.36}$$

10.2.4.1 Properties of the exponentials
These exponentials have several interesting properties [7, 89].
1. *Relation between exponentials.* Substituting $-s$ for s in the function defined in (10.36), we obtain the inverse of (10.35):

$$e_\Delta(t, s) = 1/e_\nabla(t, -s) = e_\Delta(-t, -s) \tag{10.37}$$

2. As $h \to 0$, *both exponentials converge to e^{st}*. This is obtained from the well-known relation $\lim_{h\to 0} (1 + ah)^{1/h} = e^a$.
3. Attending to the way how both exponentials were obtained, we can conclude easily that

$$D_\nabla^\alpha e_\nabla(t, s) = s^\alpha e_\nabla(t, s) \tag{10.38}$$

$$\alpha = 0.25 \qquad\qquad \alpha = 0.75$$

$$\alpha = 1.25 \qquad\qquad \alpha = 1.75$$

$$\alpha = 2.25 \qquad\qquad \alpha = 2.75$$

$$\alpha = 3.25 \qquad\qquad \alpha = 3.75$$

$$\alpha = 4.25 \qquad\qquad \alpha = 4.75$$

$$\alpha = 5.25 \qquad\qquad \alpha = 5.75$$

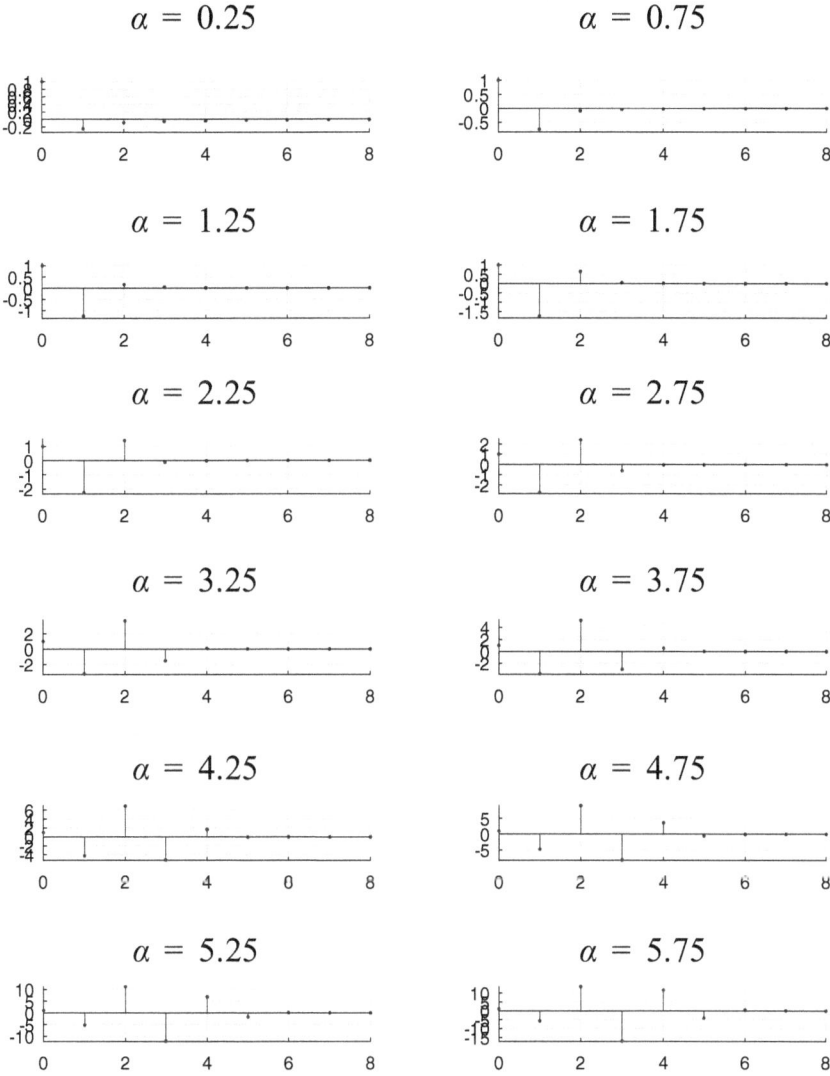

Fig. 10.1: Fractional derivatives of the impulse for $\alpha > 0$.

$$D_\Delta^\alpha e_\Delta(t, s) = s^\alpha e_\Delta(t, s) \tag{10.39}$$

This can be proved from the definitions and the properties of the binomial series. We will present one of the proofs in detail because it has important consequences. Consider the causal (nabla) situation. We have

$$D_\nabla^\alpha e_\nabla(t, s) = e_\nabla(t, s) \frac{\sum\limits_{k=0}^{\infty} (-1)^k \binom{\alpha}{k} [1 - sh]^k}{h^\alpha} \tag{10.40}$$

$$\alpha = -0.25 \qquad\qquad \alpha = -0.75$$

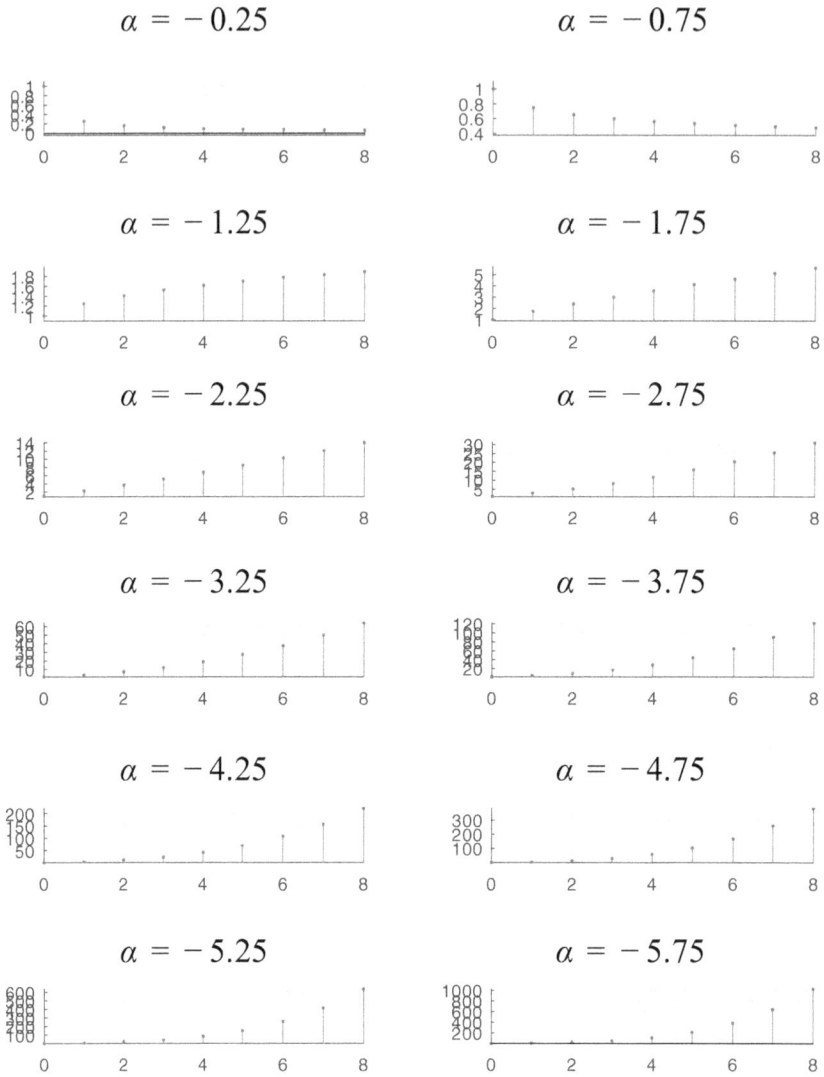

$$\alpha = -1.25 \qquad\qquad \alpha = -1.75$$

$$\alpha = -2.25 \qquad\qquad \alpha = -2.75$$

$$\alpha = -3.25 \qquad\qquad \alpha = -3.75$$

$$\alpha = -4.25 \qquad\qquad \alpha = -4.75$$

$$\alpha = -5.25 \qquad\qquad \alpha = -5.75$$

Fig. 10.2: Fractional derivatives of the impulse for $\alpha < 0$.

The series converges to $(sh)^\alpha$ if

$$|1 - sh| < 1, \tag{10.41}$$

which means that s must be inside the so-called right Hilger circle defined by

$$|1 - sh| = 1. \tag{10.42}$$

Then we have

$$s^\alpha = h^{-\alpha} \sum_{k=0}^{\infty} (-1)^k \binom{\alpha}{k} [1 - sh]^k, \tag{10.43}$$

an interesting relation that will be used later. We assume the principal branch of the multivalued expression. We can choose any branch cut line in the left half complex plane and starting at $s = 0$.

In the delta case, being anti-causal, we have the reverse situation:

$$|1 + sh| > 1 \qquad (10.44)$$

The variable s must be outside the left Hilger circle:

$$|1 + sh| = 1 \qquad (10.45)$$

To test for back compatibility, we are going to compute the limit as h goes to zero in the expression of s^α. Letting $k = t/h$, we have

$$(-1)^k \binom{\alpha}{k} = \frac{\Gamma(-\alpha + t/h)}{\Gamma(-\alpha)\Gamma(t/h + 1)}. \qquad (10.46)$$

Using the property of the quotient of gamma functions above referred, we can write

$$\frac{\Gamma(-\alpha + t/h)}{\Gamma(t/h + 1)} \approx \left(\frac{t}{h}\right)^{-\alpha - 1} \qquad (10.47)$$

valid for very small values of h. On the other hand,

$$\lim_{h \to 0} [1 - sh]^{t/h} = e^{-st}. \qquad (10.48)$$

Joining the above expressions, we obtain

$$s^\alpha = \frac{\int\limits_0^\infty t^{-\alpha - 1} e^{-st} dt}{\Gamma(-\alpha)}, \qquad (10.49)$$

which is a well-known result ([87]).

4. *Behaviour for $s \in \mathbb{C}$.* It is important to know how these exponentials increase or decrease for $s \in \mathbb{C}$. We are going to consider the nabla case. The other is similar. Concerning this exponential, we can say that
   - It is real for real s;
   - It is positive for $s = x < \frac{1}{h}$, $x \in \mathbb{R}$;
   - It oscillates for $s = x > \frac{1}{h}$, $x \in \mathbb{R}$;
   - For values of s inside the right Hilger circle it is bounded and goes to zero as $s \to \frac{1}{h}$;
   - On the Hilger circle it has absolute value equal to 1: it degenerates into a complex exponential;
   - Outside the Hilger circle its absolute value increases as $|s|$ increases and goes to infinite as $|s| \to \infty$.

5. *Delayed exponentials.* Let $n_0 > 0$. The delayed exponentials verify the following relations, easily deduced:

$$e_\nabla(t \mp n_0 h, s) = e_\nabla(t, s).e_\Delta(\pm n_0 h, -s) \tag{10.50}$$

$$e_\Delta(t \mp n_0 h, s) = e_\Delta(t, s).e_\nabla(\pm n_0 h, -s) \tag{10.51}$$

6. *Product of exponentials.*
   - *Different types, same time:*

$$e_\nabla(t, s) \cdot e_\Delta(t, v) = e_\nabla(t, \frac{s + v}{1 + vh}) \tag{10.52}$$

$$= P_\Delta(t, \frac{s + v}{1 - sh})$$

The proof is very simple:

$$\frac{1 - sh}{1 + vh} = 1 - \frac{s + v}{1 + vh} \tag{10.53}$$

$$\frac{1 + vh}{1 - sh} = 1 + \frac{s + v}{1 - sh} \tag{10.54}$$

   - *Different types, different times:*

$$e_\nabla(t, s) \cdot e_\Delta(\tau, -s) = e_\nabla(t - \tau, s) \tag{10.55}$$

$$e_\nabla(t, -s) \cdot e_\Delta(\tau, s) = e_\Delta(\tau - t, s) \tag{10.56}$$

   - *Same type:*

$$e_\nabla(t, s) \cdot e_\nabla(t, v) = e_\nabla(t, s + v - svh) \tag{10.57}$$

$$e_\Delta(t, s) \cdot e_\Delta(t, v) = e_\Delta(t, s + v + svh) \tag{10.58}$$

7. *Cross derivatives:*

$$D_\Delta e_\nabla(t, s) = s \cdot h \cdot e_\nabla(t + h, s) \tag{10.59}$$

$$D_\nabla e_\Delta(t, s) = s \cdot h \cdot e_\Delta(t - h, s) \tag{10.60}$$

10.3 Suitable transforms

10.3.1 The nabla transform

We want suitable transforms able to deal with the differential systems constructed using the above defined derivatives. The two exponentials we just introduced are eigenfunctions of the forward and backward derivatives. Therefore, it is natural that we use them to define transforms. We are going to start from the inverse transform, the so-called *synthesis formula* [89].

Definition 10.3.1. *Assume a given signal, $f(nh)$, and that the arrow of time is from past to future. Assume also that the signal has a transform $F_\nabla(s)$ (undefined for now);*

when there is no danger of confusion, we will omitt the nabla subscript. The inverse transform would serve to synthesise the function from a continuous set of elemental exponentials:

$$f(nh) = -\frac{1}{2\pi i} \oint_{\gamma} F_{\nabla}(s) ds \cdot e_{\nabla}((n+1)h, s) \tag{10.61}$$

where the integration path γ is any simple close contour in a region of analyticity of the integrand and including the point $s = \frac{1}{h}$.

The simplest case is a circle with centre at $\frac{1}{h}$ and radius $r = \frac{1}{h}$. Consider first a simple example: $F(s) = 1$. We are expecting that the inverse transform be an impulse. In fact, we obtain

$$\mathcal{N}_{\mathcal{L}}^{-1}[1] = \delta(nh), \tag{10.62}$$

where $\mathcal{N}_{\mathcal{L}}$ stands for the Nabla Laplace transform. The relation between the two exponentials expressed by (10.37) suggests the following definition.

Definition 10.3.2. *The Nabla Laplace transform (NLT) is defined by:*

$$\mathcal{N}_{\mathcal{L}} f(t) = F_{\nabla}(s) = h \sum_{n=-\infty}^{+\infty} f(nh) e_{\Delta}(nh, -s) \tag{10.63}$$

Attending to the properties of the exponential that we stated before, the limit of (10.63) as $h \to 0$ is the usual two-sided Laplace transform.

10.3.2 Main properties of the NLT

This transform enjoys several properties, described below. We will assume that s is inside the region of convergence and in most cases consider only the properties corresponding to the nabla derivative/transform case; the analog delta cases are easily obtained [89].

- *Linearity.* It is an evident property, the proof of which does not need any computation.
- *Transform of the derivative.* Attending to the previous results and to (10.38), we deduce immediately that

$$\mathcal{N}_{\mathcal{L}}\left[f_{\nabla}^{(\alpha)}(nh)\right] = s^{\alpha} F(s), \tag{10.64}$$

and thus we find again a well known result in the context of the Laplace transform. The region of convergence is the disk inside the Hilger circle. It is worth to remark here that we are dealing with a two-sided transform. Therefore, no *initial values* are required. As we can see from (10.43), s^{α} is, aside the $1/h^{\alpha}$ factor, the NLT of the causal sequence formed by binomial coefficients. Consequently, it is the transfer function of the system (differenciator) defined by the fractional derivative. The ROC is the interior of the Hilger circle.

- *Time shift.* The NLT of $f(nh - n_0 h)$ with $n_0 \in \mathbb{Z}$ is given by

$$\mathcal{N}_{\mathcal{L}}\left[f(nh - n_0 h)\right] = e_\Delta(n_0 h, -s) F(s) \tag{10.65}$$

This result comes from the definition of transform (10.52).
- *Modulation or shift in s.*

$$\mathcal{N}_{\mathcal{L}}\left[f(nh)e_\nabla(nh, -s_0)\right] = F\left(\frac{s - s_0}{1 - s_0 h}\right) \tag{10.66}$$

To obtain this result we only have to use the inverse transform (10.61). Similarly,

$$\mathcal{N}_{\mathcal{L}}\left[f(nh)e_\Delta(nh, -s_0)\right] = F(s + s_0 - s\,s_0 h) \tag{10.67}$$

- *Convolution in time.*

$$\mathcal{N}_{\mathcal{L}}\left[h\sum_{k=-\infty}^{+\infty} f(kh)g(nh - kh)\right] = F(s)G(s) \tag{10.68}$$

To prove this relation we only have to insert $F(s)G(s)$ into (10.61), substitute $F(s)$ by its expression given by (10.63), and use (10.61) again.
- *Time scaling.* Let a be a positive real number (the negative case is not interesting). We have

$$\mathcal{N}_{\mathcal{L}}\left[f(nh/a)\right] = \frac{1}{a}F(s/a) \tag{10.69}$$

Essentially this result expresses the way how we can do a rate change: we had $1/h$ samples per second, and end up with a rate of a/h samples per second.

10.3.3 Examples

We are going to compute the NLT of some interesting functions.
- *Unit steps.* Let $f(t) = \varepsilon(t)$. The NLT is given by

$$\mathcal{N}_{\mathcal{L}}\left[\varepsilon(t)\right] = h\sum_{n=0}^{+\infty} e_\Delta(nh, -s)$$

$$= \frac{h}{1 - (1 - sh)} = \frac{1}{s}, \tag{10.70}$$

provided that $|1 - sh| < 1$. For the anti-causal step,

$$\mathcal{N}_{\mathcal{L}}\left[\varepsilon(-t)\right] = h\sum_{n=-\infty}^{0} e_\Delta(nh, -s)$$

$$= h\frac{1}{1 - (1 - sh)^{-1}} = -\frac{1 - sh}{s} \tag{10.71}$$

This result is somehow different from what we were waiting for, but it can be explained by the shift property of the NLT. Let us modify the problem and compute

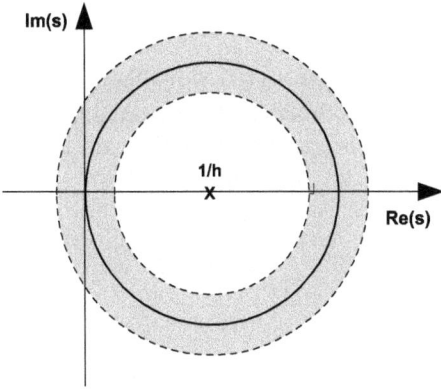

Fig. 10.3: Region of convergence for the NLT of a two-sided signal.

the transform of $\varepsilon(-t + h)$. We have now

$$\mathcal{N}_{\mathcal{L}} [\varepsilon(-t - h)] = h \sum_{n=-\infty}^{-1} e_{\Delta}(nh, -s) = \frac{1}{s} \tag{10.72}$$

Now the ROC is the region outside the right Hilger circle, $|1 - sh| > 1$.

− *Causal and anti-causal exponentials.*

$$\mathcal{N}_{\mathcal{L}} [e_{\nabla}(t + h, p)\varepsilon(t)] = \frac{1}{s - p} \tag{10.73}$$

$$\mathcal{N}_{\mathcal{L}} [-e_{\nabla}(t + h, p)\varepsilon(-t - h)] = \frac{1}{s - p} \tag{10.74}$$

To find the corresponding ROC we must test the fraction $\frac{1-sh}{1-ph}$. In the first case its absolute value must be less than 1. It is not difficult to see that we must have

$$\left| s - \frac{1}{h} \right| < \left| p - \frac{1}{h} \right| \tag{10.75}$$

So we conclude that the region of convergence is the inside of a circle centred on $1/h$ with radius $\left| p - \frac{1}{h} \right|$. We can obtain an easier criterion by imposing $\left| s - \frac{1}{h} \right| < 1$ and $\left| p - \frac{1}{h} \right| > 1$. The pole must stay outside the Hilger circle and the ROC is the disk with radius $|p|$ centred in $\frac{1}{h}$. For the anti-causal case, the situation is reversed. We can generalise the results above for multiple poles by computing successive integer order derivatives relatively to p.

− *Causal powers.* As seen above, the NLT of the unit step is given by (10.70). Using the property of the derivative with negative α and using the results of section 10.2, we obtain

$$\mathcal{N}_{\mathcal{L}} \left[h^{\alpha} \frac{(\alpha + 1)_n}{n!} \varepsilon(nh) \right] = \frac{1}{s^{\alpha+1}} \tag{10.76}$$

The region of convergence is the disk inside the Hilger circle. For the anti-causal case the situation is similar, but the region of convergence is outside the Hilger circle.

10.3.4 Existence of NLT

As for both the Laplace transform and Z transform all we can is to formulate sufficient conditions for the NLT to exist. Let us consider a function $f(nh)$ assuming finite values for every finite $n \in \mathbb{Z}$. We will assume that it is bounded asymptotically by two exponentials:

$$|f(nh)| < A\, e_\nabla(nh, a), \qquad n > n_2 \tag{10.77}$$

$$|f(nh)| < B\, e_\nabla(nh, b), \qquad n < n_1 \tag{10.78}$$

We can write

$$\left| \sum_{n=-\infty}^{+\infty} f(nh) e_\Delta(nh, -s) \right| \le \sum_{n=-\infty}^{+\infty} |f(nh) e_\Delta(nh, -s)| \tag{10.79}$$

$$< \sum_{n=n_1}^{n_2} |f(nh) e_\Delta(nh, -s)| + A \sum_{n=-\infty}^{n_1-1} |e_\nabla(nh, b) e_\Delta(nh, -s)|$$

$$+ B \sum_{n=n_2+1}^{+\infty} |e_\nabla(nh, a) e_\Delta(nh, -s)|$$

According to the results obtained in the last section, we conclude that the two infinite summations converge inside and outside two regions delimited by $|a - 1/h|$ and $|b - 1/h|$, concentric with the right Hilger circle. To have a non empty region of convergence we must have

$$|a - 1/h| > |b - 1/h|, \tag{10.80}$$

as shown in figure 10.8.

10.3.5 Unicity of the transform

Let $g(nh)$ a function defined on a given time scale $\mathbb{T} = h\mathbb{Z}$. It is not a hard task to show that the NLT given by (10.63) is a unique function inside its ROC. Now generalise the situation: let g_n, $n \in \mathbb{Z}$ be a discrete-time function that we want to transform. To do so, we must fix a time scale. This means that the above signal has infinite NLT, one for each time scale.

This is very interesting, since it can be used to change from one scale to another. Let h and q be two scales and $g(nh)$ and $g(nq)$ the corresponding functions that have equal NLT, with ROC R_h and R_q respectively, such that $R_h \subset R_q$. Choose an integration path in R_h. We then have

$$f(nh) = -q \sum_{k=-\infty}^{+\infty} f(kq) \frac{1}{2\pi i} \oint_\gamma e_\Delta(kq, -s)\, e_\nabla((n+1)h, s)\, ds, \tag{10.81}$$

where function

$$\phi(nh, kq) = -q \frac{1}{2\pi i} \oint_\gamma e_\Delta(kq, -s)\, e_\nabla((n+1)h, s)\, ds \tag{10.82}$$

is a generalisation of the well known sinc function. It can be calculated using (10.55)–(10.56).

As to the inverse NLT, the situation is not exactly the same. Even having fixed a time scale, we can have several inverse transforms accordingly to the region of convergence that we choose. In the rational transform case we can have three situations:

- one causal inverse;
- one anti-causal inverse:
- one or several non-causal inverses (we can call them *acausal*).

10.3.6 Initial value theorem

Consider the causal functions

$$g(t) = f(t)\varepsilon(t) \tag{10.83}$$

and particularise the NLT definition for this case. We obtain easily the *initial value theorem* from (10.63). We only have to remark that the summation goes from 0 to ∞,

$$G(s) = hg(0) + hg(1)e_\Delta(h, -s) + hg(2)e_\Delta(2h, -s) + \ldots$$
$$\cdots + hg(3)e_\Delta(3h, -s) + \ldots \tag{10.84}$$

and all the terms in the series, except one (viz. $h.g(0)$), become zero if $s = \frac{1}{h}$. So

$$g(0) = \lim_{s \to \frac{1}{h}} s\, G(s). \tag{10.85}$$

This theorem can be readily generalised to the case where $g(t) = 0$ for a given $t < Kh$, $K > 0$. We only have to multiply $G(s)$ by $e_\Delta(Kh, -s)$.

10.3.7 Final value theorem

To find the *final value theorem*, we define $y(nh) = h \sum_{k=0}^{n} g(kh)$. If $g(t)$ is a causal function, $y(t)$ is the convolution of $g(t)$ with the unit step. So the transform of $y(t)$ is $\frac{G(s)}{s}$ and, as can be easily verified,

$$y(\infty) = \lim_{s \to 0} G(s) = \lim_{s \to 0} s\, Y(s), \tag{10.86}$$

which is equal to the usual result.

Now consider an anti-causal signal:

$$g(t) = f(t)\varepsilon(-t - h) \tag{10.87}$$

Its NLT is given by

$$G(s) = \cdots + hg(-3h)e_\Delta(-3h, -s)$$
$$+ hg(-2h)e_\Delta(-2h, -s) + hg(-h)e_\Delta(-h, -s) \tag{10.88}$$

From this we can define the *last value* by

$$g(-h) = \lim_{s \to \infty} \left[s - \frac{1}{h} \right] G(s) \tag{10.89}$$

10.3.8 The delta transform and the correlation

We introduced and studied more carefully the transform suitable for dealing with causal systems. This does not mean that we cannot invert the roles of the exponentials and formulate another transform. It is not very difficult to do it. Now, and as before, a given function $f(nh)$ can be synthesised by

$$f(nh) = \frac{1}{2\pi i} \oint F_\Delta(s) e_\Delta((n-1)h, s) \, ds, \tag{10.90}$$

where the integration path is any simple close contour in a region of analyticity of the integrand and including the point $s = -\frac{1}{h}$.

It is again the relation between the two exponentials expressed by (10.37) that suggests that we define the delta Laplace tansform (DLT) by

$$F_\Delta(s) = h \sum_{n=-\infty}^{+\infty} f(nh) e_\nabla(nh, -s) \tag{10.91}$$

The limit as $h \to 0$ in (10.90) and in (10.91) leads once more to the usual two-sided Laplace transform. The properties the DLT are similar to those of the NLT, but the ROC and the Hilger circle are now in the left complex plane.

This transform is useful in the computation of the transform of the correlation. Accordingly to the well-known relation between the convolution and the correlation, we introduce this for two functions $f(t)$ and $g(t)$ by

$$r_{fg}(nh) = h \sum_{k=-\infty}^{+\infty} f(kh) g(nh + kh). \tag{10.92}$$

Using the shift property of the NLT, we have

$$\mathcal{N}_{\mathcal{L}} \left[r_{fg}(nh) \right] = h \sum_{k=-\infty}^{+\infty} f(kh) e_\Delta(-kh, -s) G_\nabla(s) \tag{10.93}$$

Using (10.37) and attending to (10.91), we can write

$$\mathcal{N}_{\mathcal{L}} \left[r_{fg}(nh) \right] = F_\Delta(-s) G_\nabla(s), \tag{10.94}$$

that is a somehow strange result, but that can be important when generalising this theory for other time scales. It is not a hard task to show that, in the continuous time case, the right hand side is $F(-s)G(s)$, and in the Z transform case it is $F(z^{-1})G(z)$.

10.3.9 Backward compatibility

Accordingly to what we said above, when we perform the limit $h \to 0$ we recover the classic continuous-time formulations. Applying the Euler transformation $s = \frac{1-z^{-1}}{h}$, we obtain the current discrete-time difference based equations. In figure 10.4 we can see the evolution of the Hilger circle as h decreases to zero.

So, when $h \to 0$:

- The two derivatives degenerate into generalised fractional order derivatives;
- The negative integer order gives the well-known formulation of the Riemann integral;
- The exponentials degenerate into one: e^{st};
- The nabla and delta Laplace transforms degenerate also into the classic two-sided Laplace transform;
- Putting $s = i\omega$ we obtain the Fourier transform.

In the case of a difference formulation, instead of a limit computation we make two variable transformations: $s = \frac{1-z^{-1}}{h}$ in the nabla case, and $s = \frac{z-1}{h}$ in the delta case.

- Both exponentials degenerate into the current exponential z^n (see figure 10.5);
- Both transforms recover the classic two-sided Z transform;
- Making $s = e^{i\omega}$ we obtain the discrete-time Fourier transform.

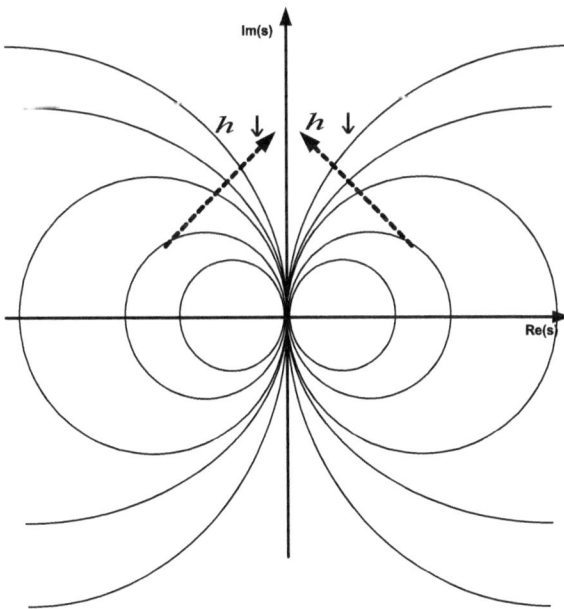

Fig. 10.4: Evolution of the Hilger circle as h decreases to zero.

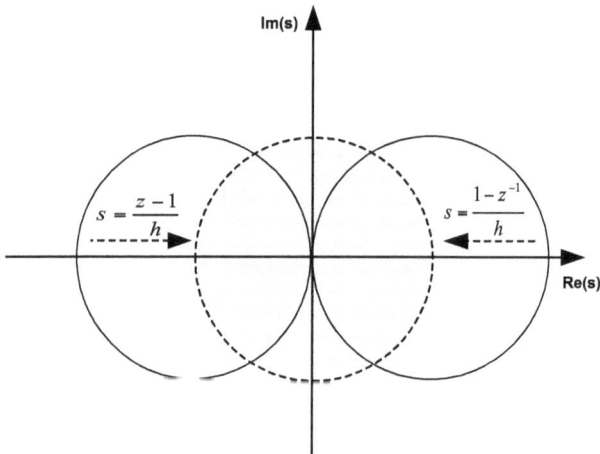

Fig. 10.5: Transformations from the Hilger circles to the unit circle.

10.4 Discrete-time fractional linear systems

We are going to consider discrete-time systems with the general format [89]

$$\sum_{k=0}^{N_0} a_k D^{\alpha_k} y(t) = \sum_{k=0}^{M_0} b_k D^{\beta_k} x(t) \tag{10.95}$$

where $a_N = 1$. Orders N_0 and M_0 are any given positive integers. Let $g(t)$ be the impulse response of the system defined by (10.95), i.e. the case $x(t) = \delta(nh)$. As in previous defied linear systems, the output is the convolution of the input and the impulse response.

If $x(t) = e_\nabla(nh, s)$ the output is given by

$$y(t) = e_\nabla(nh, s)\left[h \sum_{n=-\infty}^{\infty} g(nh)e_\Delta(nh, -s) \right]. \tag{10.96}$$

The summation expression is, as usually, the **transfer function**. We then write

$$G(s) = h \sum_{n=-\infty}^{\infty} g(nh)e_\Delta(nh, -s) \tag{10.97}$$

In other words, the transfer function is the NLT of the impulse response. With these results we can easily express the transfer function as

$$G(s) = \frac{\sum_{k=0}^{M_0} b_k s^{\beta_k}}{\sum_{k=0}^{N_0} a_k s^{\alpha_k}}. \tag{10.98}$$

In general, the systems described by (10.95) or (10.98) are IIR systems. Only when $N_0 = 0$ and all the derivative orders are positive integers do we have FIR Systems. The general case stated in (10.98) is very difficult to study due to the difficulties in obtaining the poles and zeros. Therefore, we will consider the commensurate case:

$$G(s) = \frac{\sum\limits_{k=0}^{M_0} b_k s^{\alpha k}}{\sum\limits_{k=0}^{N_0} a_k s^{\alpha k}}. \tag{10.99}$$

Systems given by (10.99) will be called fractional ARMA (FARMA) systems. Assume that the fraction in (10.99) is decomposed into a sum of a polynomial (only zeros) plus a proper fraction (with poles and zeros). We will study them separately.

10.4.1 Polynomial case

Let us consider a fractional MA (FMA) transfer function:

$$G_{MA}(s) = \sum_{k=0}^{M_0} b_k s^{\alpha k} \tag{10.100}$$

To invert this expression, we recall our previous statement that s^α is the transfer function of the differentiator. The corresponding impulse response given by the binomial coefficients is in agreement with (10.43). Therefore, the impulse response corresponding to (10.100) is

$$g_{MA}(nh) = b_0 \delta(nh) + \sum_{k=1}^{M_0} b_k h^{-\alpha-1} \frac{(-\alpha k)_n}{n!} \varepsilon(nh) \tag{10.101}$$

In figure 10.6 we depict the impulse responses of this kind of systems for $\alpha = 0.1, 0.2, \ldots, 1$. We set $h = 1$. In the computations we used as b_k, $k = 0, 1, \ldots, M$ the FIR impulse response shown in the last strip. Although we are dealing with theoretically IIR systems, the impulse responses go to zero very fast, so the behaviour in practice is that of a FIR.

10.4.2 Proper fraction case

Consider the FARMA(N_0, M_0) given by (10.99) with $N_0 > M_0$. For the sake of simplicity, we assume that the polynomial $A(w) = \sum\limits_{k=0}^{N_0} a_k w^k$ has only simple roots. In this case we can write $G(s)$ as

$$G_{ARMA}(s) = \sum_{k=1}^{N_0} \frac{R_k}{s^\alpha - p_k}, \tag{10.102}$$

$$\alpha = 0.2$$

$$\alpha = 0.4$$

$$\alpha = 0.6$$

$$\alpha = 0.8$$

$$\alpha = 1.0$$

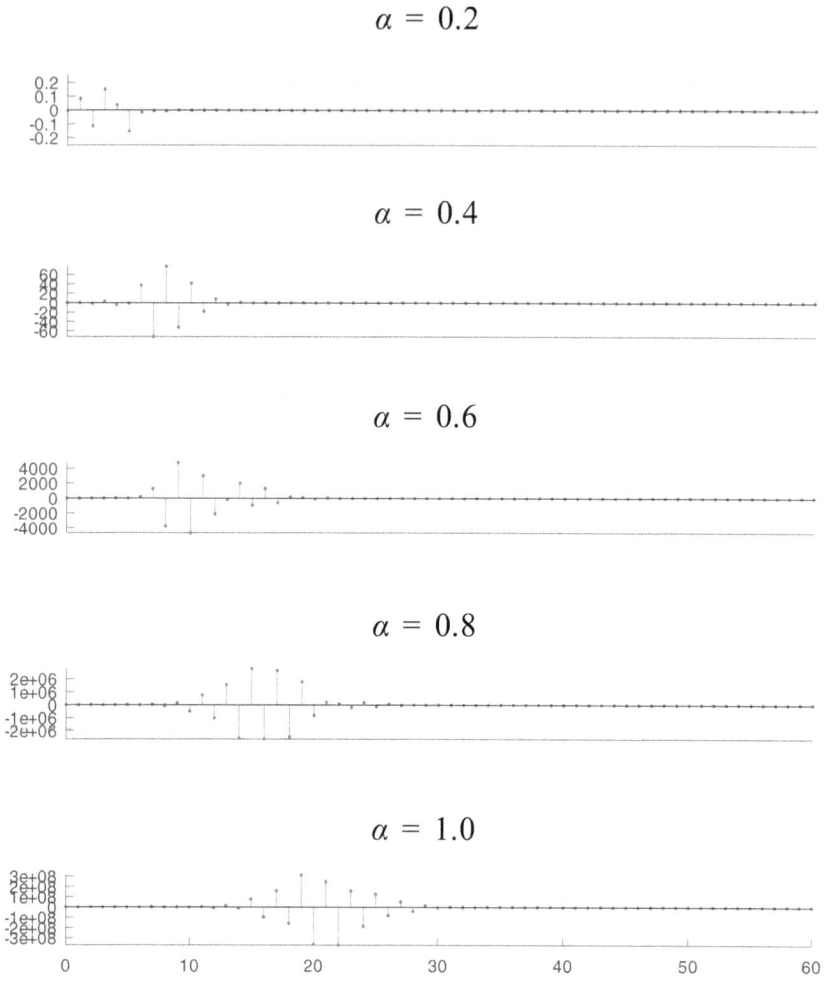

Fig. 10.6: Impulse responses of polynomial systems.

where the R_k are the residues, and p_k, $k = 1, 2, \ldots, N$ the poles of $A(w)$. The NLT inverse of $G_{ARMA}(s)$ is obtained by inverting separately the partial fractions of the form

$$F(s) = \frac{1}{s^\alpha - p}. \tag{10.103}$$

To invert (10.103), we insert $F(s)$ into the inversion integral and use the residue theorem. However, this implies the computation of the $(n + 1)$th derivative of $F(s)$ and then the substitution of $1/h$ for s. A simpler alternative is to use the geometric series. We have two situations corresponding to different regions:

1. *Intersection of the Hilger circle with the disk* $|s| < |p|^{\frac{1}{\alpha}}$.

$$\alpha = 0.2$$

$$\alpha = 0.4$$

$$\alpha = 0.6$$

$$\alpha = 0.8$$

$$\alpha = 1.0$$

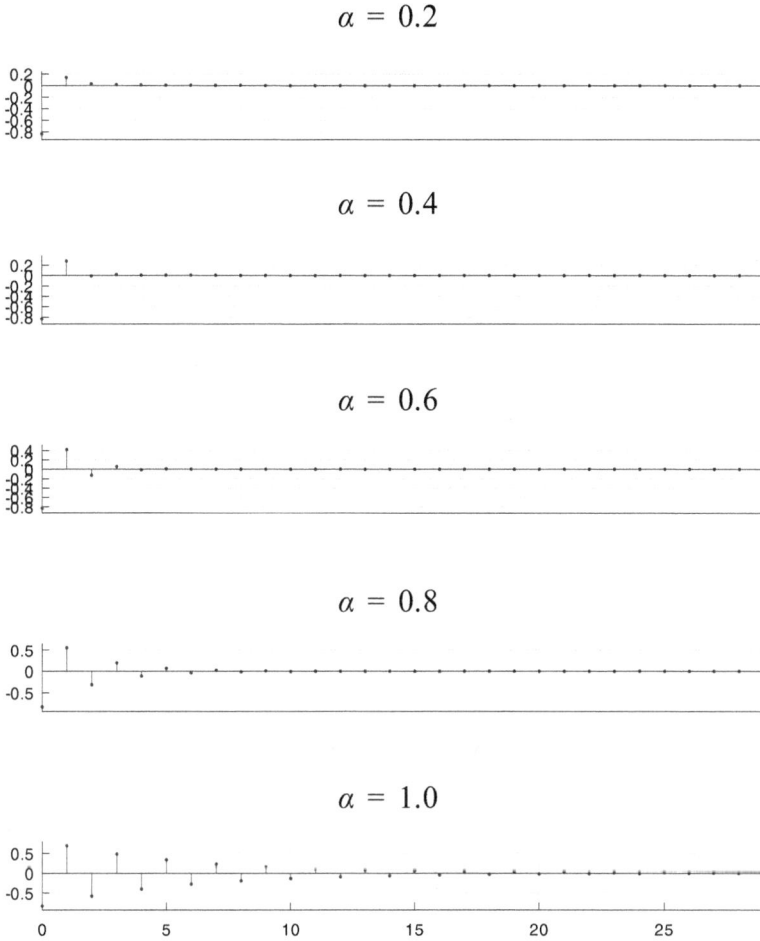

Fig. 10.7: Impulse response corresponding to $\frac{1}{s^\alpha - p}$ with $p = 2.1$.

In this case we have

$$F(s) = -\frac{1}{p}\frac{1}{1-\frac{s^\alpha}{p}} = -\frac{1}{p}\left[1 + \sum_{k=1}^{\infty} p^{-k}s^{\alpha k}\right]. \tag{10.104}$$

From the above expression it results that the impulse response $f(nh)$ corresponding to a partial fraction (10.103) is given by

$$f(nh) = -\left[\sum_{k=0}^{\infty} p^{-k-1}h^{-\alpha k-1}\frac{(-\alpha k)_n}{n!}\varepsilon(nh)\right]. \tag{10.105}$$

In figure 10.7, we present an illustration showing the impulse responses of a one-pole system for $\alpha = 0.1, 0.2, \ldots, 1$. As seen, the response goes to zero quickly, resembling a FIR system. However this will not be an interesting system, since the values of p must be high for short h.

In practice we have pairs of complex conjugate poles that introduce terms of the form:

$$V(s) = \frac{A}{s^\alpha - p} + \frac{A^*}{s^\alpha - p^*} \tag{10.106}$$

Using (10.105) we obtain:

$$v(nh) = -2.Re\left\{\frac{A}{p}\right\}\delta(nh) - 2\sum_{k=1}^{\infty} Re\left\{\frac{A}{p^{k+1}}\right\} h^{-k\alpha-1}\frac{(-\alpha k)_n}{n!}\varepsilon(nh) \tag{10.107}$$

In figure 10.8, we show the impulse responses of a complex conjugate pair of poles system for $\alpha = 0.1, 0.2, \ldots, 1$ and $p = 1.001\left(1 + e^{i\pi/2}\right)$. As it can be seen, only with multiplicity equal to 1 do we obtain a sinusoid. In the other cases, the system goes from a dumped oscillation to an undumped oscillation with decreasing order.

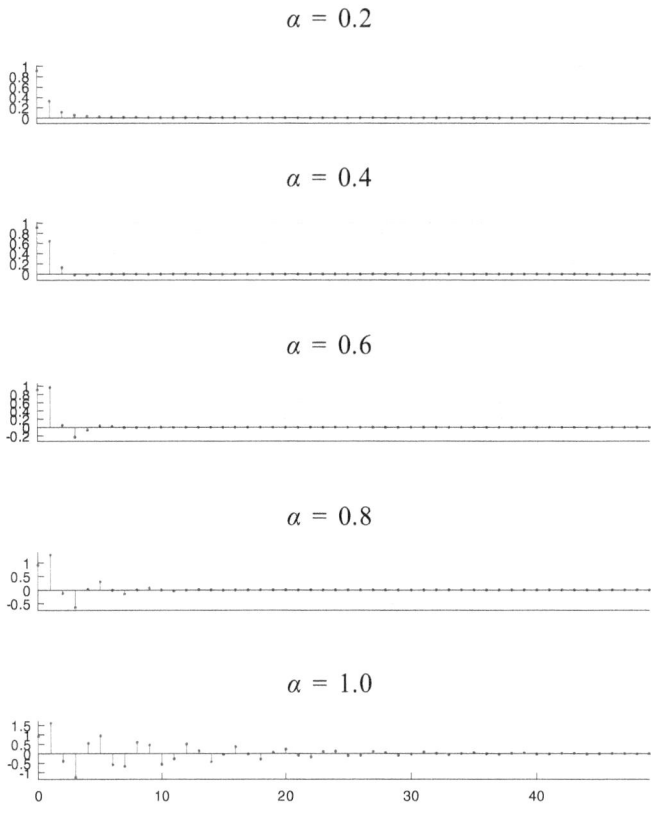

$$\alpha = 0.2$$

$$\alpha = 0.4$$

$$\alpha = 0.6$$

$$\alpha = 0.8$$

$$\alpha = 1.0$$

Fig. 10.8: Impulse response corresponding to system with a complex conjugate pair of poles for $p = 1.1(1 + i)$.

2. *Intersection of the Hilger circle with the disk $|s| > |p|^{\frac{1}{a}}$.*
   For this case, we can write

$$F(s) = \frac{s^{-\alpha}}{1 - ps^{-\alpha}} = A\left[\sum_{k=1}^{\infty} p^{k-1}s^{-\alpha k}\right]. \tag{10.108}$$

So, the corresponding impulse response $f(nh)$ is given by

$$f(nh) = A\left[\sum_{k=1}^{\infty} p^{k-1}h^{\alpha k-1}\frac{(\alpha k)_n}{n!}\varepsilon(nh)\right] \tag{10.109}$$

This expression is the discrete-time version of the alpha-exponential we found in continuous-time systems and that is related to the Mittag-Leffler function. To see that it is so, we are going to compute the limit as h goes to zero.

$$\frac{(\alpha k)_n}{n!} = \frac{\Gamma(\alpha k + t/h)}{\Gamma(\alpha k)\Gamma(t/h + 1)} \approx \frac{1}{\Gamma(\alpha k)}\left(\frac{t}{h}\right)^{\alpha k-1} \tag{10.110}$$

That is why this situation is more interesting than the one above, since it allows us to approach the continuous-time situation as much as we want, even for small values of p.

To obtain the expressions for a complex conjugate pair, we proceed as above to obtain (10.109)

$$v(nh) = 2\sum_{k=1}^{\infty} Re\left\{A.p^{k-1}\right\}h^{\alpha k-1}\frac{(\alpha k)_n}{n!}\varepsilon(nh) \tag{10.111}$$

10.4.3 The anti-causal case

For comparison purposes, we are going to compute the anti-causal inverse of the one-pole fraction $\frac{1}{s^\alpha - p}$. We proceed as above:

$$F(s) = A\frac{s^{-\alpha}}{1 - ps^{-\alpha}} = A\left[\sum_{k=1}^{\infty} p^{k-1}s^{-\alpha k}\right] \tag{10.112}$$

This leads to

$$f(nh) = -A\left[\sum_{k=1}^{\infty} p^{k-1}h^{\alpha k-1}\frac{(\alpha k)_{-n}}{(-n)!}\varepsilon(-nh)\right] \tag{10.113}$$

10.4.4 Inversion without partial fractions

The inversion of fractional transfer functions proceeds as described in chapter 2. Here we will make a brief description of the algorithm adapted to the commensurate case. The algorithm is based on a negative power series expansion of the transfer function. The term by term NLT inversion of such series leads to a series that is analogue to well-known power series: those of MacLaurin and Taylor.

Let $H(s)$ be a given transfer function. We are going to obtain a decomposition of $H(s)$ into a sum of negative power functions plus an error term:

1. Define $R_0(s) = H(s)$.
2. Let $n_0 = (N - M)\alpha$ be the integer value such that $s^{n_0\alpha}B(s)$ is of order N, and let B_0 be the coefficient of the highest power. Then let

$$R_1(s) = H(s) - B_0 s^{-n_0\alpha} \tag{10.114}$$

   The order of its denominator is $N + n_0$, and the order of the numerator $B_1(s)$ is less than N.
3. Now, repeat the process. Let n_1 be the positive integer value such that the order of $s^{n_1\alpha}B_1(s)$ is $N + n_0$, and let B_1 be the coefficient of the highest power. Then let

$$R_2(s) = H(s) - B_0 s^{-n_0\alpha} - B_1 s^{-n_1\alpha} \tag{10.115}$$

   Again, the order of its numerator $B_2(s)$ is less than N, and the order of the denominator is equal to $N + n_0 + n_1$.
4. In general, let n_n be the real value such that the order of $s^{n_n\alpha}B_{n-1}(s)$ is equal to that of the denominator, and let B_n be the coefficient of the highest power in the new numerator. Then let

$$R_n(s) = H(s) - \sum_0^{n-1} B_k s^{-n_k\alpha}, \tag{10.116}$$

   and the numerator in $R_n(s)$ has order less than N. It is interesting to remark that, while the order of the numerator remains equal to $N - 1$, the order of the denominator increases from iteration to iteration as $\sum_0^{n-1} n_k$.

We can write:

$$H(s) = \sum_0^{n-1} B_k s^{-n_k\alpha} + R_n(s) \tag{10.117}$$

It is easy to conclude that $|R_n(s)| = o\left(\left|\frac{1}{s^{n\alpha}}\right|\right)$, allowing us to write what we can consider a generalized Laurent series [92]:

$$H(s) = \sum_0^{\infty} B_k s^{-n_k\alpha} \tag{10.118}$$

About this series we can say the following:

- *Convergence.* It is easy to state the convergence of this series. We can apply the theory of Z transform [102]. In this case, if the sequence B_k is of exponential order, the series in (10.118) converges in the region that is the exterior of a circle with centre at $s = 0$.
- *Order zero term.* As we can see from the description of the algorithm, the order zero term is constant only if the orders of the numerator and denominator are equal.

– *Simplified expression.* The n_k $k = 0, \ldots, \infty$ values form a strictly increasing sequence of positive integers. So we can always write the above series in the format

$$H(s) = \sum_0^\infty B_k s^{-k\alpha}, \tag{10.119}$$

even if we have to include null coefficients.

Let $h(t)$ be the impulse response corresponding to $H(s)$. Then

$$h(t) = \sum_0^\infty B_k h^{\alpha k - 1} \frac{(\alpha k)_n}{n!} \varepsilon(t). \tag{10.120}$$

Accordingly to what we wrote before, this series can be considered as a discrete power series. The orders of the "powers" are $\alpha k - 1$, $k = 0, \ldots, \infty$. (Notice that we do not try to assign any meaning to the B_k coefficients.)

Example 10.4.1. *Consider transfer function*

$$H(s) = \frac{s^{\alpha - 1}}{s^\alpha + 1}, \tag{10.121}$$

which in the continuous-time case is the Laplace transform of the Mittag-Leffler function. We are going to proceed as pointed above. We have

$B_0 = 1$	and $\gamma_0 = 1$	(10.122)
$B_1 = -1$	and $\gamma_1 = \alpha + 1$	(10.123)

Repeating the process, we obtain:

$$R_n(s) = (-1)^n \frac{1}{s^{1+n.\alpha}(s^\alpha + 1)}, \tag{10.124}$$

with

$$B_n = (-1)^n \text{ and } \gamma_n = n.\alpha + 1. \tag{10.125}$$

This leads to

$$H(s) = s^{-1} \sum_0^\infty (-1)^k s^{-k\alpha}. \tag{10.126}$$

Its inverse is easily obtained:

$$h(t) = \sum_0^\infty (-1)^k h^{\alpha k} \frac{(\alpha k + 1)_n}{n!} . \varepsilon(t) \tag{10.127}$$

This can be considered a discrete-time Mittag-Leffler function. In figures 10.9 and 10.10 we present the results of inverting the above transform for $\beta = 1$ and $\alpha = 1$ or

0.5, respectively. To have a reference, we plotted as well the exponential $e^{-t}\varepsilon(t)$, that corresponds to the limit when $h \longrightarrow 0$, with $\alpha = 1$.

10.4.5 Some stability issues

The properties of the nabla exponential and the sequence of operations we followed to compute the NLT of the causal transform showed that, if the poles are outside the right Hilger circle, then the system is stable. Likewise, the partial fraction inversions we computed above showed that the series defining the time functions were convergent if $|p|h^\alpha > 1$ in the first case, and $|p|h^\alpha < 1$ in the second (and more interesting) case. This means that, if p is outside the Hilger circle, the system is indeed stable. Otherwise, the system can be stable even with the pole inside the Hilger circle, provided that it is outside the circle $|s| = |p|^{\frac{1}{\alpha}}$. This is similar to the situation found in the continuous-time case with the α–exponential or with the Mittag-Leffler function where can have poles in th right hand complex half-plane.

For the integer order systems we can study the pole distribution by a Routh-Hurwitz like criterion, that we will not present in detail; see [17].

10.4.6 A correspondence principle

The above development showed that the algorithm has two steps:
- Expand the transfer function in negative powers of s^α;
- Invert the series term by term.

On the other hand, the description of the algorithm showed that it can be used for both integer or fractional orders. In fact, the fractional order α does not have any role. This means that the algorithm is valid for any α, and leads us to state the possibility of obtaining the fractional solution from the integer one and vice-versa. We call *correspondence principle* to this conversion procedure [92]. More formally:

Theorem 10.4.1 (Discrete version of the correspondence principle). *Let $H(s)$ be the Laplace transform of the impulse response $h(t)$ of a causal commensurate linear system. The substitution of s^α for s in the transfer function corresponds to the substitution of $(k\alpha)$ for k in the series associated to the impulse response (and, obviously, in the differential equation).*

Proof. In terms of the series, we can do as follows: let $G(s) = \sum_{k_0}^{\infty} g_k s^{-k}$. The substitution of s^α for s to obtain the series

$$G_\alpha(s) = \sum_{k_0}^{\infty} g_k s^{-\alpha k} \qquad (10.128)$$

is equivalent to the substitution of $k\alpha$ for k in the series $g(t) = \sum_{k_0}^{\infty} g_k h^{k-1} \frac{(k)_n}{n!} \varepsilon(nh)$ to

obtain $g_\alpha(t) = \sum_{k_0}^{\infty} g_k h^{k\alpha-1} \frac{(k\alpha)_n}{n!} \varepsilon(nh)$. □

$$\alpha = 0.25$$

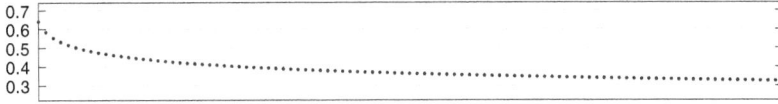

$$\alpha = 0.5$$

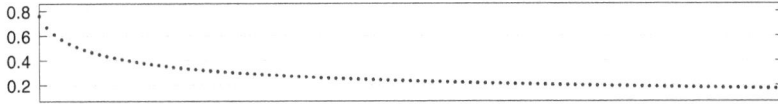

$$\alpha = 0.75$$

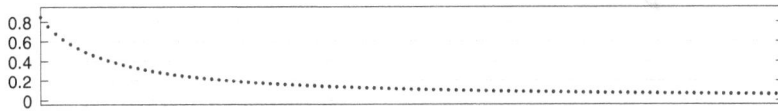

$$\alpha = 1.0$$

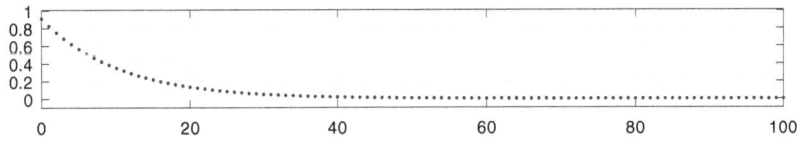

Fig. 10.9: Computation of the discrete-time Mittag-Leffler function for $\beta = 1$ and $h = 0.1$.

Example 10.4.2. *The inverse NLT of $G(s) = \frac{1}{s+a}$ is given by $g(t) = e_\nabla(t + h, -a)\varepsilon(t)$. Suppose we want to compute the inverse of $G_\alpha(s) = \frac{1}{s^\alpha + a}$ using the correspondence principle. First we rewrite $g(nh)$ in a series format:*

$$
\begin{aligned}
g(nh) &= (1 + ah)^{-(n+1)}\varepsilon(nh) \\
&= \sum_0^\infty (-1)^k (ah)^k \frac{(n+1)_k}{k!}\varepsilon(nh) \\
&= \sum_0^\infty (-ah)^k \frac{(k+1)_n}{n!}\varepsilon(nh) \\
&= \sum_1^\infty (-a)^{k-1} h^{k-1}\frac{(k)_n}{n!}\varepsilon(nh) \qquad\qquad (10.129)
\end{aligned}
$$

$$h = 0.1$$

$$h = 0.075$$

$$h = 0.05$$

$$h = 0.025$$

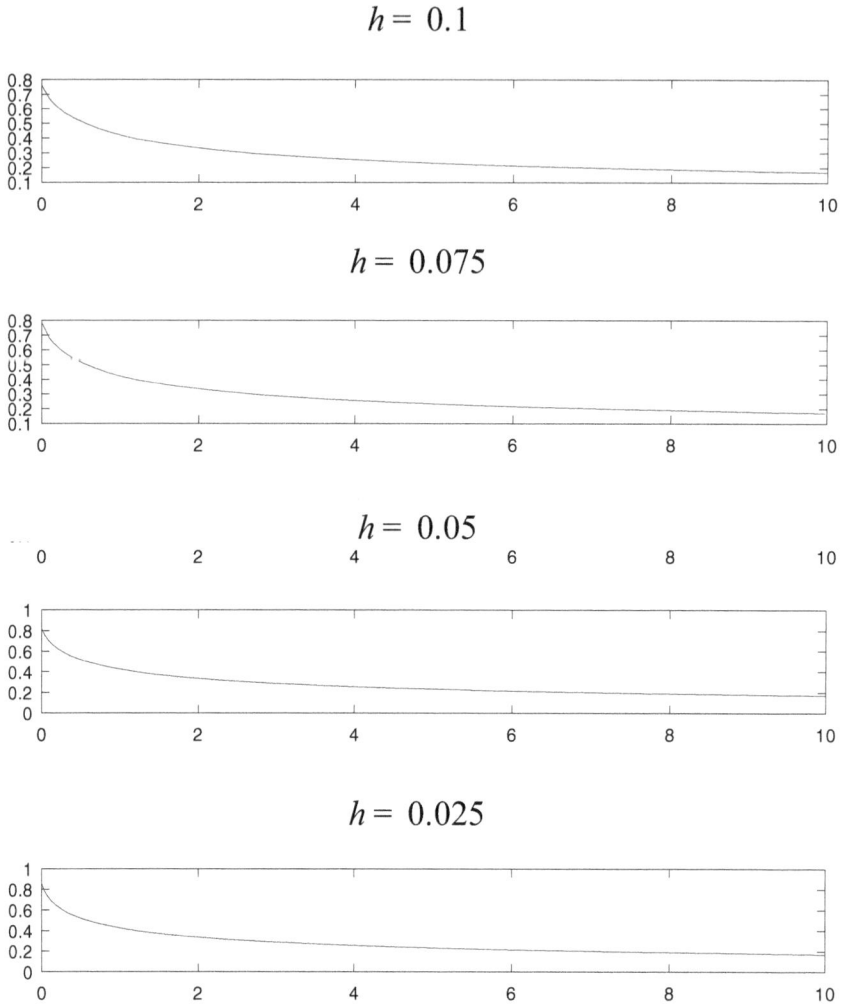

Fig. 10.10: Computation of the discrete-time Mittag-Leffler function for $\alpha = 0.5$ and $\beta = 1$.

Then we perform the above substitution to obtain

$$g_\alpha(nh) = \sum_{1}^{\infty} (-a)^{k-1} h^{\alpha k-1} \frac{(\alpha k)_n}{n!} \varepsilon(nh). \tag{10.130}$$

Computing the NLT of this series term by term, we obtain

$$G_\alpha(s) = \sum_{1}^{\infty} (-a)^{n-1} s^{-n\alpha}. \tag{10.131}$$

Finally we assume that $|s| > |a|^{\frac{1}{\alpha}}$ and sum the series to obtain

$$G_\alpha(s) = s^{-\alpha} \frac{1}{1 + as^{-\alpha}} = \frac{1}{s^\alpha + a} = H(s^\alpha), \qquad (10.132)$$

confirming the above statement.

10.4.7 On initial conditions

In many problems we are mainly interested in transient responses of the systems, obtained for instants after the initial one (that we assume to be $t = 0$). This may be dangerous, because we may be forgetting the past. In current literature the one-sided Laplace or Z transforms are frequently used. In the fractional case the results may be incorrect due to the unsuitableness of the initial conditions employed [75, 87, 89].

Let us begin by considering a simple problem: compute the output of a first order system

$$y'(t) + ay(t) = x(t) \qquad (10.133)$$

under a suitable initial condition with zero input. Consider that $x(t) \neq 0$ for $t < 0$. Obviously, $y(t)$ is a two-sided function. If we multiply it by the unit step, we obtain a causal signal that is not a solution of the above system, unless we remove the effect of multiplying $y(t)$ by the step when computing the derivative. So the equation for the causal $y(t)$ will be

$$[y(t) - y(0)\varepsilon(t)]' + ay(t) = 0 \qquad (10.134)$$

that leads to

$$y'(t) - y(0)\delta(t) + ay(t) = 0 \qquad (10.135)$$

In the general integer order case, we have, from (10.11),

$$g_\nabla^{(N)}(t)\varepsilon(t) = D_\nabla^N[g(t)\varepsilon(t)] - \sum_{k=0}^{N-1} g_\nabla^{(k)}(-h)\delta_\nabla^{(N-1-k)}(t)$$

$$= D_\nabla^N[g(t)\varepsilon(t)] - \sum_{k=0}^{N-1} g_\nabla^{(k)}(-h)\varepsilon^{(N-k)}(t) \qquad (10.136)$$

Substitute $N\alpha$ for N and $k\alpha$ for k to obtain

$$g_\nabla^{(N\alpha)}(t)\varepsilon(t) = D_\nabla^{N\alpha}[g(t)\varepsilon(t)] - \sum_{k=0}^{N-1} g_\nabla^{(k\alpha)}(-h)\varepsilon^{(N-k)\alpha}(t) \qquad (10.137)$$

It is easy to verify the coherence with the above formulae and also the results presented in [75]. To be coherent with the correspondence principle, we rewrite the above relation in the format

$$g_\nabla^{(N\alpha)}(t)\varepsilon(t) = D_\nabla^{N\alpha}[g(t)\varepsilon(t)] -$$

$$- \sum_{k=0}^{N-1} g_\nabla^{(k\alpha)}(-h)h^{-(N-k)\alpha} \frac{(-(N-k)\alpha + 1)_n}{n!} \varepsilon(nh) \qquad (10.138)$$

Consider (10.11) and apply the NLT to both members. We obtain

$$\mathcal{N}_{\mathcal{L}}\left[g_\nabla^{(N\alpha)}(t)\varepsilon(t)\right] = s^{N\alpha}\mathcal{N}_{\mathcal{L}}\left[f(t)\varepsilon(t)\right] - \sum_{k=0}^{N-1} g_\nabla^{(k\alpha)}(-h)s^{(N-k)\alpha-1} \tag{10.139}$$

This result did not require defining a new transform. To solve the general integer order initial condition problem, we only have to substitute the input and output transforms for the corresponding expressions given by (10.11).

So, with the above expression, we can insert the initial conditions in any system [75, 87]. We must call the attention to an extremely important fact: initial conditions must be taken at instants before $t = 0$; in particular, in the continuous-time case case, initial conditions must be at $t = 0^-$, as we conclude from (10.139). This is too frequently forgotten.

10.5 The Fourier transform and the frequency response

The nabla and delta Laplace transforms are suitable for dealing with discrete differential systems. When the exponential degenerates into a complex sinusoid, its absolute value is equal to 1, and the transforms degenerate into a discrete-time Fourier transform. From definition (10.35), we make $|1 - sh| = 1$, defining a circle centred on $1/h$ and with radius equal to $1/h$ (right hand Hilger circle). From definition (10.36), with $|1 + sh| = 1$ we obtain a circle centred on $-1/h$ and with radius also equal to $1/h$ (left hand Hilger circle). With the variable change $s = \frac{1-e^{-i\omega h}}{h}$ in (10.35), (10.61), and (10.63), we obtain (see figure 10.5)

$$e_\nabla(nh, \omega) = e^{i\omega h n} \tag{10.140}$$

$$f(nh) = \frac{1}{2\pi} \int_{-\pi/h}^{\pi/h} F(e^{i\omega})e^{i\omega h n}d\omega \tag{10.141}$$

$$F(e^{i\omega}) = h \sum_{n=-\infty}^{\infty} f(nh)e^{-i\omega h n} \tag{10.142}$$

With substitution $\omega h \Rightarrow \omega$, we obtain expressions that are independent of the graininess h:

$$e_\nabla(n, \omega) = e^{i\omega n} \tag{10.143}$$

$$f(nh) = \frac{1}{2\pi} \int_{-\pi}^{\pi} F(e^{i\omega})e^{i\omega n}d\omega \tag{10.144}$$

$$F(e^{i\omega}) = h \sum_{n=-\infty}^{\infty} f(nh)e^{-i\omega n} \tag{10.145}$$

This means that we can use these expressions for any discrete-time signal independently of the underlying time scale. Expressions (10.144) and (10.145) define the discrete-time Fourier transform pair (DTFT), respectively the inverse and direct. Considering the delta case instead, we arrive at the same expressions for the Fourier transform, provided that we make the substitution $s = \frac{e^{i\omega h}-1}{h}$.

Now we are in conditions of defining the *frequency response* of a linear system. We only have to make $s = \frac{1-e^{-i\omega h}}{h}$ in (10.98). This involves the transformation of the parameters of the transfer function using the binomial coefficients. We can transform separately the numerator and denominator coefficients. For each one, we obtain new coefficients given by (10.101). We observed in section 10.4.1 that we can truncate the sequence. We can verify by means of simulations that, for $\alpha < 1$, the number of such coefficients is less than twice the length of the coefficient sequence corresponding to $\alpha = 1$.

From these considerations we come to the conclusion that transfer function (10.99) has a frequency response given by

$$G(e^{i\omega}) = \frac{\sum\limits_{k=0}^{M_0} B_k e^{-i\omega k}}{\sum\limits_{k=0}^{N_0} R_k e^{-i\omega k}}, \tag{10.146}$$

where the coefficients B_k are given from (10.101) by

$$B_k = \sum_{l=1}^{M_0} b_l h^{-\alpha l-1} \frac{(-\alpha l)_k}{k!} \tag{10.147}$$

for $k = 0, 1, 2, \ldots, M_0$. For the R_k, the situation is similar.

Computing the inverse Fourier transform of $G(e^{i\omega})$, we obtain a difference equation equivalent to differential equation (10.95).

Part III: **Advanced topics**

11 Fractional stochastic processes and two-sided derivatives

11.1 Stochastic input

In previous chapters we studied linear systems, and in particular their responses to some special deterministic inputs: steps, ramps, and sinusoids. Here, we will use random signals as inputs [65, 97]. We will only consider the continuous-time case; the theory is easily adapted for discrete time.

Take a continuous-time linear system with TF $H(s)$, having no poles on the imaginary axis (regular system). Assume that the input $x(t)$ to the system is a stationary stochastic process with autocorrelation function

$$R_{xx}(t) = E[x(\tau + t)x(\tau)] \tag{11.1}$$

where $E[\cdot]$ stands for the expected value. The output is given by $y(t) = h(t) * x(t)$, and the corresponding autocorrelation is

$$R_{yy}(t) = h(t) * h(-t) * R_{xx}(t). \tag{11.2}$$

Let $S_{xx}(s) = \mathcal{L}[R_{xx}(t)]$ represent the LT of the autocorrelation, and let us define the *power spectral density* (or simply the *spectrum*) of $x(t)$ as

$$S_{xx}(i\omega) = \mathcal{F}[R_{xx}(t)], \tag{11.3}$$

obtained restricting s to the imaginary axis, i.e. $s = i\omega$. The relation (11.3) states the Wiener-Khintchin Einstein theorem [25]. We get

$$S_{yy}(s) = H(s)H(-s)S_{xx}(s), \tag{11.4}$$

In the integer order case, $S_{yy}(s)$ has a non empty ROC that includes the imaginary axis, but, in general, the ROC is empty, since $H(s)$ exists only for $\mathbb{R}(s) > 0$. This leads us to define

$$S_{yy}(i\omega) = \lim_{s \to i\omega} H(s)H(-s) \cdot S_{xx}(i\omega) = |H(i\omega)|^2 S_{xx}(i\omega), \tag{11.5}$$

that relates the input with the corresponding output power spectral densities, $S_{xx}(i\omega)$ and $S_{yy}(i\omega)$, respectively.

Remark 11.1.1. *In applications, mainly in modelling real data, we assume that the input is white noise, $w(t)$. This is considered as a generalised derivative of the Brownian motion. It is assumed to have zero mean and variance σ^2. Although not physically realisable, white noise is a useful concept, and is in fact approximated by thermal noises in a frequency band that extends up to frequencies around 10^{10} Hz. Its autocorrelation is an impulse, usually written as*

$$R_{ww} = \sigma^2 \delta(t). \tag{11.6}$$

DOI 10.1515/9783110624588-011

Therefore, the output spectrum of a linear system is

$$S_{yy}(i\omega) = \sigma^2 \, |H(i\omega)|^2 \,, \tag{11.7}$$

stating an important relation suitable for stochastic modeling and identification. To create a model for a given realisation of a stochastic process, $x(t)$, we proceed as follows:
1. *Estimate the power spectral density;*
2. *Compute the parameters defining the transfer function $H(s)$ of a linear system such that $\sigma^2 \, |H(i\omega)|^2$ best approximates $S_{xx}(i\omega)$.*

Let us go back to equation (11.4) and substitute there the expression of the transfer function. We have successively

$$S_{yy}(s) = H(s)H(-s)S_{xx}(s) = \frac{\displaystyle\sum_{k=0}^{M} b_k s^{k\alpha} \ \sum_{k=0}^{M} b_k(-s)^{k\alpha}}{\displaystyle\sum_{k=0}^{N} a_k s^{k\alpha} \ \sum_{k=0}^{N} a_k(-s)^{k\alpha}} \cdot S_{xx}(s), \tag{11.8}$$

or, equivalently,

$$\sum_{k=0}^{N} a_k s^{k\alpha} \sum_{k=0}^{N} a_k(-s)^{k\alpha} S_{yy}(s) = \sum_{k=0}^{M} b_k s^{k\alpha} \sum_{k=0}^{M} b_k(-s)^{k\alpha} S_{xx}(s), \tag{11.9}$$

that we can write as

$$\sum_{k=0}^{N} \sum_{m=0}^{N} a_k a_m s^{k\alpha}(-s)^{m\alpha} S_{yy}(s) = \sum_{k=0}^{M} \sum_{m=0}^{M} b_k b_m s^{m\alpha}(-s)^{k\alpha} S_{xx}(s). \tag{11.10}$$

However, we must take into account that, if $\alpha \neq 1$, these relations are only valid in the limit when $s \to i\omega$. Consequently, we have

$$\sum_{k=0}^{N} \sum_{m=0}^{N} a_k a_m (i\omega)^{k\alpha}(-i\omega)^{m\alpha} S_{yy}(i\omega) = \sum_{k=0}^{M} \sum_{m=0}^{M} b_k b_m (i\omega)^{m\alpha}(-i\omega)^{k\alpha} S_{xx}(i\omega). \tag{11.11}$$

With the inverse Fourier transform and introducing a (centred) two-sided derivative D_θ^γ

$$D_{\alpha-\beta}^{\alpha+\beta} f(t) = \mathcal{F}^{-1}\left[(i\omega)^\alpha(-i\omega)^\beta F(i\omega) \right], \tag{11.12}$$

we obtain a new (two-sided) differential equation

$$\sum_{k=0}^{N} \sum_{m=0}^{N} a_k a_m D_{k\alpha-m\alpha}^{k\alpha+m\alpha} R_{yy}(t) = \sum_{k=0}^{M} \sum_{m=0}^{M} b_k b_m D_{k\alpha-m\alpha}^{k\alpha+m\alpha} R_{xx}(t), \tag{11.13}$$

that defines a new linear system relating the autocorrelation functions of input and output signals. Noting that

$$\Psi_{\alpha-\beta}^{\alpha+\beta}(i\omega) = (i\omega)^\alpha(-i\omega)^\beta = |\omega|^{\alpha+\beta} \, e^{i(\alpha-\beta)\frac{\pi}{2}\,\mathrm{sgn}(\omega)}, \tag{11.14}$$

the frequency response of such system is

$$|H(i\omega)|^2 = \frac{\sum_{k=0}^{M}\sum_{m=0}^{M} b_k b_m |\omega|^{(k+m)\alpha} e^{i(k-m)\frac{\alpha\pi}{2}\operatorname{sgn}(\omega)}}{\sum_{k=0}^{N}\sum_{m=0}^{N} a_k a_m |\omega|^{(k+m)\alpha} e^{i(k-m)\frac{\alpha\pi}{2}\operatorname{sgn}(\omega)}} \tag{11.15}$$

The next section will be dedicated to the study of the two-sided derivatives [70, 86].

11.2 Two-sided derivatives

11.2.1 Derivative of white noise

Consider that we want to study the fractional derivative of white noise, $D^\alpha w(t)$. Assume that we choose the GL derivative (though another could be used instead). The autocorrelation of the output of the derivative operator, $D^\alpha w(t)$, is

$$R_f^\alpha(t_1, t_2) = \lim_{h \to 0^+} \frac{\sum_{k=0}^{\infty}\sum_{n=0}^{\infty} \binom{\alpha}{k}(-1)^{k-n}\binom{\alpha}{n} R_f[t_1 - t_2 - (k-n)h]}{h^{2\alpha}}. \tag{11.16}$$

With a change in the summation variable, it is not hard to show, using the properties of Gauss hypergeometric function, that

$$R_f^\alpha(t_1, t_2) = \lim_{h \to 0} \frac{\Gamma(2\alpha+1)}{h^{2\alpha}} \sum_{k=-\infty}^{+\infty} \frac{(-1)^k}{\Gamma(\alpha-k+1)\Gamma(\alpha+k+1)} R_f(t_1 - t_2 - kh), \tag{11.17}$$

provided that $\alpha > -1$. Therefore, the autocorrelation function of the forward α-order derivative of a stationary stochastic process is a new derivative of order 2α ($\alpha > -1/2$) of the impulse. This suggests we ought to introduce yet another type of derivatives [87].

11.2.2 GL type two-sided derivative

Two centred fractional derivatives were formally introduced by M. Ortigueira [70, 85–87]. They were unified into a formulation that included the one-sided (forward/backward) Grünwald-Letnikov (GL) derivatives.

Definition 11.2.1. *Let $f(t)$, $t \in \mathbb{R}$, be a real function and $\gamma, \theta \in \mathbb{R}$ two real parameters. We define a two-sided GL type fractional derivative of $f(t)$ by*

$$D_\theta^\gamma f(t) = \lim_{h \to 0^+} h^{-\gamma} \sum_{n=-\infty}^{+\infty} \frac{(-1)^n \Gamma(\gamma+1)}{\Gamma(\frac{\gamma+\theta}{2} - n + 1)\Gamma(\frac{\gamma-\theta}{2} + n + 1)} f(t - nh), \tag{11.18}$$

where $\gamma > -1$ is the derivative order, and θ, sometimes called skewness, is in fact an asymmetry parameter.

Remark 11.2.1. *If the function is delayed by $h/2$, another two-sided derivative is obtained, that we will not consider here [68, 69, 87].*

For functions in $L_1(\mathbb{R})$ or in $L_2(\mathbb{R})$, the FT of (11.18) is given by

$$\mathcal{F}\left[D_\theta^\gamma f(t)\right] = \Psi_\theta^\gamma(\omega)\mathcal{F}\left[f(t)\right],\qquad(11.19)$$

where

$$\Psi_\theta^\gamma(\omega) = |\omega|^\gamma e^{i\theta\frac{\pi}{2}\operatorname{sgn}(\omega)}.\qquad(11.20)$$

Some particular cases of (11.18) are very interesting:
1. GL derivatives:

$$D_{\pm\gamma}^\gamma f(t) = \lim_{h\to 0^+} h^{-\gamma} \sum_{n=0}^{+\infty} \frac{(-1)^n \Gamma(\gamma+1)}{\Gamma(\gamma-n+1)\Gamma(n+1)} f(t\mp h).\qquad(11.21)$$

2. GL-type Riesz derivative:

$$D_0^\gamma f(t) = \lim_{h\to 0^+} h^{-\gamma} \sum_{n=-\infty}^{+\infty} \frac{(-1)^n \Gamma(\gamma+1)}{\Gamma(\frac{\gamma}{2}-n+1)\Gamma(\frac{\gamma}{2}+n+1)} f(t-nh).\qquad(11.22)$$

3. GL-type Feller derivative:

$$D_1^\gamma f(t) = \lim_{h\to 0^+} h^{-\gamma} \sum_{n=-\infty}^{+\infty} \frac{(-1)^n \Gamma(\gamma+1)}{\Gamma(\frac{\gamma}{2}+\frac{1}{2}-n+1)\Gamma(\frac{\gamma}{2}-\frac{1}{2}+n+1)} f(t-nh),\qquad(11.23)$$

4. GL-type Hilbert transform:

$$D_\theta^0 f(t) = \lim_{h\to 0^+} \sum_{n=-\infty}^{+\infty} (-1)^n \cdot \frac{1}{\Gamma\left(\frac{\theta}{2}-n+1\right)\Gamma\left(\frac{-\theta}{2}+n+1\right)} f(t-nh).\qquad(11.24)$$

With $\theta = 1$ we obtain the usual discrete-time formulation of the Hilbert transform [100].

11.2.3 The general two-sided fractional derivative

Definition 11.2.2. *We introduce formally a general two-sided fractional derivative (TSFD), D_θ^γ, through its Fourier transform [77, 82]*

$$\mathcal{F}\left[D_\theta^\gamma f(t)\right] = |\omega|^\gamma e^{i\frac{\pi}{2}\theta\cdot\operatorname{sgn}(\omega)} F(\omega),\qquad(11.25)$$

where γ and θ are any real numbers, that we will call derivative order and asymmetry parameter, respectively.

Some properties of this derivative are:
1. **Eigenfunctions.** Let $f(t) = e^{i\omega t}$, $\omega, t \in \mathbb{R}$. Then

$$D_\theta^\gamma e^{i\omega t} = |\omega|^\gamma e^{i\frac{\pi}{2}\theta\cdot\operatorname{sgn}(\omega)} e^{i\omega t},\qquad(11.26)$$

meaning that the sinusoids are the eigenfunctions of the TSFD with eigenvalue $\Psi_\theta^\gamma(\omega) = |\omega|^\gamma e^{i\frac{\pi}{2}\theta \cdot \text{sgn}(\omega)} e^{i\omega t}$. Notice that $\Psi_\theta^\gamma(\omega)$ is the *frequency response* of the TSFD system.

2. **Relation to Riesz and Feller derivatives.** As

$$\Psi_\theta^\gamma(\omega) = \cos(\theta\frac{\pi}{2})\Psi_0^\beta(\omega) + \sin(\theta\frac{\pi}{2})\Psi_1^\beta(\omega), \tag{11.27}$$

we can write

$$D_\theta^\gamma f(t) = \cos(\theta\frac{\pi}{2})D_0^\gamma f(t) + \sin(\theta\frac{\pi}{2})D_1^\gamma f(t). \tag{11.28}$$

3. **Periodicity in θ.** As we observe from (11.25), the TSFD is periodic and antiperiodic in θ as follows:

$$D_\theta^\gamma f(t) = D_{\theta+4n}^\gamma f(t), \quad n \in \mathbb{Z}, \tag{11.29}$$

$$D_\theta^\gamma f(t) = -D_{\theta+4n+2}^\gamma f(t), \quad n \in \mathbb{Z}. \tag{11.30}$$

4. **Additivity and commutativity of the orders.**

$$D_{\theta_1}^{\gamma_1} D_{\theta_2}^{\gamma_2} f(t) = D_{\theta_1+\theta_2}^{\gamma_1+\gamma_2} f(t) \tag{11.31}$$

In particular, letting $\gamma_1 = \gamma_2 = \gamma$ and $\theta_1 = \theta_2 = \theta$ gives

$$D_\theta^\gamma D_\theta^\gamma f(t) = D_{2\theta}^{2\gamma} f(t), \tag{11.32}$$

and, in general, if $n \in \mathbb{Z}$,

$$D_\theta^\gamma D_{(n-1)\theta}^{(n-1)\gamma} f(t) = D_{n\theta}^{n\gamma} f(t). \tag{11.33}$$

As it is easy to observe, operation (11.31) is commutative. Three other relations that can be obtained from (11.31) are

$$D_\theta^\gamma D_0^{(n-1)\gamma} f(t) = D_\theta^{n\gamma} f(t), \tag{11.34}$$

$$D_\theta^{\frac{\gamma}{2}} D_{-\theta}^{\frac{\gamma}{2}} f(t) = D_0^\gamma f(t), \tag{11.35}$$

$$D_\theta^{\frac{\gamma}{2}} D_{1-\theta}^{\frac{\gamma}{2}} f(t) = D_1^\gamma f(t). \tag{11.36}$$

This shows that the Riesz and Feller derivatives can be considered as the composition of left and right one-sided derivatives, GL or Liouville. In the following, we will assume that $\theta \geq 0$.

5. **Existence of inverse derivative.** From (11.31), the anti-derivative exists when $\gamma_2 = -\gamma_1$ and $\theta_1 = -\theta_2$. Therefore,

$$D_\theta^\gamma D_{-\theta}^{-\gamma} f(t) = D_{-\theta}^{-\gamma} D_\theta^\gamma f(t) = f(t). \tag{11.37}$$

6. **Identity operator.** According to (11.31) and (11.37), the identity operator is defined by

$$D_0^0 f(t) = f(t). \tag{11.38}$$

7. **Symmetric and anti-symmetric inverses.** It can be shown that, if γ is not an integer number, D_0^γ and D_1^γ derivatives of the impulse

$$\mathcal{F}\left[D_0^\gamma\delta(t)\right] = |\omega|^\gamma F(\omega) \tag{11.39}$$

$$\mathcal{F}\left[D_1^\gamma\delta(t)\right] = i|\omega|^\gamma \mathrm{sgn}(\omega)F(\omega) \tag{11.40}$$

are given respectively by

$$\mathcal{F}^{-1}\left[|\omega|^\gamma\right] = \frac{|t|^{-\gamma-1}}{2\Gamma(-\gamma)\cos(\gamma\frac{\pi}{2})}, \tag{11.41}$$

$$\mathcal{F}^{-1}\left[|\omega|^\gamma\mathrm{sgn}(\omega)\right] - -i\frac{|t|^{-\gamma-1}\mathrm{sgn}(x)}{2\Gamma(-\gamma)\sin(\gamma\frac{\pi}{2})} \tag{11.42}$$

(the latter only if γ is not an odd integer value).

8. **General derivative.**

Theorem 11.2.1. *Let $\gamma \in \mathbb{R} \setminus \mathbb{Z}$. The inverse Fourier transform of $\Psi_\theta^\gamma(\omega)$,*

$$\psi_\theta^\gamma(t) = \mathcal{F}^{-1}\left[\Psi_\theta^\gamma(\omega)\right],$$

is

$$\psi_\theta^\gamma(t) = \frac{\sin\left[(\gamma+\theta\cdot\mathrm{sgn}(t))\pi/2\right]}{\sin(\gamma\pi)\Gamma(-\gamma)}|t|^{-\gamma-1}. \tag{11.43}$$

The case $\gamma \in \mathbb{Z}$ will be considered in section 11.2.5.

Relation (11.25) can be rewritten as

$$D_\theta^\gamma f(t) = \mathcal{F}^{-1}\left[|\omega|^\gamma e^{i\frac{\pi}{2}\theta\cdot\mathrm{sgn}(\omega)}F(\omega)\right], \tag{11.44}$$

and the use of (11.43) in the convolution theorem of the Fourier transform allows introducing an integral formulation for the TSFD.

Definition 11.2.3. *Let $\gamma \in \mathbb{R} \setminus \mathbb{Z}$ and $f(t)$ in $L_1(\mathbb{R})$ or in $L_2(\mathbb{R})$. The generalised TSFD is defined by*

$$D_\theta^\gamma f(t) = \frac{1}{\sin(\gamma\pi)\Gamma(-\gamma)} \int_{\mathbb{R}} f(t-y)\sin\left[(\gamma+\theta\cdot\mathrm{sgn}(y))\pi/2\right]|y|^{-\gamma-1}\,dy. \tag{11.45}$$

Some particular cases are interesting:

1. With $\gamma < 0$ and $\theta = 0$, we obtain the (symmetric) Riesz potential

$$^R\mathcal{I}^\gamma f(t) = \frac{1}{2\cos(\gamma\frac{\pi}{2})\Gamma(\gamma)} \int_{\mathbb{R}} f(t-y)|y|^{\gamma-1}\,dy, \tag{11.46}$$

where $f(t)$ is a real function and $\gamma > 0$, $\gamma \neq 1, 3, 5, \ldots$

Tab. 11.1: Table with integer order inverses

$N > 0$	$|\omega|^n$	$i|\omega|^n \, \text{sgn}(\omega)$
$n = 2N$	$(-1)^N \delta^{(2N)}(t)$	$-\frac{(-1)^N (2N)!}{\pi} |t|^{-2N-1} \, \text{sgn}(t)$
$n = 2N + 1$	$-\frac{(-1)^N (2N+1)!}{\pi} |t|^{-2N-2}$	$(-1)^N \delta^{(2N+1)}(t)$
$n = -2N$	$(-1)^N \frac{1}{2(2N-1)!} |t|^{2N-1}$	$-\frac{(-1)^N x^{2N-1}}{\pi(2N)!} \left[\ln|t| + \gamma \right]$
$n = -2N - 1$	$\frac{(-1)^N x^{2N}}{\pi(2N)!} \left[\ln|t| + \gamma \right]$	$-\frac{(-1)^N}{2(2N)!} |t|^{2N} \, \text{sgn}(t)$

2. With $\gamma < 0$ and $\theta = 1$, we get the (anti-symmetric) Feller potential

$$^R \mathcal{J}^\gamma f(t) = \frac{1}{2 \sin\left(\gamma \frac{\pi}{2}\right) \Gamma(\gamma)} \int_{\mathbb{R}} f(t - y)|y|^{\gamma - 1} \text{sgn}(y) \, dy, \qquad (11.47)$$

where $f(t)$ is a real function and $\gamma > 0$, $\gamma \neq 2, 4, 6, \ldots$

3. If $\gamma < 0$ and $\theta = \pm\gamma$, we obtain the Liouville integral formulae for the causal/anti-causal anti-derivative:

$$D^\gamma_{\pm\gamma} f(t) = \frac{1}{\Gamma(-\gamma)} \int_0^\infty f(t \mp \tau) \tau^{-\gamma - 1} \, d\tau. \qquad (11.48)$$

4. If $\gamma > 0$ and $\theta = \pm\gamma$, we obtain the Liouville integral formulae for the causal/anti-causal non regularised derivative, also given by (11.48). The regularised versions of Liouville derivatives that were presented in Chapter 7 will be used to obtain regularised Riesz and Feller derivatives.

11.2.4 The integer order cases

Here we will present the rules for the inverse Fourier transform of

$$\Psi^\gamma_\theta(\omega) = |\omega|^\gamma e^{i \frac{\pi}{2} \theta \cdot \text{sgn}(\omega)}, \quad \omega \in \mathbb{R} \qquad (11.49)$$

when γ is integer [82]. Attending to (11.28), we only need to consider the cases $\theta = \pm 1$. The inverses are shown in Table 11.1. With such formulae, we are able to compute the TSFD for any real order, using (11.46) and (11.47) and their regularised versions that we will present next

11.2.5 Other properties and regularisations

The substitution of $-\gamma$ for γ in the expressions of Riesz (11.46) and Feller potentials (11.47) leads to singular integrals that will be regularised [82]. We start by presenting some important results.

1. **Relations involving the Liouville derivatives.** Let $\omega, \gamma \in \mathbb{R}$. It is a simple task to show that

$$|\omega|^\gamma = \frac{(i\omega)^\gamma + (-i\omega)^\gamma}{2\cos(\gamma\frac{\pi}{2})} \qquad \gamma \neq 1, 3, 5\cdots \tag{11.50}$$

$$i|\omega|^\gamma \mathrm{sgn}(\omega) = \frac{(i\omega)^\gamma - (-i\omega)^\gamma}{2\sin(\gamma\frac{\pi}{2})} \qquad \gamma \neq 2, 4, 6\cdots \tag{11.51}$$

which means that the Riesz derivative is, aside a constant, equal to the sum of the left and right Liouville derivatives. Similarly, the Feller derivative is the difference.

2. **Relation among the TSFD and the Liouville derivatives.** We only have to insert the results in (11.50) and (11.51) into (11.28) to obtain

$$D_\theta^\gamma f(t) = \frac{\cos\left[(\gamma+\theta)\frac{\pi}{2}\right]}{\sin(\gamma\frac{\pi}{2})} D_\gamma^\gamma f(t) + \frac{\cos\left[(\gamma-\theta)\frac{\pi}{2}\right]}{\sin(\gamma\frac{\pi}{2})} D_{-\gamma}^\gamma f(t). \tag{11.52}$$

This relation shows that we can compute the TSFD directly from the Liouville derivatives.

3. **Regularised Liouville derivatives (see chapter 7).** Let $\gamma \in \mathbb{R}$ and $\varepsilon(\cdot)$ be the Heaviside unit step. The regularised Liouville derivatives are given by

$$D_{\pm\gamma}^\gamma f(t) = \frac{1}{\Gamma(-\gamma)} \int_0^\infty \tau^{-\gamma-1} \left[f(t \mp \tau) - \varepsilon(\gamma) \sum_0^N \frac{(\mp)^m f^{(m)}(t)}{m!} \tau^m \right] d\tau, \tag{11.53}$$

where $N = \lfloor \gamma \rfloor$.

11.2.5.1 Riesz derivative

In the following, we are going to present the exact inverse of (11.46). We start be rewriting the first relation in (11.50) in the form:

$$D_0^\gamma f(t) = \frac{1}{2\cos(\gamma\frac{\pi}{2})} \left[D_\gamma^\gamma f(t) + D_{-\gamma}^\gamma f(t) \right]. \tag{11.54}$$

As

$$\sum_0^N \frac{(-1)^m f^{(m)}(t)}{m!} \tau^m + \sum_0^N \frac{f^{(m)}(t)}{m!} \tau^m = 2\sum_0^M \frac{f^{(2m)}(t)}{(2m)!} \tau^{2m}, \tag{11.55}$$

where $2M \leq \gamma < 2M + 2$, $M \in \mathbb{N}_0$, we are led to the following definition.

Definition 11.2.4. *We define the Riesz derivative by*

$$^R D_0^\gamma f(t) = \frac{1}{2\cos\left(\gamma\frac{\pi}{2}\right)\Gamma(-\gamma)} \int_{-\infty}^\infty \left[f(t-y) - \sum_{k=0}^M \frac{f^{(2k)}(t)}{(2k)!} y^{2k} \right] |y|^{-\gamma-1} dy, \tag{11.56}$$

for $2M < \gamma < 2M + 2$.

Example 11.2.1. *With $M = 0$ and with $0 < \gamma < 2$, we obtain*

$$^R D_0^\gamma f(t) = \frac{1}{2\cos\left(\gamma\frac{\pi}{2}\right)\Gamma(-\gamma)} \int_0^\infty [f(t-y) + f(t+y) - 2f(t)]\, y^{-\gamma-1}\, dy. \tag{11.57}$$

Remark 11.2.2. *In agreement with the previous subsection, we may extend the validity of (11.56) to odd integers. For positive even orders, we obtain, from (11.18), a classic centred derivative.*

11.2.6 Feller derivative

Similarly to above, we start by rewriting the second relation in (11.50) in the form:

$$D_1^\gamma f(t) = \frac{1}{2\sin(\gamma\frac{\pi}{2})} \left[D_\gamma^\gamma f(t) - D_{-\gamma}^\gamma f(t) \right] \tag{11.58}$$

as

$$\sum_0^N \frac{(-1)^m f^{(m)}(t)}{m!} \tau^m - \sum_0^N \frac{f^{(m)}(t)}{m!} \tau^m = 2 \sum_0^M \frac{f^{(2m+1)}(t)}{(2m+1)!} \tau^{2m+1} \tag{11.59}$$

where $2M + 1 \le \gamma < 2M + 3$, $M \in \mathbb{N}_0$, we proceed as in the Riesz case and apply this result to the Feller potential.

Definition 11.2.5. *We define the Feller derivative by*

$$^F D_1^\gamma f(t) - \frac{1}{2\sin\left(\gamma\frac{\pi}{2}\right)\Gamma(-\gamma)} \cdot \int_{-\infty}^\infty \left[f(t-y) - 2 \sum_{k=0}^M \frac{f^{(2k+1)}(t)}{(2k+1)!} y^{2k+1} \right] |y|^{-\gamma-1} \operatorname{sgn}(y)\, dy,$$

$$\tag{11.60}$$

for $2M + 1 < \gamma < 2M + 3$.

It is simple to show that

$$^F D_1^\gamma f(t) =$$

$$\frac{1}{2\sin\left(\gamma\frac{\pi}{2}\right)\Gamma(-\gamma)} \cdot \int_0^\infty \left[f(t-y) - f(t+y) + 2 \sum_{k=0}^M \frac{f^{(2k+1)}(t)}{(2k+1)!} y^{2k+1} \right] |y|^{-\gamma-1} \operatorname{sgn}(y)\, dy,$$

$$\tag{11.61}$$

agreeing with (11.53).

Remark 11.2.3. *Again, in agreement with previous subsection, we may extend the validity of (11.60) to even integers. For positive odd orders, we obtain, from (11.18), a classic centred derivative.*

11.3 The fractional Brownian motion

11.3.1 Definition

As an application of the theory we developed, we are going to study the fractional Brownian motion (fBm). This process was studied first by Mandelbrot and Van Ness [56] that suggested it as a model for non-stationary signals, but with stationary increments. These are suitable to understand phenomena exhibiting long range or $1/f^\alpha$ dependences.

Definition 11.3.1. *Let $H \in (0, 1)$ be the so-called Hurst parameter and let $b_0 \in \mathbb{R}$. The fBm $B_H(t)$ with parameter H is defined by*

$$
B_H(t) - B_H(0) = \frac{1}{\Gamma(H + 1/2)} \left\{ \int_{-\infty}^{0} \left[(t - \tau)^{H-1/2} - (-\tau)^{H-1/2} \right] dB(\tau) \right.
$$
$$
\left. + \int_{0}^{t} (t - \tau)^{H-1/2} dB(\tau) \right\} \tag{11.62}
$$

where $B_H(0) = b_0$, and $B(t)$ is the standard Brownian motion.

$B(t)$ is not differentiable, but, as referred above, we can assign it a generalised derivative, $w(t)$, $t \in \mathbb{R}$, the *white noise*, so that $dB(t) = w(t) dt$. Then

$$
B_H(t) - B_H(0) =
$$
$$
\frac{1}{\Gamma(H + 1/2)} \left\{ \int_{-\infty}^{0} w(\tau) \left[(t - \tau)^{H-1/2} - (-\tau)^{H-1/2} \right] d\tau + \int_{0}^{t} w(\tau)(t - \tau)^{H-1/2} d\tau \right\},
$$

that we can rewrite as

$$
B_H(t) - B_H(0) = \frac{1}{\Gamma(H + 1/2)} \left\{ \int_{-\infty}^{+\infty} w(\tau) \left[(t - \tau)^{H-1/2} \varepsilon(t - \tau) - (-\tau)^{H-1/2} \varepsilon(-\tau) \right] \right\} d\tau, \tag{11.63}
$$

The inner expression can be considered as the result of applying Barrow formula to $(t - \tau)^{H-1/2} \varepsilon(t - \tau)$. A straightforward computation leads to

$$
\frac{1}{\Gamma(H - 1/2)} \int_{0}^{t} (s - \tau)^{H-3/2} \varepsilon(s - \tau) ds = \tag{11.64}
$$
$$
\frac{1}{\Gamma(H + 1/2)} \left[(t - \tau)^{H-1/2} \varepsilon(t - \tau) - (-\tau)^{H-1/2} \varepsilon(-\tau) \right]
$$

This relation allows us to write heuristically

$$B_H(t) - B_H(0) = \frac{1}{\Gamma(H-1/2)} \int_{-\infty}^{+\infty} w(\tau) \int_0^t (s-\tau)^{H-3/2} \varepsilon(s-\tau) \, ds \, d\tau \tag{11.65}$$

$$= \frac{1}{\Gamma(H-1/2)} \int_0^t \int_{-\infty}^s w(\tau)(s-\tau)^{H-3/2} \, d\tau \, ds \tag{11.66}$$

The inner integral is the Liouville forward derivative of order $-H+1/2$:

$$D_f^\alpha w(t) = \frac{1}{\Gamma(-\alpha)} \int_{-\infty}^t w(\tau)(t-\tau)^{-\alpha-1} \, d\tau. \tag{11.67}$$

This means that

$$v_H(t) = B_H(t) - B_H(0) = \int_0^t D_f^\alpha w(\tau) \, d\tau, \tag{11.68}$$

where $\alpha = -H + 1/2$ is the derivative order. As $H \in (0,1)$, then $\alpha \in (-1/2, 1/2)$. Expression (11.68) is similar to the current definition of Brownian motion, provided that $\alpha = 0$ (i.e. that $H = 1/2$). This formula suggests us the use of other fractional derivative definitions alternative to the Liouville definition, as the Grünwald-Letnikov derivative.

Definition 11.3.2. *Let $w(t)$ be a continuous-time stationary white noise with variance σ^2. We call α–order fractional noise, $r_\alpha(t)$, to the output of a differintegrator of order α with $|\alpha| < 1$, when the input is white noise:*

$$r_\alpha(t) = \frac{1}{\Gamma(\alpha)} \int_{-\infty}^t (t-\tau)^{-\alpha-1} w(\tau) \, d\tau. \tag{11.69}$$

As expected, $r_0(t) = w(t)$. If the white noise is gaussian, $r_\alpha(t)$ is a fractional Gaussian noise *(FGN).*

Although the signal has an infinite power, its mean value is constant (and null) and the autocorrelation function $R_r(t, \tau) = E[r_\alpha(t+\tau)r_\alpha(\tau)]$ depends only on t, not on τ. Therefore, fractional noise $r_\alpha(t)$ is a wide sense stationary stochastic process and its autocorrelation function is

$$R_\alpha(t) = \sigma^2 \frac{|t|^{-2\alpha-1}}{2\Gamma(-2\alpha)\cos(\alpha\pi)}. \tag{11.70}$$

Relation (11.70) shows that we only have a (wide sense) stationary (hyperbolic) noise if

$$2\alpha + 1 > 0 \quad \text{and} \quad \Gamma(-2\alpha)\cos(\alpha\pi) > 0 \tag{11.71}$$

The other cases do not lead to a valid autocorrelation function of a stationary stochastic process, since it does not have a maximum at the origin. Then, for $-1/2 < \alpha < 0$ and $\alpha \in (2n, 2n+1), n \in \mathbb{Z}^+$, we obtain valid autocorrelation functions. We conclude that, if $|\alpha| < 1/2$, we obtain a stationary process in the anti-derivative case ($\alpha < 0$), and a nonstationary process in the derivative case ($\alpha > 0$) [72–74].

The above-defined process is a somehow strange process with infinite power. However, the power inside any finite frequency band is always finite. Its spectrum is

$$S_\alpha(\omega) = \frac{\sigma^2}{|\omega|^{2\alpha}}, \tag{11.72}$$

and thus a "$1/f$ noise". With this fractional noise, we can generate a fractional Brownian motion, using (11.68).

Remark 11.3.1. *In defining the fractional noise (11.69), we can use the Grünwald-Letnikov derivative also. Although not so useful from an analytical point of view, it is better for numerical simulations and for a discrete-time definition of fractional noise.*

Definition 11.3.3. *Let $r_\alpha(t)$ be a continuous-time stationary fractional noise, as introduced above. In agreement with (11.68), define another process $v_\alpha(t)$, $t \geq 0$, by*

$$v_\alpha(t) = \int_0^t r_\alpha(\tau) \, d\tau. \tag{11.73}$$

If $|\alpha| < 1/2$, we will call this process a fractional Brownian motion (or generalised Wiener–Lévy process). This formulation is a generalisation of the former procedure for obtaining the Brownian motion from white noise ($\alpha = 0$).

11.3.2 Properties

The process introduced in (11.73) enjoys the properties usually attributed to fBm [72–74]. Meanwhile, note that, if $w(t)$ is Gaussian, so are $r_\alpha(t)$ and $v_\alpha(t)$.
1. $v_\alpha(0) = 0$ and $E[v_\alpha(t)] = 0$ for every $t \geq 0$.
2. The covariance is

$$E[v_\alpha(t)v_\alpha(s)] = E\left[\int_0^t r_\alpha(\tau) \, d\tau \int_0^s r_\alpha(t') \, d\tau'\right] \tag{11.74}$$

$$= \int_0^t \int_0^s E\left[r_\alpha(\tau)r_\alpha(\tau')\right] d\tau \, d\tau'$$

$$= \sigma^2 \frac{1}{2\Gamma(-2\alpha)\cos\alpha\pi} \int_0^t \int_0^s |\tau - \tau'|^{-2\alpha-1} \, d\tau \, d\tau'$$

$$= \frac{\sigma^2}{2\Gamma(-2\alpha+2)\cos\alpha\pi} \left[|t|^{-2\alpha+1} + |s|^{-2\alpha+1} - |t - s|^{-2\alpha+1}\right]$$

With $-2\alpha + 1 = 2H$, the usual formulation is obtained:

$$E\left[v_\alpha(t)v_\alpha(s)\right] = \frac{V_H}{2}\left[|t|^{2H} + |s|^{2H} - |t-s|^{2\alpha+1}\right], \tag{11.75}$$

where

$$V_H = \frac{\sigma^2}{\Gamma(2H+1)\sin H\pi}. \tag{11.76}$$

The variance is

$$E\left[v_\alpha^2(t)\right] = V_H|t|^{2H}. \tag{11.77}$$

3. The process has stationary increments.
    In fact, letting the increments be defined by

$$\Delta v_\alpha(t,s) = v_\alpha(t) - v_\alpha(s) = \int_s^t r_\alpha(\tau)\,d\tau, \tag{11.78}$$

their variance is

$$\text{Var}\{\Delta v_\alpha(t,s)\} = E\left[\int_s^t r_\alpha(\tau)\,d\tau\int_s^t r_\alpha\left(\tau'\right)d\tau'\right] = \int_s^t\int_s^t E\left[r_\alpha(\tau)r_\alpha\left(\tau'\right)\right]d\tau\,d\tau', \tag{11.79}$$

that can be transformed by using the third relation in (11.74) to give

$$\text{Var}\{\Delta v_\alpha(t,s)\} = \sigma^2\frac{1}{2\Gamma(-2\alpha)\cos\alpha\pi}\int_s^t\int_s^t\left|\tau-\tau'\right|^{-2\alpha-1}d\tau\,d\tau'. \tag{11.80}$$

To compute the integral, we observe that $D\frac{|u|^{\alpha+1}\text{sgn}(u)}{\alpha+1} - |u|^\alpha$, that leads to

$$\text{Var}\{\Delta v_\alpha(t,s)\} = \sigma^2\frac{|t-s|^{-2\alpha+1}}{2\Gamma(2\alpha+2)\cos\alpha\pi}, \tag{11.81}$$

confirming that the increments are stationary. Letting $t = s + T$, above we get

$$\text{Var}\{\Delta v_\alpha(s+T,s)\} = \sigma^2\frac{T^{-2\alpha+1}}{2\Gamma(-2\alpha+2)\cos\alpha\pi}$$

$$= \sigma^2\frac{T^{2H}}{2\Gamma(2H+1)\sin H\pi}, \tag{11.82}$$

where we set $H = -\alpha + 1/2$. Since $\alpha \in (-1/2, 1/2)$, we have $H \in (0, 1)$. Remark that, contrarily to the ordinary Brownian motion, $v_\alpha(t)$ does not have independent increments.

4. The process is self similar.
    To see this, consider again (11.75) to obtain

$$E\left[v_\alpha(at)v_\alpha(as)\right] = \frac{V_H}{2}|a|^{2H}\left[|t|^{2H} + |s|^{2H} - |t-s|^{2H}\right] \tag{11.83}$$

that shows the self-similarity of the process.
5. The incremental process has a $1/f^\beta$ spectrum.
   Consider the process corresponding $v_{H+1/2}(t)$ and its increments obtained for t
   and $t - T$ with $t \in \mathbb{R}$ and $T \in \mathbb{R}^+$. The incremental process defined by

   $$d_h(t) = v_{H+1/2}(t) - v_{H+1/2}(t - T) \tag{11.84}$$

   has the following autocorrelation function:

   $$R_d(t) = \frac{V_H}{2} \left[|t + T|^{2H} + |t - T|^{2H} - 2|t|^{2H} \right]. \tag{11.85}$$

   It can be shown that, if $\beta > 0$,

   $$\mathcal{F} \left[\frac{1}{2\Gamma(\beta)\cos(\beta\pi/2)} |t|^{\beta-1} \right] = \frac{1}{|\omega|^\beta}, \tag{11.86}$$

   so that the Fourier transform of $R_d(t)$ leads to the spectrum of the incremental
   process:

   $$S_d(\omega) = \sigma^2 \frac{\sin^2(\omega T/2)}{|\omega|^{2H+1}}. \tag{11.87}$$

   For $|\omega| \ll \pi/T$, the spectrum can be approximated by

   $$S_d(\omega) \sim \frac{\sigma^2 T^2}{4} \frac{1}{|\omega|^{2H-1}}. \tag{11.88}$$

   This result shows that:
   – If $0 < H < 1/2$, the spectrum is parabolic and corresponds to an antipersistent
     fBm, because the increments tend to have opposite signs; this case corre-
     sponds to the integration of a stationary fractional noise.
   – If $1/2 < H < 1$, the spectrum has a hyperbolic character and corresponds to a
     persistent fBm, because the increments tend to have the same sign; this case
     corresponds to the integration of a nonstationary fractional noise.

Exercises

1. Consider the centred derivatives $D^{\alpha+\beta}$ from section 11.2.1, given by (11.2.1). Study
   the case $\alpha + \beta = 0$, $\alpha - \beta = \pm 1$, and relate it to the Hilbert transform.
2. Calculate numerically $D^{\alpha+\beta} \sin t$, $t \in [-\pi, +\pi]$, according to (11.2.1), for $\alpha + \beta = 1$
   and $\alpha = 0, 0.1, 0.2, 0.3, \ldots, 1$.
3. Repeat this for $\alpha + \beta = 2$ and $\alpha = 0, 0.1, 0.2, 0.3, \ldots, 2$.
4. Consider the Riesz-Feller potential given by (11.45). Study the case $a + b = 0$ and
   $a - b = \pm 1$, as in exercise 1.

5. Using (11.18) and (11.28) show that, for $\gamma > -1$, $\theta \in \mathbb{R}$, and $n \in \mathbb{Z}$, then

$$\frac{1}{\Gamma(\frac{\beta+\theta}{2} - n + 1)\Gamma(\frac{\beta-\theta}{2} + n + 1)} =$$

$$\frac{\cos(\frac{\pi}{2}\theta)}{\Gamma(\frac{\beta}{2} - n + 1)\Gamma(\frac{\beta}{2} + n + 1)} + \frac{\sin(\frac{\pi}{2}\theta)}{\Gamma(\frac{\beta+1}{2} - n + 1)\Gamma(\frac{\beta-1}{2} + n + 1)}.$$

6. Find numerically the derivatives of a unit step and of a unit-slope ramp, defined for $t > 0$, and for orders $\alpha = \pm\frac{1}{2}$, according to the several definitions presented in this chapter. In the case of two-sided derivatives, evaluate their existence; use the six combinations $\gamma = \pm\frac{1}{2}$, $\theta = 0, \pm\frac{1}{2}$. Try the signum function also.

12 Fractional delay discrete-time linear systems

12.1 Introduction

In chapter 8, we studied discrete-time systems defined by difference equations with the form

$$\sum_{k=0}^{N_0} a_k y(n-k) = \sum_{k=0}^{M_0} b_k x(n-k) \tag{12.1}$$

with $a_0 = 1$. The system is defined on the time scale $\mathbb{T} = T\mathbb{Z}$, $T > 0$.

Here, we will generalise these systems to the case of fractional delays. Basically, we work on two scales, $\mathbb{T}$ and $\mathbb{T}_\alpha = \mathbb{T} - \alpha T\mathbb{Z}$, $0 < \alpha \le 1$. The first is used to establish the input/output relation, while the second serves to define past values.

Definition 12.1.1. *These considerations give a support for introducing the fractional delay systems through the fractional difference equation [83, 84, 87]*

$$\sum_{k=0}^{N_0} a_k y(n-k\alpha) = \sum_{k=0}^{M_0} b_k x(n-k\alpha) \tag{12.2}$$

with $a_0 = 1$. The orders N_0 and M_0 are any given positive integers and the parameters a_k and b_k are real numbers.

As we referred before, we do not need to explicitly indicate the value of the sampling interval, T. In the following we will show how to solve this kind of equations.

12.2 Fractional delays

Consider a discrete-time signal $x(n)$, defined on $\mathbb{T}$. Its discrete-time Fourier transform is

$$X(e^{i\omega}) = \sum_{-\infty}^{+\infty} x(n)e^{-i\omega n}, \tag{12.3}$$

where $\omega = 2\pi f$, f being the frequency, $|f| \le 1/2$. We are going to show how we can produce a fractional delay to obtain a new signal defined on $\mathbb{T}_\alpha$.

Definition 12.2.1. *Let $\alpha \in \mathbb{R}$. A fractional delay operator [79, 83, 84] is a system with frequency response given by*

$$H_\alpha(e^{i\omega}) = e^{-i\omega\alpha}, \qquad |\omega| \le \pi \tag{12.4}$$

and impulse response given by

$$h_\alpha(n) = \frac{1}{2\pi} \int_{-\pi}^{\pi} e^{i\omega\alpha} e^{i\omega n}\, d\omega = \frac{\sin[\pi(n-\alpha)]}{\pi(n-\alpha)}, \qquad n \in \mathbb{Z}. \tag{12.5}$$

DOI 10.1515/9783110624588-012

When acting over a given signal $x(n)$ this operator produces a (fractional) delay:

$$x(n - \alpha) = \frac{1}{2\pi} \int_{-\pi}^{\pi} e^{-i\omega\alpha} X(e^{i\omega}) e^{i\omega n} d\omega \qquad (12.6)$$

This is a generalisation of the shift property of the discrete-time Fourier transform.

Using the expression of the Fourier transform, we obtain an interesting relation:

$$x(n - \alpha) = h_\alpha(n) * x(n) = \sum_{-\infty}^{+\infty} x(k) \frac{\sin[\pi(n - k - \alpha)]}{\pi(n - k - \alpha)} \qquad (12.7)$$

Therefore, the linear system with impulse response $h_\alpha(n)$ produces a fractional delay, that is to say, a delay which is not given by an integer number of sampling intervals. With this delay we are led to consider systems with the general form given by (12.2).

12.3 Impulse responses

Contrarily to what happens with the (integer) delay operator, the general exponential function z^n may not be an eigenfunction of the fractional delay operator. In fact, if we insert z^n in (12.7) the output is $H(z) z^n$ iff $|z| = 1$. This means that we cannot use the Z-transform for studying this kind of systems. However, for $z = e^{i\omega}$ we obtain the frequency response of the system:

$$H(e^{i\omega}) = \frac{\sum_{k=0}^{M_0} b_k e^{-i\omega a k}}{\sum_{k=0}^{N_0} a_k e^{-i\omega a k}}. \qquad (12.8)$$

The inverse discrete-time Fourier transform of $H(e^{i\omega})$ gives the acausal impulse response.

Remark 12.3.1. *As we cannot use directly the Z-transform, we have to use other tools to get the transfer functions, corresponding to causal and anti-causal systems. Such tools are the Cauchy integrals that are used for projecting $H(z)$, above defined on the unit circle, to the spaces outside and inside it [27].*

Assume that $M_0 < N_0$. If this is not the case, we know already how to proceed. Setting $A(e^{i\omega}) = \sum_{k=0}^{N_0} a_k e^{-i\omega a k}$ and $B(e^{i\omega}) = \sum_{k=0}^{M_0} b_k e^{-i\omega a k}$, we put

$$B(e^{i\omega}) = A(e^{i\omega}) Q(e^{i\omega}) + R(e^{i\omega})$$

and treat separately the quotient and the remainder. For a proper fraction, we follow the steps pointed out in the study of other systems:

1. Consider function $H(w)$, by substitution of w for z^α.
2. The polynomial denominator $H(w)$ is again the indicial polynomial or characteristic pseudo-polynomial. Perform the expansion of $H(w)$ into partial fractions of the form

$$F(w) = \frac{1}{(1 - pw^{-1})^k},$$
(12.9)

where we represented by p a generic root of the indicial polynomial (pole).
3. Substitute back z^α for w to obtain $H(z)$, expanded as a linear combination of fractions of the form

$$F(z) = \frac{1}{(1 - pz^{-\alpha})^k}.$$
(12.10)

4. Compute the impulse responses corresponding to each partial fraction.
5. Add the impulse responses.

In agreement to the above developments and the definition of the fractional delay operator, this kind of systems is intrinsically acausal. However, it is possible obtain causal responses with the help of Cauchy integrals, that we will not study here [84]. The impulse response $f(n)$ corresponding to $F(z) = \frac{1}{(1-p.z^{-\alpha})}$ is given by

$$f(n) = \sum_{k=0}^{\infty} p^k \frac{\sin\left[\pi(n - k - \alpha)\right]}{\pi(n - k - \alpha)}, \quad n \in \mathbb{Z}.$$
(12.11)

As referred above, these systems are interesting because they allow us to relate signals defined on two different time scales.

12.4 Scale conversion

As the system defined in (12.2) relates two signals defined on two different time scales, this suggests us to find a way of relating directly such sequences. Here, we will see how we can use above results to produce a scale conversion. What we really want is to obtain a direct conversion from one time scale, say $\mathbb{T}$, to another one, $\alpha\mathbb{T}$. This means that we are going from one time grid $t_n = nT$ (T will be assumed equal to 1 for simplicity) to $\tau_n = n\alpha T$, $(0 < \alpha < 1)$.

The procedure we will describe is equivalent to an ideal discrete to continuous-time conversion followed by a new sampling, now with αT as the new sampling interval. However, our procedure is direct in the sense that it proceeds without passing through the continuous-time domain. In spectral terms, this conversion maintains the shape of the spectrum, but spreads it [79]. Consider a signal x_n, $n \in \mathbb{Z}$, with Fourier transform $X(e^{i\omega})$, and a real constant α such that $0 < \alpha < 1$. Define a new function $X_\alpha(e^{i\omega})$ by

$$X_\alpha\left(e^{i\omega}\right) = \begin{cases} X\left(e^{i\omega}\right), & \text{if } |\omega| \le \pi \\ 0, & \text{if } \pi < |\omega| < \frac{\pi}{\alpha}, \end{cases}$$
(12.12)

and construct a periodic function, by repeating it with period $\frac{2\pi}{\alpha}$. This periodic function has an associated Fourier series, and the corresponding coefficients are given by

$$c_n = \frac{\alpha}{2\pi} \int_{-\pi}^{\pi} X\left(e^{i\omega}\right) e^{i\omega n} \, d\omega, \tag{12.13}$$

where we used (12.12). (12.13) suggests we set $c_n = x(\alpha n)$. As

$$X\left(e^{i\omega}\right) = \sum_{m=-\infty}^{+\infty} x_n e^{-i\omega n}, \tag{12.14}$$

we obtain

$$x(\alpha n) = \sum_{k=-\infty}^{+\infty} x(k) \frac{\sin[\pi(n\alpha - k)]}{\pi(n\alpha - k)}, \tag{12.15}$$

that solves our problem. We must note that the scale converted of a pulse does not have a finite duration.

Now, we note that for $|\omega| \leq \pi$ the Fourier series associated to $X(e^{i\omega})$ and $X_\alpha(e^{i\omega})$ represent the same function:

$$\sum_{m=-\infty}^{+\infty} x(n) e^{-i\omega n} = \alpha \sum_{m=-\infty}^{+\infty} x(\alpha n) e^{-i\omega\alpha n}, \quad |\omega| < \pi \tag{12.16}$$

Then

$$\mathcal{F}[x(k\alpha)] = \frac{1}{\alpha} X\left(e^{i\omega/\alpha}\right), \tag{12.17}$$

which expresses a generalisation of a well-known property of the discrete-time Fourier transform.

12.5 Linear prediction

Consider a stochastic process $x(n)$ that was observed during a certain time interval until time k [79]. It is intended to predict its value at a later time $k + d$, $d > 0$. For this we will have to take advantage of the information provided by known values, so the value to predict will be a function thereof:

$$\bar{x}(k + d) = f\left(x(k), x(k - 1), \dots, x(k - N)\right) \tag{12.18}$$

A difficulty may arise in the choice of the predictor function, $f(.)$. It is clear that such a function depends on the criterion to be used to measure the precision of $\bar{x}(k + d)$ as an estimator of $x(k + d)$. The simplest and most widely used criterion is the mean square error:

$$P_N(d) = E\left\{[\bar{x}(k + d) - x(k + d)]^2\right\} \tag{12.19}$$

So we choose $f(.)$ to minimise $P_N(d)$. To d we give the name of *prediction step*. We will only treat the cases $0 < d \leq 1$. The optimum predictor is the expected conditional value of $x(k+d)$ assuming $x(k-N), \ldots, x(k)$. Although this result is very important from the theoretical point of view, it has, however, little practical interest, since, in general, one does not have such a detailed knowledge of the structure of the process — and in particular of the conditional probability. However, in the Gaussian case, the predictor takes a linear form:

$$\bar{x}(k+d) = \sum_{i=1}^{N} a_i x(k-i) \tag{12.20}$$

Here the a_i, $i = 0, 1, 2, \ldots, N$ are the predictor coefficients of the d-step predictor.

Remark 12.5.1. *Since P_d is a quadratic function of $x(n)$ values, predictor coefficients a_i can be determined from a second-order statistic of the process (autocorrelation or spectrum). In the non-Gaussian case, it may happen that the optimal predictor is still linear. But, in general, this is not the case. Of course, we can decide to consider only and always linear predictors, whether the process is Gaussian or not. In this case, the linear predictor, not being optimal, is certainly easy to implement and often a good estimator. Wiener and Kolmogorov studied the problem and concluded that, in the limiting case where $n \to \infty$, which corresponds to the situation in which the entire process has been observed up to time k, the linear predictor is still optimal. This, of course, is an unrealistic assumption, which, however, greatly facilitates calculations. Notwithstanding, this is not generally very critical, because it will be expected that, from a certain order, the coefficients will become small enough to be neglected; that is to say, it is to be expected that, in general, $x(k+m)$ does not depend very much on a value too far apart in time $(m \gg 1)$.*

The minimisation of the error power (12.19) is as follows.

Theorem 12.5.1. *According to definition (12.20), and assuming that the correlation matrix of $x(n)$ has, at least, rank N, the optimum d-step predictor is given by the solution of the following set of normal equations:*

$$\sum_{i=1}^{N} a_i R_x(k-i) = R_x(-k-d-1), \quad k = 1, 2, \ldots, N \tag{12.21}$$

The matrix in (12.21) is Toeplitz, which makes easier the inversion through the Levinson recursion [100].

Inserting (12.21) into (12.19), the minimum error power is obtained:

$$P_{d\min} = R(0) + \sum_{i=1}^{N} a_i R(-i-d+1) \tag{12.22}$$

To solve (12.21), we need to know $R_x(n)$ and, from it, $R_x(n+d)$. There are many ways of computing $R_x(n)$. The most reliable is given by the *maximum entropy method* or AR modelling, that first obtains the spectrum $S_x(\omega)$, and then

$$R(k) = \mathcal{F}^{-1}[S_x(\omega)] \qquad R(k+d) = \mathcal{F}^{-1}\left[e^{i\omega d} S_x(\omega)\right] \tag{12.23}$$

As alternatives, other well-known methods are Burg's, modified Burg's, modified covariance, and Marple's [81].

A practical algorithm can be the following:

1. Compute the $N-1$ consecutive 1-step linear predictors using a suitable algorithm. The most interesting ones are based on the Levinson recursion [31, 65].
2. Use the $(N-1)$th linear 1-step predictor to estimate the spectrum, $S_x(\omega)$, using

$$S_x(\omega) = \frac{P_{N-1}}{\left|\sum_{n=0}^{N-1} p_i^{N-1} e^{-i\omega n}\right|^2}, \tag{12.24}$$

    where P_{N-1} is the corresponding error power.
3. Invert $S_x(\omega)$ to obtain the autocorrelation.
4. Compute the shifted autocorrelation

$$R(k+d) = \mathcal{F}^{-1}\left[e^{i\omega d} S_x(\omega)\right]$$

5. Use equation (12.21) to obtain the coefficients of the fractional predictor.

Remark 12.5.2. *The relation between $R(k+d)$ and $R(k)$ is, from (12.6),*

$$R(-k-d+1) = R(k-1+d) = \frac{\sin(\pi d)}{\pi d} \sum_{-\infty}^{\infty} R(n)\frac{(-1)^{k-1-n}}{\alpha+k-1-n} \tag{12.25}$$

As the autocorrelation function is an even function, we can transform the previous relation into

$$R(-k-d+1) = R(k-1+d) = \frac{(-1)^{k-1}\sin(\pi d)}{\pi(\alpha+k-1)}\left[R(0) + 2\sum_{n=1}^{\infty}\frac{(-1)^n R(n)}{1-\left(\frac{n}{\alpha+k-1}\right)^2}\right] \tag{12.26}$$

Observe that the coefficients go to zero, at least quadratically; so, the series converges quickly, allowing its truncation.

Exercises

1. Consider the scale conversion problem, discussed in section 12.4. Let β be another real constant such that $\beta \neq \alpha$ and $0 < \beta < 1$. Show that

$$\frac{\sin[\pi(n\alpha - k)]}{\pi(n\alpha - k)} = \beta \sum_{m=-\infty}^{+\infty} \frac{\sin[\pi(\beta m - k)]}{\pi(\beta m - k)}\frac{\sin[\pi(n\alpha - \beta m)]}{\pi(n\alpha - \beta m)} \tag{12.27}$$

2. Prove that, if $\alpha > 1$, the same procedure leads to

$$x(\alpha n) = \sum_{k=-\infty}^{+\infty} x(k)\frac{\sin[\pi(n - k/\alpha)]}{\pi(n - k/\alpha)} \tag{12.28}$$

which corresponds to an ideal low-pass filtering followed by a down-sampling. Show that, in this case, the spectrum of the new signal has a different shape.

3. Let x_n be each of the functions below, with $n = 1, 2, \ldots, 600$. Resample these signals for $\alpha = 0.1$. Compute the error committed.

a) $x_n = \cos \frac{n\pi}{20}$

b) $x_n = \cos \frac{n\pi}{5}$

c) $x_n = 0.6 \cos \frac{n\pi}{23} + 0.3 \cos \frac{n\pi}{31}$

d) $x_n = \log_{10} \frac{n}{20}$

4. Find d-step predictors for the signals above, with $d = 1, 2, 3, 4, 5$. Compute the error committed.

13 Fractional derivatives with variable orders

13.1 Introduction

Variable order (VO) fractional derivatives are derivative operators where the order will no longer restricted to assume a constant value $\alpha \in \mathbb{R}$, but will be considered as a bounded function, $\alpha(t) \in \mathbb{R}$. We consider the derivative order as function of the independent variable, that is, we have $\alpha(t) = f(t)$, $t \in \mathbb{R}$, since this leads to the simplest theoretical formulation of operator $D^{\alpha(t)}$. The variable $t \in \mathbb{R}$ will be called "time", but it can have any other physical meaning (e.g. position). Furthermore, the order can even depend on some other variable, as in

$$y(t) = g\left(D^{b(y(t))}u(t)\right) \tag{13.1}$$

where the order of the derivative $b(y(t))$ is now function of the output $y(t)$ of a differential equation depending on the derivative itself.

The VO fractional derivative is not a new theme. In fact, the first proposal was developed by Ross and Samko [103, 104, 106]. Their approach was based on the VO RL integral and the Marchaud (M) derivative. Lorenzo and Hartley proposed several versions of the VO RL derivative and integral [51], but observed that the VO derivative is not the left inverse of the integral. This problem was also found by Ross and Samko in their first formulation, and that limitation justified the use of the M derivative. Nonetheless, the M derivative is not the left inverse of the VO RL integral [2, 104]. Coimbra proposed a Caputo based VO derivative [9]. Later, Valério and Sá da Costa presented a review of the state of the art including expressions based on the GL derivative [118, 119, 122] that were generalised in [109, 110]. Several applications using VO derivatives were considered in [9, 62, 63, 101, 112]. Bohannan presented a critical view of the VO based on physical considerations [6].

13.2 Past proposals

There are several directions for extending the current FD definitions [118, 119]. For the sake of parsimony we will present them only for the forward (left) case, since the backward formulation involves a similar reasoning.
1. **Ross–Samko approach.** Ross and Samko [103, 104, 106] considered the integral

$$_cI_t^{\alpha(t)}f(t) = \int_c^t \frac{(t-\tau)^{\alpha(t)-1}}{\Gamma(\alpha(t))}f(\tau)\,d\tau \tag{13.2}$$

and the derivative

$$_cD_t^{\alpha(t)}f(t) = \frac{1}{\Gamma(\alpha(t))}\frac{d}{dt}\int_c^t (t-\tau)^{\alpha(t)-1}f(\tau)\,d\tau, \tag{13.3}$$

DOI 10.1515/9783110624588-013

assuming that $0 < \alpha(t) < 1$. However, as noted by S. Samko [2, 106], this derivative does not provide the left inverse operator to (13.2). This problem motivated the formulation based on the M derivative:

$$_cD_t^{\alpha(t)}f(t) = \frac{f(t)}{\Gamma(1-\alpha)(t)(t-c)^{\alpha(t)}} + \frac{\alpha(t)}{\Gamma(1-\alpha(t))}\int_c^t \frac{f(t)-f(\tau)}{(t-\tau)^{\alpha(t)+1}}\,d\tau,$$

$$0 < \alpha(t) < 1 \tag{13.4}$$

This was considered more appropriate than (13.3). Nevertheless, (13.4) is not the left inverse of the integral (13.2), as it can be seen by direct computation.

2. **Lorenzo–Hartley approach.** Lorenzo and Hartley [51] adopted the usual procedure: the RL derivative is obtained by integer order differentiation of the RL integral. They constructed the VO RL derivative of order $\alpha(t)$ in two steps:
   (a) Start from a RL FI as in (13.2), but with order $\lceil\alpha(t)\rceil - \alpha(t)$, where $\lceil\alpha(t)\rceil$ is the least positive integer greater than $\alpha(t)$;
   (b) The VO RL FD is obtained by differentiating the VO RL FI. This derivative has order $\lceil\alpha(t)\rceil$.

Three different VO FI were proposed:

Definition 13.2.1 (VO RL FI, type A). *The integral is the same as in (13.2):*

$$_cI_t^{\beta(t)}f(t) - \int_c^t \frac{(t-\tau)^{\beta(t)-1}}{\Gamma(\beta(t))}f(\tau)\,d\tau, \quad \text{if } \beta(t) \in \mathbb{R}^+ \tag{13.5}$$

Definition 13.2.2 (VO RL FI, type B).

$$_cI_t^{\beta(t)}f(t) = \int_c^t \frac{(t-\tau)^{\beta(\tau)-1}}{\Gamma(\beta(\tau))}f(\tau)\,d\tau, \quad \text{if } \beta(t) \in \mathbb{R}^+ \tag{13.6}$$

Definition 13.2.3 (VO RL FI, type C).

$$_cI_t^{\beta(t)}f(t) = \int_c^t \frac{(t-\tau)^{\beta(t-\tau)-1}}{\Gamma(\beta(t-\tau))}f(\tau)\,d\tau, \quad \text{if } \beta(t) \in \mathbb{R}^+ \tag{13.7}$$

The VO RL FD is defined as an integer order derivative of the RL FI:

$$_cD_t^{\alpha(t)}f(t) = \frac{d^{\lceil\alpha(t)\rceil}}{dt^{\lceil\alpha(t)\rceil}}\, _cI_t^{\lceil\alpha(t)\rceil-\alpha(t)}f(t). \tag{13.8}$$

This derivative was studied in [51, 118, 119, 122]. Nevertheless, the procedure does not seem suitable for implementing a VO FD, since we are simultaneously differentiating both $f(t)$ and $\alpha(t)$.

3. **Coimbra and Valério–Sá da Costa approach.** Coimbra proposed a VO derivative that is essentially a VO Caputo derivative, but including a term depending on the initial conditions that is null if the function is continuous [9]. Valério and

Sá da Costa started from (13.2)–(13.7) to obtain three VO derivatives based on the Caputo approach [62, 63, 118, 119, 122]. The VO FI is defined as in the case of the Lorenzo-Hartley approach. The main difference lies in the derivative definition: the sequence of operations, integral–derivative, is reversed, and becomes derivative–integral. This VO derivative is written as

$$ {}_cD_t^{\alpha(t)}f(t) = {}_cI_t^{\lceil\alpha(t)\rceil-\alpha(t)}\frac{d^{\lceil\alpha(t)\rceil}}{dt^{\lceil\alpha(t)\rceil}}f(t). \tag{13.9}$$

This formulation has disadvantages similar to those of the VO RL definition. However, this problem can be alleviated making

$$ {}_cD_t^{\alpha(t)}f(t) = \frac{d^m}{dt^m}\,{}_cI_t^{m-\alpha(t)}f(t), \tag{13.10}$$

where m is a positive integer such that $m \geq \alpha(t)$, $t \in \mathbb{R}$

4. **Grünwald–Letnikov type formulations.** The Grünwald–Letnikov definition was modified to lead to alternative VO derivatives. The most straightforward expressions can be written as follows, with $h \in \mathbb{R}^+$ [118, 119, 122]:
Definition 13.2.4 (VO GL FD, type A).

$$ {}_cD_t^{\alpha(t)}f(t) = \lim_{h\to 0^+} h^{-\alpha(t)} \sum_{k=0}^{\lfloor\frac{t-c}{h}\rfloor} (-1)^k \binom{\alpha(t)}{k} f(t-kh) \tag{13.11}$$

Definition 13.2.5 (VOGL FD, type B).

$$ {}_cD_t^{\alpha(t)}f(t) = \lim_{h\to 0^+} \sum_{k=0}^{\lfloor\frac{t-c}{h}\rfloor} h^{-\alpha(kh)}(-1)^k \binom{\alpha(kh)}{k} f(t-kh) \tag{13.12}$$

Definition 13.2.6 (VO GL FD, type C).

$$ {}_cD_t^{\alpha(t)}f(t) = \lim_{h\to 0^+} \sum_{k=0}^{\lfloor\frac{t-c}{h}\rfloor} h^{\alpha(t-kh)}(-1)^k \binom{\alpha(t-kh)}{k} f(t-kh) \tag{13.13}$$

A type D derivative definition, based upon a recursive relation, was introduced in [109, 110]
Definition 13.2.7 (VO GL FD, type D).

$$ {}_cD_t^{\alpha(t)}f(t) = \lim_{h\to 0^+} \left[\frac{f(t)}{h^{\alpha(t)}} - \sum_{k=1}^{\lfloor\frac{t-c}{h}\rfloor} (-1)^k \binom{-\alpha(t)}{k} {}_cD_{t-kh}^{\alpha(t)}f(t) \right] \tag{13.14}$$

Analytical solutions for differential equations using definitions (13.11), (13.12) and (13.14) were studied in [55].

Remark 13.2.1. *None of the VO FD outlined in sub-section 13.2 satisfies all the properties required to be a derivative in agreement with the criterion presented in* *chapter 7. In particular, satisfying the index law entails the existence of left inverse [106]; that is to say, the FD has to obey the condition*

$$D^{\alpha(t)} D^{-\alpha(t)} f(t) = f(t).$$ (13.15)

The direct analytical computation of (13.15) for the above derivative–integral pairs shows that none of the above definitions satisfies this requirement, not even if we enlarge the integral limits in (13.2)–(13.7) to c = −∞, or set the upper limit of the summation in (13.11)–(13.14) to +∞. This means that, to be coherent with [80], we should not use the term "derivative" to designate the operators above.

13.3 An approach to VO derivatives based on the GL derivative

The considerations in the previous section motivate a definition of VO FD [93], obtained by proceeding in agreement with the criterion for fractional derivatives described in chapter 7.

Definition 13.3.1. *Let f(t) be a function defined on ℝ, with Laplace transform, and α a real number. An operator T [f(t), α] verifying the properties of the above referred criterion is said to be a fractional derivative of order α.*

This definition is the base for introducing the notion of VO FD:

Definition 13.3.2. *In the conditions of definition 13.3.1, let us assume that α(t) = g(t), where g(t), t ∈ ℝ, is any bounded real function. We define as VO FD of order α any operator*

$$T [f(t), \alpha(t)]$$ (13.16)

that becomes a FD when α(t) is constant.

This definition is the base for the approach to VO FD. Consider the constant order forward (causal) GL derivative we presented in chapter 7.

$$D_f^\alpha f(t) = \lim_{h \to 0^+} h^{-\alpha} \sum_{k=0}^\infty \frac{(-\alpha)_k}{k!} f(t - kh),$$ (13.17)

where $f(t)$, $t \in \mathbb{R}$, is any real function. For a given non zero time increment, h, the right hand side of (13.17) represents a weighted sum that is essentially a moving average model of infinite order, where the weights are given by $h^{-\alpha} \frac{(-\alpha)_k}{k!}$, $k = 0, 1, 2, \ldots$, thus depending only the parameter α. Therefore, following definition 13.3.2, to change (13.17) into a VO we have the following formulation:

Definition 13.3.3. *The VO forward GL derivative is given by:*

$$D_f^{\alpha(t)} f(t) = \lim_{h \to 0^+} h^{-\alpha(t)} \sum_{k=0}^\infty \frac{(-\alpha(t))_k}{k!} f(t - kh).$$ (13.18)

Definition (13.18) preserves a very important property of the GL FD. Let $f(t) = e^{st}$, $s \in \mathbb{C}$. Then

$$D_f^{\alpha(t)} e^{st} = \lim_{h \to 0^+} h^{-\alpha(t)} \sum_{k=0}^{\infty} \frac{(-\alpha(t))_k}{k!} e^{s(t-kh)} = s^{\alpha(t)} e^{st}, \qquad \text{for } Re(s) > 0. \qquad (13.19)$$

Remark 13.3.1. *If $f(t) = e^{i\omega t}$, then $D_f^{\alpha(t)} f(t) = (i\omega)^{\alpha(t)} e^{i\omega t}$, and the derivative of a co-sine (or sine) is an amplitude-phase modulated co-sine (or sine):*

$$D_f^{\alpha(t)} \cos(\omega t) = \omega^{\alpha(t)} \cos\left[\omega t + \alpha(t)\frac{\pi}{2}\right], \quad \omega > 0. \qquad (13.20)$$

Remark 13.3.2. *If $Re(s) < 0$, then we have a backward FD, that we will not discuss, due to the similarities with the forward case.*

Let us consider a special case of a function that can be expressed as a sum of exponentials:

$$f(t) = \sum a_k e^{s_k t}, \qquad t \in \mathbb{R}, Re(s_k) > 0, \ k = 1, 2, \dots \qquad (13.21)$$

We obtain

$$D_f^{\alpha(t)} f(t) = \sum a_k s_k^{\alpha(t)} e^{s_k t}. \qquad (13.22)$$

This result can be generalised for functions with LT using the inversion integral (Bromwich integral).

Definition 13.3.4. *If $f(t) = 0$, $t < t_0$, the ROC is defined by $Re(s) > a$, with $a \leq 0$. Then the VO derivative of $f(t)$ is given by*

$$D_f^{\alpha(t)} f(t) = \frac{1}{2\pi i} \int_{\sigma - i\infty}^{\sigma + i\infty} s^{\alpha(t)} F(s) e^{st} \, ds, \quad t \in \mathbb{R} \qquad (13.23)$$

where $\sigma > a$.

Substituting

$$F(s) = \int_{\mathbb{R}} f(t) e^{-st} \, ds$$

into (13.23), we can write

$$D_f^{\alpha(t)} f(t) = \int_{\mathbb{R}} f(\tau) \frac{1}{2\pi i} \int_{\sigma - i\infty}^{\sigma + i\infty} s^{\alpha(t)} e^{s(t-\tau)} \, ds \, d\tau, \qquad (13.24)$$

since the inversion integral is uniformly convergent. If $\alpha(t) < 0$, the calculation of the inner inverse LT integral leads to

$$\frac{1}{2\pi i} \int_{\sigma - i\infty}^{\sigma + i\infty} s^{\alpha(t)} e^{s(t-\tau)} \, ds = \frac{(t-\tau)^{-\alpha(t)-1}}{\Gamma(-\alpha(t))} \varepsilon(t-\tau) \qquad (13.25)$$

where $\varepsilon(t)$ is the Heaviside unit step. Then, it yields

$$D_f^{\alpha(t)} f(t) = \frac{1}{\Gamma(-\alpha(t))} \int_{-\infty}^{t} f(\tau)(t-\tau)^{-\alpha(t)-1} \, d\tau. \tag{13.26}$$

Expression (13.26) can be written as

$$D_f^{\alpha(t)} f(t) = \frac{1}{\Gamma(-\alpha(t))} \int_{0}^{\infty} f(t-\tau)\tau^{-\alpha(t)-1} \, d\tau \tag{13.27}$$

that constitutes the VO Liouville integral [12, 76]. If $\alpha(t) < 0$, the right hand side in (13.27) is in general a singular integral. However, it is possible to regularize it to get the regularised VO Liouville derivative

$$D_f^{\alpha} f(t) = \frac{1}{\Gamma(-\alpha(t))} \int_{0}^{\infty} \tau^{-\alpha(t)-1} \left[f(t-\tau) - \varepsilon(\alpha(t)) \sum_{0}^{N(t)} \frac{(-1)^m f^{(m)}(t)}{m!} \tau^m \right] d\tau, \tag{13.28}$$

where $N(t) = \lfloor \alpha(t) \rfloor$. This expression includes both positive and negative values of α. Therefore, we can interpret (13.28) as a VO differintegral.

In particular, if $0 < \alpha(t) < 1$, then we obtain

$$D_f^{\alpha} f(t) = \frac{1}{\Gamma(-\alpha(t))} \int_{0}^{\infty} \tau^{-\alpha(t)-1} [f(t-\tau) - f(t)] \, d\tau, \tag{13.29}$$

recovering the VO M derivative, above referred.

To illustrate the use of (13.28), consider two examples.

Example 13.3.1. *Consider the Heaviside unit step, $f(t) = \varepsilon(t)$. We have*

$$D_f^{\alpha(t)} \varepsilon(t) = \frac{1}{\Gamma(-\alpha(t))} \int_{0}^{t} (t-\tau)^{-\alpha(t)-1} d\tau = \frac{t^{-\alpha(t)}}{\Gamma(-\alpha(t)+1)} \varepsilon(t), \tag{13.30}$$

a result similar to the constant order case.

Example 13.3.2. *Consider the impulse, $f(t) = \delta(t)$. For $\alpha(t) < 0$, its VO FI is given by (13.25), while for $\alpha(t) > 0$ we use (13.28), obtaining an infinite term which requires a Hadamard regularisation [30]. We have then*

$$D_f^{\alpha(t)} \delta(t) = \frac{1}{\Gamma(-\alpha(t))} \int_{0}^{\infty} f(t-\tau)\tau^{-\alpha(t)-1} d\tau = \frac{t^{-\alpha(t)-1}}{\Gamma(-\alpha(t))} \varepsilon(t), \tag{13.31}$$

that can be considered as the impulse response of the VO differintegrator, for any VO. With a constant α, we obtain the standard result $D^{\alpha} \delta(t) = \frac{t^{-\alpha-1}}{\Gamma(-\alpha)} \varepsilon(t)$.

13.4 Variable order linear systems

We are going introduce a general framework for dealing with VO linear systems that extends the results described in chapter 3

Definition 13.4.1. *Let $x(t)$ and $y(t)$ be the input and output functions of a VO fractional linear system defined by*

$$\sum_{k=0}^{N} a_k D^{\alpha_k(t)} y(t) = \sum_{k=0}^{M} b_k D^{\beta_k(t)} x(t) \tag{13.32}$$

with $t \in \mathbb{R}$. The symbol D represents the derivative operator defined previously in (13.18), $\alpha_k(t)$ and $\beta_k(t)$ are variable orders, and $N, M \in \mathbb{N}_0$. We assume also that $x(t)$ and $y(t)$ have LT denoted by $X(s)$ and $Y(s)$ with suitable ROC.

To simplify the mathematical treatment, let us consider the commensurate case

$$\sum_{k=0}^{N} a_k D^{k\alpha(t)} y(t) = \sum_{k=0}^{M} b_k D^{k\alpha(t)} x(t) \tag{13.33}$$

with $0 < \alpha(t) \leq 1$. Using the VO derivative definition (13.23), we can rewrite (13.33) as

$$\frac{1}{2\pi i} \int_{\sigma-i\infty}^{\sigma+i\infty} \left[\sum_{k=0}^{N} a_k s^{k\alpha(t)} Y(s) - \sum_{k=0}^{M} b_k s^{k\alpha(t)} X(s) \right] e^{st} \, ds = 0, \tag{13.34}$$

where we made the permutation of the summations and integrals. These operations are valid because the LT inversion integral is uniformly convergent.

Definition 13.4.2. *The VO transfer function corresponding to (13.33) is given by*

$$H(s, \alpha(t)) = \frac{\sum\limits_{k=0}^{M} b_k s^{k\alpha(t)}}{\sum\limits_{k=0}^{N} a_k s^{k\alpha(t)}}, \quad s \in \mathbb{C}, \, t \in \mathbb{R}. \tag{13.35}$$

With this tool, for any input signal $x(t)$, we obtain the system output through

$$y(t) = \frac{1}{2\pi i} \int_{\sigma-i\infty}^{\sigma+i\infty} H(s, \alpha(t)) X(s) e^{st} ds. \tag{13.36}$$

This relation generalises for VO the classic formula of the integer case.

Notice that s is independent of t, which has the role of parameterising the transfer function.

13.5 The VO Mittag-Leffler function

With the transfer function defined in (13.35), we can invert (13.36) directly, by means of the usual following steps:

1. Transform $H(s, \alpha(t))$ into $H(z)$, by substitution of $s^{\alpha}(t)$ for z;
2. Perform the expansion of $H(z)$ in partial fractions;
3. Substitute z back for $s^{\alpha}(t)$, to obtain the partial fraction decomposition:

$$H(s) = \sum_{k=0}^{N_p} \frac{A_k}{(s^{\alpha(t)} - p_k)^{n_k}} \tag{13.37}$$

where $p_k \in \mathbb{C}$ are called pseudo-poles, N_p is the number of parcels, and $n_k \in \mathbb{N}_0$ are the corresponding multiplicities.

Remark 13.5.1. *Observe that, having solved the problem of multiplicity one, the solution for other multiplicities is immediate. In fact, we note that, if p is a generic pseudo-pole with order 2, we have*

$$\frac{1}{(s^{\alpha(t)} - p)^2} = \frac{d}{dp}\left[\frac{1}{s^{\alpha(t)} - p}\right]. \tag{13.38}$$

For other multiplicities we repeat the procedure.

To simplify things, and since the multiplicity is not relevant, in the following we consider $n_k = 1$, $k = 1, 2, \ldots$. Let $F(s) = \frac{1}{s^{\alpha(t)} - p}$, where p is any pseudo-pole, be a generic partial fraction. Using the geometric series we can write

$$\frac{1}{s^{\alpha(t)} - p} = s^{-\alpha(t)} \sum_{n=0}^{\infty} p^n s^{-n\alpha(t)} = \sum_{n=1}^{\infty} p^{n-1} s^{-n\alpha(t)} \tag{13.39}$$

valid for $|ps^{-\alpha(t)}| < 1$. Substituting (13.37) and (13.39) into (4.15), we can compute N_p output terms given by

$$y_k(t) = \frac{1}{2\pi i} \int_{\sigma-i\infty}^{\sigma+i\infty} \sum_{n=1}^{\infty} p_k^{n-1} s^{-n\alpha(t)} X(s) e^{st} ds = \sum_{n=1}^{\infty} p_k^{n-1} \frac{1}{2\pi i} \int_{\sigma-i\infty}^{\sigma+i\infty} s^{-n\alpha(t)} X(s) e^{st} ds \tag{13.40}$$

Proceeding as in section 13.3, with (13.25) we can write

$$\frac{1}{2\pi i} \int_{\sigma-i\infty}^{\sigma+i\infty} s^{-n\alpha(t)} e^{s(t-\tau)} ds = \frac{(t-\tau)^{n\alpha(t)-1}}{\Gamma(n\alpha(t))} \varepsilon(t-\tau) \tag{13.41}$$

which allows us to obtain the output corresponding to the k-th partial fraction:

$$y_k(t) = \sum_{n=1}^{\infty} p_k^{n-1} \frac{1}{\Gamma(n\alpha(t))} \int_{-\infty}^{t} x(\tau)(t-\tau)^{n\alpha(t)-1} d\tau \tag{13.42}$$

$$= \sum_{n=0}^{\infty} p_k^n \frac{1}{\Gamma(n\alpha(t) + \alpha(t))} \int_{-\infty}^{t} x(\tau)(t - \tau)^{n\alpha(t)+\alpha(t)-1} d\tau. \tag{13.43}$$

So, we to obtain a result similar to the one for the constant order case.

Equation (13.42) points toward the following VO Mittag-Leffler function definition, similar as well to that of the constant order case.

Definition 13.5.1. *The VO Mittag-Leffler function is given by*

$$E_{\alpha(t),\beta(t)}(pt^{\alpha(t)}) = \sum_{n=0}^{\infty} p^n \frac{t^{n\alpha(t)}}{\Gamma(n\alpha(t) + \beta(t))}, \quad t \ge 0. \tag{13.44}$$

With (13.44) we can rewrite (13.42) as

$$y_k(t) = \int_{-\infty}^{t} x(\tau)(t - \tau)^{\alpha(t)-1} E_{\alpha(t),\alpha(t)}(p_k(t - \tau)^{\alpha(t)}) d\tau, \tag{13.45}$$

or

$$y_k(t) = \int_{0}^{\infty} x(t - \tau)\tau^{\alpha(t)-1} E_{\alpha(t),\alpha(t)}(p_k\tau^{\alpha(t)}) d\tau. \tag{13.46}$$

With (13.46) it is possible to obtain the output of a VO fractional linear system when the input is $x(t)$. In what follows we present some examples.

Remark 13.5.2. *The computation of the output under non null initial conditions follows the rules described in [75]. We will not detail the calculation procedure, since it employs standard signal processing techniques only.*

Example 13.5.1. *The particular case of the unit step response gives*

$$y_k(t) = \int_{0}^{t} \tau^{\alpha(t)-1} E_{\alpha(t),\alpha(t)}(p_k\tau^{\alpha(t)}) d\tau. \tag{13.47}$$

Example 13.5.2. *The unit step response of*

$$F(s) = \frac{1}{0.1 + s^{\alpha(t)}}, \tag{13.48}$$

calculated with (13.47) for a variable order $\alpha(t)$ given by a sequence of steps, is shown in Figure 13.1. (The figure shows the variation of $\alpha(t)$ as well.)

Example 13.5.3. *The response of (13.48) when the input is a unit slope ramp $x(t) = t$, calculated with (13.46), is shown in Figure 13.2 when the order varies with time as the sequence of steps of the previous example.*

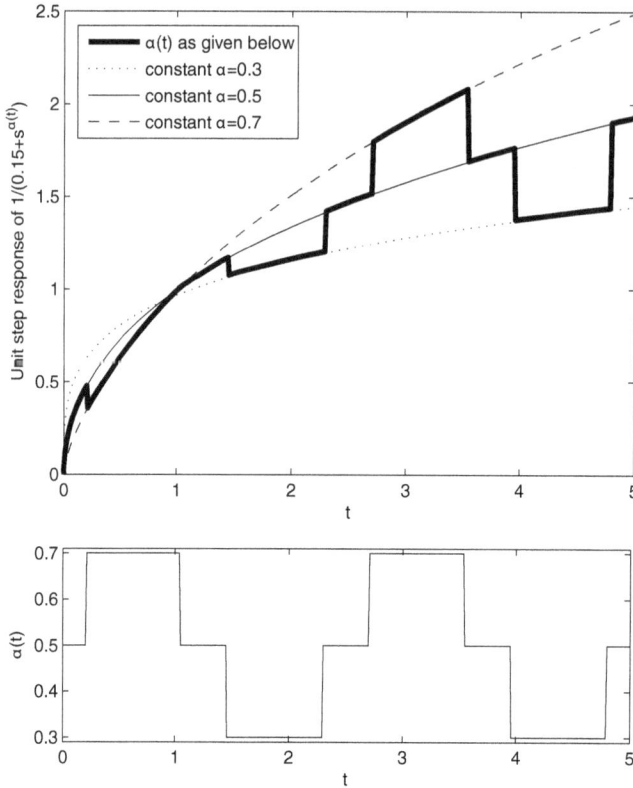

Fig. 13.1: Unit step response of (13.48) according to (13.47) when the order varies with time as a sequence of steps; see Example 13.5.2.

13.6 Variable order two-sided derivatives

Two constant-order centred derivatives based on the GL formulation were introduced in chapter 11. They can be considered as the composition of a left and a right derivative and expressed in a constant order unified general two-sided derivative given by

$$D_\theta^\alpha f(t) = \lim_{h \to 0^+} h^{-\alpha} \sum_{n=-\infty}^{\infty} \frac{(-1)^n \, \Gamma(\alpha + 1)}{\Gamma(\frac{\alpha+\theta}{2} - n + 1)\Gamma(\frac{\alpha-\theta}{2} + n + 1)} f(t - nh), \qquad (13.49)$$

where α is the derivative order and θ is the dissymmetry parameter. The order α cannot be a negative integer unless $\alpha - \theta$ is an even integer.

Definition 13.6.1. *Considering that α and θ are now variable functions, we define the VO two-sided derivative by*

$$D_{\theta(t)}^{\alpha(t)} f(t) = \lim_{h \to 0^+} h^{-\alpha(t)} \sum_{n=-\infty}^{\infty} \frac{(-1)^n \, \Gamma(\alpha(t) + 1)}{\Gamma\left(\frac{\alpha(t)+\theta(t)}{2} - n + 1\right)\Gamma\left(\frac{\alpha(t)-\theta(t)}{2} + n + 1\right)} f(t - nh) \qquad (13.50)$$

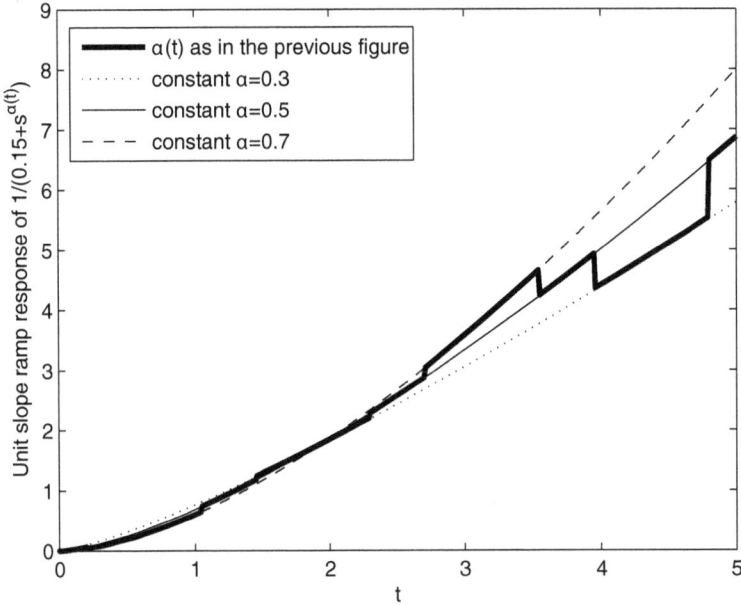

Fig. 13.2: Unit slope ramp response of (13.48) according to (13.46) when the order varies with time as a sequence of steps; see Example 13.5.3.

If $f(t) = e^{i\omega t}$, then we obtain

$$D_{\theta(t)}^{\alpha(t)} e^{i\omega t} = |\omega|^{\alpha(t)} e^{i\frac{\pi}{2}\theta(t)\cdot\text{sgn}(\omega)} e^{i\omega t}. \tag{13.51}$$

Let us assume that a given function $f(t)$ has FT given by $F(\omega)$. The inverse FT (C.2) allows us to write

$$D_{\theta(t)}^{\alpha(t)} f(t) = \frac{1}{2\pi} \int_{-\infty}^{\infty} |\omega|^{\alpha(t)} e^{i\frac{\pi}{2}\theta(t)\cdot\text{sgn}(\omega)} F(\omega) e^{i\omega t} dt. \tag{13.52}$$

Substituting the FT (See appendix C) in (13.52), we have

$$D_{\theta(t)}^{\alpha(t)} f(t) = \int_{-\infty}^{\infty} f(\tau) \frac{1}{2\pi} \int_{-\infty}^{\infty} |\omega|^{\alpha(t)} e^{i\frac{\pi}{2}\theta(t)\cdot\text{sgn}(\omega)} e^{i\omega(t-\tau)} d\omega \, d\tau. \tag{13.53}$$

To compute the inner integral, we use the relations

$$|\omega|^{\alpha(t)} e^{i\frac{\pi}{2}\theta(t)\cdot\text{sgn}(\omega)} = |\omega|^{\alpha(t)} \left[\cos\left(\frac{\pi}{2}\theta(t)\right) + i\sin\left(\frac{\pi}{2}\theta(t)\right) \text{sgn}(\omega) \right] \tag{13.54}$$

$$\frac{1}{2\pi} \int_{-\infty}^{\infty} |\omega|^{\alpha(t)} e^{i\omega(t-\tau)} d\omega = \frac{1}{2\Gamma(-\alpha(t))\cos\left(\frac{\pi}{2}\alpha(t)\right)} |t - \tau|^{-\alpha(t)-1} \tag{13.55}$$

$$i\frac{1}{2\pi}\int_{-\infty}^{\infty}|\omega|^{\alpha(t)}\,\mathrm{sgn}(\omega)e^{i\omega(t-\tau)}d\omega = \frac{\mathrm{sgn}(t-\tau)}{2\Gamma(-\alpha(t))\sin\left(\frac{\pi}{2}\alpha(t)\right)}|t-\tau|^{-\alpha(t)-1} \tag{13.56}$$

to obtain

$$D_{\theta(t)}^{\alpha(t)}f(t) = \cos\left(\frac{\pi}{2}\theta(t)\right)\frac{1}{2\Gamma(-\alpha(t))\cos\left(\frac{\pi}{2}\alpha(t)\right)}\int_{-\infty}^{\infty}f(\tau)|t-\tau|^{-\alpha(t)-1}\,d\tau \tag{13.57}$$

$$+\sin\left(\frac{\pi}{2}\theta(t)\right)\frac{1}{2\Gamma(-\alpha(t))\sin\left(\frac{\pi}{2}\alpha(t)\right)}\int_{-\infty}^{\infty}f(\tau)|t-\tau|^{-\alpha(t)-1}\mathrm{sgn}(t-\tau)\,d\tau,$$

that expresses an important result: the VO two-sided derivative is a linear combination of the VO Riesz, $R_R(t)$, and the VO Riesz-Feller, $R_{RF}(t)$, 1-D potentials, given by

$$R_R(t) = \frac{1}{2\Gamma(-\alpha(t))\cos\left(\frac{\pi}{2}\alpha(t)\right)}\int_{-\infty}^{\infty}f(\tau)|t-\tau|^{-\alpha(t)-1}\,d\tau \tag{13.58}$$

$$R_{RF}(t) = \frac{1}{2\Gamma(-\alpha(t))\sin\left(\frac{\pi}{2}\alpha(t)\right)}\int_{-\infty}^{\infty}f(\tau)|t-\tau|^{-\alpha(t)-1}\mathrm{sgn}(t-\tau)\,d\tau \tag{13.59}$$

With these results, and those of section 13.3, we can solve an equation of the type

$$D_f^{\alpha(t)}f(t,x) = D_{\theta(x)}^{\beta(x)}f(t,x) \text{ for } t\in\mathbb{R},\ x\in\mathbb{R}, \tag{13.60}$$

where the orders of the two-sided derivatives are now function of the space.

Exercises

1. Implement the VO Mittag-Leffler function (13.44) numerically, and use your implementation to find $E_{\alpha(t),\beta(t)}(t)$, $t\in[0,1]$ when:
   a) $\alpha(t) = 1$ and $\beta(t) = 0.3$
   b) $\alpha(t) = 0.5$ and $\beta(t) = 0.3$
   c) $\alpha(t) = 1$ and $\beta(t) = 0.6$
   d) $\alpha(t) = 0.5$ and $\beta(t) = 0.6$
   e)

$$\alpha = \begin{cases} 1, & \text{if } x < 0.7 \\ 0.5, & \text{if } x \geq 0.7 \end{cases} \tag{13.61}$$

$$\beta = \begin{cases} 0.3, & \text{if } x < 0.5 \vee x > 0.8 \\ 0.6, & \text{if } 0.5 \leq x \leq 0.8 \end{cases} \tag{13.62}$$

2. Use your VO Mittag-Leffler function to implement (13.47) numerically, and then use your implementation to:

a) Reproduce the results of example 13.5.2.
b) Repeat example 13.5.2 for constant orders $\alpha(t) = 0.25$, $\alpha(t) = 0.5$, and $\alpha(t) = 0.75$.
c) Repeat example 13.5.2 for a sinusoidal variation of the order:

$$\alpha(t) = 0.5 + 0.25\sin\left(\frac{4\pi}{5}t\right) \hspace{3cm} (13.63)$$

3. Repeat the last exercise, calculating the responses with (13.46) instead.
4. Use (13.46) to:
   a) Reproduce the results of example 13.5.3.
   b) Repeat example 13.5.3 for constant orders $\alpha(t) = 0.25$, $\alpha(t) = 0.5$, and $\alpha(t) = 0.75$.
   c) Repeat example 13.5.2 for the sinusoidal order (13.63).
5. Recalculate the derivatives above using the two-sided definition.

Appendices

A On distributions

A.1 Introduction

In the following, we will present a brief introduction to the *axiomatic theory of distributions*. This theory was developed by J. Sebastião and Silva [111] in the fifties and is almost unknown in the engineering community, even though it is the most intuitive and direct of the approaches to distributions. It intends to enlarge the class of functions in order to make possible to differentiate indefinitely any continuous function [18].

Let I be an interval in $\mathbb{R}$, let $\mathcal{C}(I)$ be the set of continuous functions on I, and let $\mathcal{C}^p(I)$ be the set of functions p times continuously differentiable in the usual sense. A function $f(t) \in \mathcal{C}(I)$ can be represented by

$$f(t) = D^p g(t), \tag{A.1}$$

where $g(t) \in \mathcal{C}^p$. The right inverse operator of D (primitive operator) is represented here by S and satisfies

$$S^p f(t) = g(t). \tag{A.2}$$

As the primitive of a given function is defined aside from a constant, the set $\mathcal{C}^1(I) = Sf + K$, where K is a constant, is contained in $C(I)$, i.e. $\mathcal{C}(I) \supset \mathcal{C}^1(I)$. Defining the powers of the operator S by the recursion as $S^n f = S(S^{n-1})f$, with $S^0 f = f$, and denoting by P_{n-1} the set of polynomials of order $n - 1$, we will have

$$C^n(I) = \{S^n f + p : f \in C(I) \text{ and } p \in P_{n-1}\}, \tag{A.3}$$

where $\mathcal{C}^n(I)$ is the set of the n times continuously differentiable functions in $\mathbb{R}$. If we now denote by $\mathcal{C}^\infty(I)$ the intersection of all the $\mathcal{C}^n(I)$, $n = 1, 2, \cdots, \infty$, we conclude that:

$$\mathcal{C}(I) \supset \mathcal{C}^1(I) \supset \mathcal{C}^2(I) \supset \cdots \supset \mathcal{C}^n(I) \supset \cdots \supset \mathcal{C}^\infty(I) \tag{A.4}$$

For every $n \in \mathbb{N}$, the derivation can be viewed as an application of $\mathcal{C}^{n+1}(I)$ in $\mathcal{C}^n(I)$. Its generalisation consists essentially in prolonging the previous sequence to the left in order to obtain

$$\mathcal{C}_\infty(I) \supset \cdots \supset \mathcal{C}_{n+1}(I) \supset \mathcal{C}_n(I) \supset \cdots \supset \mathcal{C}_1(I) \supset \mathcal{C}(I) \tag{A.5}$$

We represent by $\mathcal{C}_\infty(I)$ the reunion of all the $\mathbb{C}_n(I)$, $n = 1, 2, \cdots, \infty$, i.e. the set of all the "functions" φ that result from the repeated application of the operator D to $f \in \mathcal{C}(I)$.

Definition A.1.1. *The elements $\phi \in \mathcal{C}_\infty(I)$ such that*

$$\varphi(t) = D^p g(t), \ p \in \mathbb{N}_0, \ g \in \mathcal{C}^p(I), \tag{A.6}$$

where $\varphi \in C(I)$, will be called generalised functions *(GF) or distributions. Therefore, any distribution is a derivative of a continuous function.*

DOI 10.1515/9783110624588-014

Example A.1.1. *The second order derivative of the function $x(t) = |t|$ is twice the Dirac impulse, $\delta(t)$.*

Remark A.1.1. *It is important to note that the operation S does not coincide with the operator D^{-1}, which is the anti-derivative.*

Example A.1.2. *According to Example A.1.1 $D^{-2}\delta(t) = |t|$, while $S^2\delta(t) = |t| + at + b$, $a, b \in \mathbb{R}$.*

A.2 The axioms

In agreement with the definitions above, any continuous function is a GF. The formal framework for the GF theory begins by introducing two primitive terms, *distribution* (or GF) and (generalised) *derivative*, and is based upon four axioms, which will be presented, interpreted, and their consequences discussed.

Axiom 1. *If $f \in \mathcal{C}(I)$, then $f \in \mathcal{C}_\infty(I)$.*

Thus, $\mathcal{C}_\infty(I)$ is a nonempty set, since it contains, at least, the continuous functions.

Axiom 2. *For each $\varphi \in \mathcal{C}(I)$, there is an element $Df \in \mathcal{C}_\infty(I)$ such that, if f has a derivative $f' \in \mathcal{C}(I)$ in the usual sense, then Df coincides with f'.*

Axiom 3. *For each $\varphi \in \mathbb{C}(I)$, there is a $n \in \mathbb{N}_0$ and a function $f \in \mathcal{C}(I)$ such that*

$$\varphi(t) = D^n f(t), \tag{A.7}$$

We conclude that any distribution is defined by sets of pairs (f, n).

Example A.2.1. *The Dirac delta function can be defined by the pairs $(r, 2)$ and $(\varepsilon, 1)$, where $r(t)$ and $\varepsilon(t)$ are the ramp and Heaviside unit step functions, respectively. In general we write $\delta(t) = D\varepsilon(t)$.*

This case must be analysed carefully. The derivative of the unit step $\varepsilon(t)$ is null for $t \neq 0$, but it is ∞ at $t = 0$. This seems strange, but it is useful and we do not need to look at the problem from this point of view. We can simply say that there is an entity $\delta(t)$ that verifies:

- $D\varepsilon(t) = \delta(t)$
- $\varepsilon(t) = \int_a^t \delta(\tau)\, d\tau$, $a < 0$,
  and, as it can be proved
- $f(t)\delta(t - a) = f(a)\delta(t - a)$ (ideal sampling)
- $f(t) * \delta(t - a) = f(a)$

Axiom 4. *If $n \in \mathbb{N}_0, f \in \mathbb{C}(I)$, and $g \in \mathbb{C}(I)$, then we will have $D^n f = D^n g$ if and only if $f - g$ has the form $f - g = P_{n-1}(t)$, where P_{n-1} represents a polynomial in t of order $n - 1$ or less.*

This axiom specifies the equality of generalised functions (FG), or the conditions such that 2 pairs (f, n) and (g, m) represent the same GF.

Example A.2.2. *If $f(t) = t^3 u(t)$, the pairs $(f, 2)$, $(f', 1)$ and $(f'', 0)$ represent the same distribution.*

Consider the simple case of two continuous functions defined in $\mathbb{R}$. In this case, $Df = f'$ and $Dg = g'$, $Df = Dg$ being equivalent to $(f - g)' = 0$, or $f - g = $ constant $\in \mathbb{R}$. Axiom 4 accepts and generalises this fact.

Example A.2.3. *Let $f(t) = 2e^{at}\varepsilon(t)$. It has a jump at $t = 0$. Notice that function $g(t) = f(t) - 2\varepsilon(t)$ is continuous and has a derivative in classic sense. So, $g'(t) = 2ae^{at}u(t)$, and $f'(t) = 2ae^{at}u(t) + 2\delta(t)$, since $\delta(t) = D\varepsilon(t)$. This example can be generalised:*

When a function has a jump at a given point, its derivative has a delta at the same point.

Example A.2.4. *Let $f(t) = |t|$. Notice that we can write $f(t) = t \cdot \text{sgn}(t)$, meaning that the absolute value is the product of a continuous function and a generalised function, the signum function. The generalised derivative of $|t|$ is given by $D|t| = \text{sgn}(t) + t \cdot \delta(t)$. But, as it is easy to verify, $t \cdot \delta(t) \equiv 0$, and then $D|t| = \text{sgn}(t)$. Continuing the derivation, we obtain $D^2|t| = 2\delta(t)$, $D^3|t| = 2\delta'(t), \cdots$*

It is interesting to remark the sequence "even – odd – even – odd ..." that remains valid here.

Having these axioms in hands, we can show that, with suitable definitions of sum of distributions and multiplication by a constant, the space of distributions is a linear vectorial space. This means that, in particular, the sum of two GF is a GF, and the addition enjoys the usual properties. The same does not happen with multiplication. In fact, it is not possible to define the product of two GF in order to guarantee that it enjoys the usual properties (namely, to be associative, to satisfy the product derivative rule, and to coincide with usual product when both factors are continuous functions).

The case of tempered distributions is very important in applications [45].

Definition A.2.1. *A generalised function φ is said to be tempered or of polynomial type if and only if $\exists \alpha \in \mathbb{R}$ such that $\varphi(t)/t^\alpha$ is bounded when $t \to +\infty$.*

Exercises

1. Show that $\int_{-\infty}^{+\infty} \cos(t)\delta(t)dt = 1$.
2. Prove that the space of distributions is a linear vectorial space.
3. Show that $\lim_{t\to\infty} \cos(\omega t) = \lim_{t\to\infty} \sin(\omega t) = 0$, $\omega \in \mathbb{R}$.
4. Show that $\lim_{t\to\infty} \cos^2(\omega t) = \lim_{t\to\infty} \sin^2(\omega t) = \frac{1}{2}$, $\omega \in \mathbb{R}$

B The Gamma function and binomial coefficients

The factorial is defined for non-negative integers only. The Γ function generalises the factorial for all complex numbers.

For $z \in \mathbb{C}$, $Re(z) \in \mathbb{R}^+$, function $\Gamma(z)$ is defined as

$$\Gamma(z) = \int_0^{+\infty} e^{-y} y^{z-1} \, dy \tag{B.1}$$

with $\Gamma(0) = 1$, as can be verified by direct computation. This function has the property

$$\Gamma(z+1) = z\Gamma(z) \tag{B.2}$$

that can be used to analytically continue $\Gamma(z)$ to the left complex plane. We only have to use a reformulation of (B.2):

$$\Gamma(z) = \frac{\Gamma(z+1)}{z}. \tag{B.3}$$

Notice that $\Gamma(z)$ is not defined for non-positive integers, and, in fact, from (B.3),

$$\Gamma(0) = \frac{\Gamma(1)}{0} \tag{B.4}$$

and similarly for other values. With this extension the Γ function becomes a meromorphic function with simple poles at the the points $z = -k$, $k \in \mathbb{Z}_0^+$. The residues at these poles are given by

$$R_k = \frac{(-1)^k}{k!}. \tag{B.5}$$

It is possible to regularise the integral in (B.1) in order to give the values of the function at the whole complex plane excepting the negative integers or zero. This integral is given by:

$$\Gamma(z) = \int_0^{+\infty} \left[e^{-y} - \sum_{k=0}^{N} \frac{(-1)^k}{k!} y^k \right] y^{z-1} \, dy \tag{B.6}$$

where $N = \lfloor -z \rfloor$. If $N < 0$, the summation is null.

This function has several interesting properties that we are going to describe. It is convenient to introduce first the Pochhamer symbol defined by

$$(z)_n = z(z+1)(z+2)\cdots(z+n-1), \tag{B.7}$$

with $(z)_0 = 1$. Using this symbol we can refer some properties of the gamma function:

$$\Gamma(n+1) = n!, \ n \in \mathbb{N}_0 \tag{B.8}$$

$$\Gamma(z+n) = (z)_n \Gamma(z) \ n \in \mathbb{N}_0 \tag{B.9}$$

$$\Gamma(z)\Gamma(-z+1) = (-1)^n \Gamma(-z-n+1)\Gamma(z+n), \ n \in \mathbb{Z} \tag{B.10}$$

DOI 10.1515/9783110624588-015

$$\frac{\Gamma(z-n)}{\Gamma(z)} = \frac{(-1)^n}{(1-z)_n}, \quad n \in \mathbb{Z} \tag{B.11}$$

$$\frac{\Gamma(z+n)}{\Gamma(z)} = (-1)^n \frac{\Gamma(-z+1)}{\Gamma(-z-n+1)} \tag{B.12}$$

$$(-1)^n \frac{\Gamma(-z+n)}{\Gamma(-z)} = \prod_{k=0}^{n-1}(z-k) \tag{B.13}$$

$$(-1)^n \frac{\Gamma(-z+n)}{\Gamma(-z)} = \frac{\Gamma(z+1)}{\Gamma(z-n+1)} \tag{B.14}$$

$$\Gamma(z)\Gamma(1-z) = \frac{\pi}{\sin(\pi z)}, \quad 0 < z < 1 \tag{B.15}$$

$$\Gamma(0.5) = \sqrt{\pi} \tag{B.16}$$

The Gamma function allows us to generalise the expression of binomial coefficients to the complex plane.

Let $m, n \in \mathbb{Z}_0^+$. The binomial coefficients or combinations of m things, n at a time are defined as

$$\binom{m}{n} = \frac{m!}{n!(m-n)!} \tag{B.17}$$

The Γ function allows us to extend this by continuity, as follows:

$$\binom{a}{b} = \frac{\Gamma(a+1)}{\Gamma(b+1)\Gamma(a-b+1)}, \quad a, b, a-b \in \mathbb{R}\backslash\mathbb{Z}^- \tag{B.18}$$

$$\binom{a}{b} = \frac{(-1)^b \Gamma(b-a)}{\Gamma(b+1)\Gamma(-a)}, \quad b-a \notin \mathbb{Z}^- \wedge a \neq 0 \tag{B.19}$$

$$\binom{a}{b} = 0, \quad b \in \mathbb{Z}^- \wedge a \notin \mathbb{Z}^- \tag{B.20}$$

$$\binom{a}{b} = 0, \quad n \in \mathbb{Z}^- \wedge a \notin \mathbb{Z}^- \tag{B.21}$$

$$\binom{a}{b} = 0, \quad m, n \in \mathbb{Z}^- \wedge |m| > |n| \tag{B.22}$$

If none of the above apply (i.e. if $a \in \mathbb{Z}^- \wedge b \in \mathbb{Z}^- \wedge b \leq a$), we have to use another procedure to extend combinations. This can be done with the Pochhamer symbol. We only have to verify that

$$\binom{a}{n} = (-1)^n \frac{(-a)_n}{n!} \tag{B.23}$$

where a can be any complex and, in particular, any integer.

Exercises

1. Verify that $\Gamma(0) = 1$.
2. Prove (B.2) using integration by parts.

3. Use the two results above to prove (B.8) by mathematical induction.
4. Prove each of the results (B.9)–(B.14) from the definitions.

C The continuous-Time Fourier Transform

C.1 Definition

The Continuous-Time Fourier transform (CTFT, or, when there is no risk of confusion, FT) can be considered as a particular case of the BLT obtained when the ROC includes the imaginary axis. However, it can be used with functions that do not have LT.

Let $x(t)$ be a function defined on $\mathbb{R}$. The analysis and synthesis formulae for the FT are given by

$$\mathcal{F}[x(t)] = X(i\omega) = \int_{-\infty}^{\infty} x(t)e^{-i\omega t}\,dt, \ \omega \in \mathbb{R}, \tag{C.1}$$

and

$$x(t) = \mathcal{F}^{-1}[X(i\omega)] = \frac{1}{2\pi}\int_{-\infty}^{\infty} X(i\omega)e^{i\omega t}\,d\omega, \ t \in \mathbb{R}. \tag{C.2}$$

Frequently we write $\omega = 2\pi f$, which leads to

$$\mathcal{F}[x(t)] = X(f) = \int_{-\infty}^{\infty} x(t)e^{-i2\pi ft}\,dt, \ f \in \mathbb{R}, \tag{C.3}$$

and

$$x(t) = \mathcal{F}^{-1}[X(f)] = \int_{-\infty}^{\infty} X(f)e^{i2\pi ft}\,df, \ t \in \mathbb{R}. \tag{C.4}$$

If $x(t)$ is
- almost everywhere continuous,
- with bounded variation,
- absolutely integrable (AI),

then the FT exists and the convergence of the integral in (C.1) is uniform on $]-\infty, +\infty[$.

These conditions are not necessary, so that there are signals that do not verify them and still have a FT. For example, signals with finite energy $\int_{-\infty}^{+\infty} |x(t)|^2\,dt < \infty$ may not be absolutely integrable and even so they have FT. In this case, convergence is not uniform but in quadratic mean. If the signal is a continuous function and tends to zero with at least $1/|t|^a$, with $a > 1$, when $|t|$ tends to infinity, the convergence will be uniform and the TF will be a continuous function. In general we can say that *if $x(t)$ is differentiable until order N and its derivatives until order $N-1$ are continuous, then $X(\omega)$ tends to zero, at least with $1/|\omega|^{N+1}$ when $|\omega| \to \infty$.* The extreme and very important case is that of functions that tend to zero exponentially. They are called analytical of exponential type. The corresponding FTs are indefinitely differentiable.

DOI 10.1515/9783110624588-016

Tab. C.1: Main properties of the FT.

Linearity/homogeneity	$\mathcal{F}[a x(t) + b y(t)] = a X(\omega) + b . Y(\omega)$
Hermitean symmetry	$x(t) \in \mathbb{R} \Rightarrow X(-\omega) = X^*(\omega)$
Time reversion	$\mathcal{F}[x(-t)] = X(-\omega)$
Duality	$\mathcal{F}[X(t)] = x(-f) \Leftrightarrow \mathcal{F}[X(t)] = 2\pi x(-\omega)$
Scale change	$\mathcal{F}[x(a t)] = \dfrac{1}{|a|} X(\omega/a)$
Time shift	$\mathcal{F}[x(t - \tau)] = e^{-i\omega\tau} X(\omega)$
Frequency shift, modulation	$\mathcal{F}^{-1}[X(\omega - \omega_0)] = e^{i\omega_0 t} x(t)$
Frequency differentiation	$\mathcal{F}[t^n x(t)] = i \dfrac{d^n X(\omega)}{d\omega^n}$
Time convolution	$\mathcal{F}[x(t) * y(t)] = X(\omega) Y(\omega)$
Frequency convolution	$\mathcal{F}[x(t) y(t)] = \dfrac{1}{2\pi} X(\omega) * Y(\omega)$
Correlation	$\mathcal{F}\left[\displaystyle\int_{-\infty}^{\infty} x(\tau) y(\tau + t) \, d\tau\right] = X(\omega) Y^*(\omega)$
Parseval relation	$\displaystyle\int_{-\infty}^{\infty} |x(\tau)|^2 \, d\tau = \dfrac{1}{2\pi} \int_{-\infty}^{\infty} |X(\omega)|^2 \, d\omega$

The main properties of the FT are introduced in table C.1.

C.2 Differentiation and integration in the context of FT

The FT of the Nth order derivative of a function is given by

$$\mathcal{F}\left[\frac{d^n x(t)}{dt^n}\right] = (i\omega)^n X(\omega). \tag{C.5}$$

We want to generalise it for the fractional case:

$$\mathcal{F}[D^\alpha x(t)] = (i\omega)^\alpha X(\omega). \tag{C.6}$$

This means that we must have a derivative operator such that the derivative of a sinusoid is a sinusoid:

$$D^\alpha e^{i\omega_0 t} = (i\omega_0)^\alpha e^{i\omega_0 t}, \text{ if } t, \omega_0 \in \mathbb{R}. \tag{C.7}$$

Assume that there is such an operator. We will leave it undefined for now. Letting $x(t) = \cos(\omega_0 t) = \frac{1}{2} e^{i\omega_0 t} + \frac{1}{2} e^{-i\omega_0 t}$, we obtain

$$D^\alpha \cos(\omega_0 t) = \omega_0^\alpha \cos\left(\omega_0 t + \alpha \frac{\pi}{2}\right). \tag{C.8}$$

For the sine function the result is similar:

$$D^\alpha \sin(\omega_0 t) = \omega_0^\alpha \sin\left(\omega_0 t + \alpha \frac{\pi}{2}\right), \tag{C.9}$$

Tab. C.2: Examples of FTs.

$f(t)$	$\mathcal{F}[f(t)]$
$\delta(t)$	1
1	$2\pi\,\delta(\omega)$
$e^{-|t|}$	$\frac{2}{1+\omega^2}$
$\mathrm{sgn}(t)e^{-|t|}$	$\frac{-2i\omega}{1+\omega^2}$
$\frac{1}{1+t^2}$	$\pi e^{-|\omega|}$
$e^{-at}\varepsilon(t)$	$\frac{1}{a+i\omega}$
$\begin{cases} 1 & |t| \le T \\ 0 & |t| > T \end{cases}$	$2T\frac{\sin(\omega T)}{\omega T}$
$\frac{\sin\Omega t}{\pi t}$	$\begin{cases} 1 & |\omega| < \Omega \\ 0 & |\omega| > \Omega \end{cases}$
$\frac{\sin\Omega t}{|t|}$	$-i\,\mathrm{sgn}(\omega)\ln\left|\frac{\omega+\Omega}{\omega-\Omega}\right|$
$\begin{cases} \mathrm{sgn}(t) & 0 \le |t| \le T/2 \\ 0 & |t| > T/2 \end{cases}$	$-iT\frac{\sin^2\left(\omega\frac{T}{4}\right)}{\omega\frac{T}{4}}$
$\varepsilon(t)$	$\frac{1}{i\omega} + \pi\delta(\omega)$
$\mathrm{sgn}(t)$	$\frac{2}{i\omega}$
$\sum\limits_{m=-\infty}^{+\infty}\delta(t-mT_0)$	$\frac{T_0}{2\pi}\sum\limits_{m=-\infty}^{+\infty}\delta\left(\omega-m\frac{2\pi}{T_0}\right)$
$\begin{cases} 1-2\frac{|t|}{T} & |t| \le T/2 \\ 0 & |t| > T/2 \end{cases}$	$2T\mathrm{sinc}^2\left(\omega T/4\right)$
e^{-t^2}	$\sqrt{\pi}e^{-\frac{1}{4}t^2}$
$\begin{cases} \cos\left(\frac{\pi}{T}\right) & |t| \le T/2 \\ 0 & |t| > T/2 \end{cases}$	$\frac{2T\pi}{\pi^2-T^2\omega^2}\cos\left(\frac{\omega T}{2}\right)$
$\frac{1-e^{-\beta|t|}}{t}$	$-i\pi\,\mathrm{sgn}(\omega) + 2i\arctan\left(\frac{\omega}{\beta}\right)$
$\delta^{(n)}(t)$	$(i\omega)^n$
$\frac{(-it)^n}{2\pi}$	$\delta^{(n)}(\omega)$
$t\varepsilon(t)$	$i\pi\delta'(\omega) - \frac{1}{\omega^2}$
$\frac{1}{t}$	$-i\pi\,\mathrm{sgn}(\omega)$
$|t|$	$-\frac{2}{\omega^2}$
$\cos(\omega_0 t)$	$\pi\delta(\omega-\omega_0) + \pi\delta(\omega+\omega_0)$
$\sin(\omega_0 t)$	$-i\pi\delta(\omega-\omega_0) + i\pi\delta(\omega+\omega_0)$
$\sin\left(at^2\right)$	$-\sqrt{\frac{\pi}{a}}\sin\left[\frac{(\omega^2-a\pi)}{4a}\right]$
$\cos\left(at^2\right)$	$\sqrt{\frac{\pi}{a}}\cos\left[\frac{(\omega^2-a\pi)}{4a}\right]$
$|\sin(at)|$	$\sum\limits_{-\infty}^{+\infty}\frac{4}{1-4k^2}\delta(\omega-2ak)$
$|\cos(at)|$	$\sum\limits_{-\infty}^{+\infty}(-1)^k\frac{4}{1-4k^2}\delta(\omega-2ak)$

so that the result in 4.12 is fully justified.

Example C.2.1. *Let* $x(t) = e^{i\pi t}$. *Consider the differential equation*

$$y^{3\alpha} + y^{2\alpha} - 4y^{\alpha} + 2y = x^{2\alpha} - 4x. \tag{C.10}$$

Then

$$y(t) = \frac{\pi^{2\alpha}e^{i2\alpha\pi} + 4}{\pi^{3\alpha}e^{i3\alpha\pi} + \pi^{2\alpha}e^{i2\alpha\pi} + 4\pi^{\alpha}e^{i\alpha\pi} - 2}e^{i\pi t}. \tag{C.11}$$

The usual time integration property of the FT reads

$$\mathcal{F}\left[\int_{-\infty}^{t} x(\tau)d\tau\right] = x(t) * \varepsilon(t) = \frac{X(\omega)}{i\omega} + \pi X(0)\delta(\omega) \tag{C.12}$$

that seems to be in contradiction with similar property of the LT, $\mathcal{L}\left[\int_{-\infty}^{t} x(\tau)d\tau\right] = \frac{X(s)}{s}$, and with (C.6). The contradiction is only apparent. We note that the results are equal for $\omega \neq 0$. Only at the origin do we find a difficulty, that is nothing else than a singular case of the eigenfunction property. We will accept (C.6) as correct, except for $\omega = 0$, which is the case of a constant input.

Exercises

1. Prove the convolution properties of the FT given in Table C.1.
2. Prove the FTs of $\varepsilon(t)$ and sgn(t) given in Table C.2.

D The discrete-time Fourier Transform

D.1 Definition

The discrete-time Fourier Transform results from the Z transform, just as the continuous-time Fourier transform comes from the Laplace transform. In fact, making the transformation $z \longrightarrow e^{i\omega}$, we obtain

$$X(e^{i\omega}) = \sum_{n=-\infty}^{+\infty} x(n)e^{-in\omega} \tag{D.1}$$

and

$$x(n) = \frac{1}{2\pi} \int_{-\pi}^{\pi} X(e^{i\omega})e^{in\omega}\,d\omega. \tag{D.2}$$

The above equations admit the following interpretation: *a discrete signal can be expressed as a continuous over-positioning of elementary sinusoids with infinitesimal complex amplitudes $\frac{1}{2\pi}X(e^{i\omega})\,d\omega$, and frequencies $\omega \in (-\pi, \pi]$ infinitely close.*

With this interpretation, (D.1) and (D.2) define a discrete signal analysis/synthesis pair and are called discrete-time Fourier transform (DTFT or FTd)[1] and its inverse. As is evident, the DTFT is periodic with period 2π (and so usually we will represent only one period). We may also write $X(\omega)$ instead of $X(e^{i\omega})$.

The way how we obtained the DTFT from the Z transform allows us to extract an important conclusion: *any signal having Z-transform with a convergence region that contains the unit circumference has also a Fourier transform that can be obtained by a simple change of variable $z = e^{i\omega}$.*

To emphasise this fact, we will often write $X(e^{i\omega})$.

Example D.1.1 (Causal exponential)**.** *The Fourier transform of signal $x(n) = a^n \epsilon(n)$, $|a| < 1$ is*

$$X(e^{i\omega}) = \sum_{n=-\infty}^{\infty} a^n \epsilon(n)e^{-i\omega n} = \sum_{n=0}^{\infty} \left(ae^{-i\omega}\right)^n \tag{D.3}$$

If $|a| < 1$, then $|ae^{-i\omega} < 1$, and so the summation can be calculated using the usual summation rule of the geometric series to obtain

$$\mathscr{F}\left[a^n \epsilon(n)\right] = \frac{1}{1 - ae^{-i\omega}} \tag{D.4}$$

Example D.1.2 (Two-sided exponential)**.** *Let $x(n) = a^{|n|}$, $|a| < 1$. Insert $x(n)$ in (D.1) and decompose the summation in two, corresponding to the values of n from 0 to $+\infty$*

---

1 Whenever there is no danger of confusion with the continuous case, we will speak of a Fourier transform only, and simply say FT.

DOI 10.1515/9783110624588-017

and from $-\infty$ to -1. We then have

$$X(e^{i\omega}) = \sum_{n=-\infty}^{\infty} a^{|n|} e^{-i\omega n}$$

$$= \sum_{n=0}^{\infty} \left(ae^{-i\omega}\right)^n + \sum_{n=-\infty}^{-1} \left(ae^{-i\omega}\right)^{-n}$$

$$= \sum_{n=0}^{\infty} \left(ae^{-i\omega}\right)^n + \sum_{n=1}^{\infty} \left(ae^{-i\omega}\right)^n$$

$$= \frac{1}{1 - ae^{-i\omega}} + \frac{ae^{i\omega}}{1 - ae^{i\omega}} \tag{D.5}$$

Consequently,

$$\mathcal{F}\left[a^{|n|}\right] = \frac{1 - a^2}{1 - 2a\cos\omega + a^2} \tag{D.6}$$

We could solve the problem in another way. Note that

$$a^{|n|} = a^{|n|} \left[\varepsilon(n) + \varepsilon(-n-1)\right] = a^n \varepsilon(n) + a^{-n}\varepsilon(-n) - \delta(n), \tag{D.7}$$

and, as we shall see later, the FT of a sum is equal the sum of the transforms, so we can use (D.4). We still need to know what is the FT of $\delta(n)$. It is not hard to verify that

$$\mathcal{F}[\delta(n)] = 1 \tag{D.8}$$

Example D.1.3 (Rectangular pulse). *The FT of the rectangular pulse*

$$p(n) = \begin{cases} 1 & |n| \le N \\ 0 & |n| > N \end{cases} \tag{D.9}$$

is

$$P(e^{i\omega}) = \sum_{n=-N}^{N} e^{-i\omega n} = \sum_{n=0}^{N} e^{-i\omega n} + \sum_{n=-N}^{-1} e^{-i\omega n}$$

$$= \sum_{n=0}^{N} e^{-i\omega n} + \sum_{n=1}^{N} e^{i\omega n}$$

$$= \sum_{n=0}^{N} e^{-i\omega n} + \sum_{n=0}^{N} e^{i\omega n} - 1. \tag{D.10}$$

Summing up the geometric series, we obtain:

$$P(e^{i\omega}) = \frac{1 - e^{-i\omega(N+1)}}{1 - e^{-i\omega}} + \frac{1 - e^{i\omega(N+1)}}{1 - e^{i\omega}} - 1, \tag{D.11}$$

from where

$$P(e^{i\omega}) = 2\frac{e^{i\omega\frac{N+1}{2}} - e^{-i\omega\frac{N+1}{2}}}{e^{i\omega/2} - e^{-i\omega/2}} - 1$$

$$= e^{-i\omega\frac{N}{2}}\frac{\sin\left(\omega\frac{N+1}{2}\right)}{\sin(\omega/2)} + e^{i\omega\frac{N}{2}}\frac{\sin\left(\omega\frac{N+1}{2}\right)}{\sin(\omega/2)} - 1$$

$$= \frac{\sin\left(\omega\frac{N+1}{2}\right)}{\sin(\omega/2)}\cos\left(\omega\frac{N}{2}\right) - 1. \tag{D.12}$$

Using the well-known trigonometric identity $\sin(a \pm b) = \sin a \cos b \pm \cos a \sin b$, *we get*

$$P(e^{i\omega}) = \frac{\sin\left[\omega\left(N+\frac{1}{2}\right)\right] + \sin(\omega/2)}{\sin(\omega/2)} - 1 \tag{D.13}$$

and finally

$$\mathcal{F}[p(n)] = \frac{\sin\left[\omega\left(N+\frac{1}{2}\right)\right]}{\sin(\omega/2)}. \tag{D.14}$$

Example D.1.4 (Ramp pulse). *Consider now the signal* $r(n) = n + 1$ *for* $0 \le n \le N$. *We have*

$$R(e^{i\omega}) = \sum_{n=0}^{N} e^{-i\omega n} + \sum_{n=0}^{N} n e^{-i\omega n}$$

$$= R_1(e^{i\omega}) - i\frac{dR_1(e^{i\omega})}{d\omega} \tag{D.15}$$

where

$$R_1(e^{i\omega}) = \frac{1 - e^{-i\omega(N+1)}}{1 - e^{-i\omega}} \tag{D.16}$$

Example D.1.5 (Triangular pulse). *Consider signal* $t(n) = [0\ 1\ 2\ 3\ 2\ 1]$, *for which*

$$T(e^{i\omega}) = \sum_{n=0}^{N} n e^{-i\omega n} + \sum_{n=4}^{5} (6 - n)e^{-i\omega n}$$

$$= -i\frac{dR_1(e^{i\omega})}{d\omega} + 2e^{-4i\omega} + e^{-5i\omega}, \tag{D.17}$$

where $R_1(e^{i\omega})$ *was calculated in the previous example.*

D.2 Existence

Since, in general, the Fourier transform results from the addition of an infinite number of sinusoids, it is natural to wonder in what conditions the previous summation converges. It should be noted that, contrary to what happens for the Z transform, where convergence was ensured by a convenient choice of the variation domain for the independent variable z, in the FT we have a previously set domain. Thus, for a certain signal $x(n)$, either the summation converges or not. In the first case, the result is the Fourier transform of the signal, while in the second it is said that the signal has no Fourier transform. If the signal has finite duration and finite amplitudes, the FT always exists. In the general case, we have to study the convergence of the series. To derive a sufficient condition of convergence, simply take the absolute value in (D.1). We have

$$|X(\omega)| \le \sum_{n=-\infty}^{+\infty} |x(n)| < \infty, \tag{D.18}$$

since $\left|e^{in\omega}\right| = 1$. So, if the signal is absolutely summable, its FT exists and the convergence of the series in (D.1) is uniform in the interval $[-\pi, \pi]$. This condition is not necessary; there are signals that do not verify it and still have FT. Just as for the continuous FT, sequences with finite energy, $\sum_{n=-\infty}^{+\infty} |x(n)|^2 < \infty$, may not be absolutely summable and even so have FT. Again, in this case the convergence is not uniform but in quadratic mean.

If the signal tends to zero with at least $1/n^a$, with $a > 1$, when $|n|$ tends to infinity, the convergence will be uniform and FT will be a continuous function (the extreme and very important case is that of functions that tend to zero exponentially; the corresponding FTs are indefinitely differentiable). However, if that condition does not occur, the FT will necessarily be discontinuous. In this situation, convergence in quadratic mean should be used. Let us study this case using an interesting example.

Example D.2.1 (The ideal low-pass filter). *Consider the ideal low-pass filter*

$$H(\omega) = \begin{cases} 1 & |\omega| < \Omega \\ 0 & \Omega < |\omega| < \pi \end{cases} \tag{D.19}$$

Let us calculate its inverse FT, using the inversion integral (D.2). We have then

$$h(n) = \frac{1}{2\pi} \int_{-\Omega}^{\Omega} 1 \cdot e^{in\omega} d\omega$$

$$= \frac{1}{2\pi} \left[\frac{e^{in\omega}}{in} \right]_{-\Omega}^{+\Omega} = \frac{\sin \Omega n}{\pi n}, \tag{D.20}$$

which is a signal that tends to zero with $1/|n|$ and whose transform is a discontinuous function. Note an important fact: the ideal low-pass system is not causal, as its impulse response is bilateral.

When summing the series, we observe "peaks" that we could not be waiting for. This is because, as mentioned above, the series does not converge uniformly throughout the interval. At the discontinuity points the said peaks arise, consisting in the so-called Gibbs effect which is a consequence of the nonuniform convergence and the finite

Tab. D.1: Main properties of the DTFT

Linearity/homogeneity	$\mathcal{F}[a.x(n) + b.y(n)] = aX(e^{i\omega}) + b.Y(e^{i\omega})$
Hermitean symmetry	$x(n) \in \mathbb{R} \Rightarrow X(e^{-i\omega}) = X^*(e^{i\omega})$
Time reversion	$\mathcal{F}[x(-n)] = X(e^{-i\omega})$
Scale change	$\mathcal{F}\left[\begin{cases} x(an) & an \in \mathbb{Z} \\ 0 & an \notin \mathbb{Z} \end{cases}\right] = \frac{1}{|a|}X(e^{-i\omega/a})$
Time shift	$\mathcal{F}[x(n-k)] = e^{-i\omega k}X(e^{i\omega})$
Frequency shift, modulation	$\mathcal{F}^{-1}\left[X\left(e^{i(\omega-\omega_0)}\right)\right] = e^{i\omega_0 n}x(n)$
Time differentiation	$\mathcal{F}[x(n) - x(n-1)] = \left[1 - e^{-i\omega}\right]X(e^{i\omega})$
Frequency differentiation	$\mathcal{F}[nx(n)] = i\frac{dX(e^{i\omega})}{d\omega}$
Time Convolution	$\mathcal{F}[x(n) * y(n)] = X(e^{i\omega})Y(e^{i\omega})$
Frequency Convolution	$\mathcal{F}[x(n)y(n)] = \frac{1}{2\pi}X(e^{i\omega}) * Y(e^{i\omega})$
Accumulation	$TF\left[\sum\limits_{-\infty}^{n} x(k)\right] = \frac{X(e^{i\omega})}{[1-e^{-i\omega}]} + \pi X(e^{i0})\sum\limits_{-\infty}^{+\infty} \delta(\omega - 2k\pi)$
Correlation	$\mathcal{F}\left[\int\limits_{-\infty}^{\infty} x(k).y(n+k)d\tau\right] = X(e^{i\omega}).Y^*(e^{i\omega})$
Parseval relation	$\sum\limits_{-\infty}^{\infty} |x(n)|^2 d\tau = \frac{1}{2\pi}\int_{-\pi}^{\pi}|X(e^{i\omega})|^2 d\omega$

number of terms used in the calculation. However, at the point of discontinuity, all curves pass in the 1/2 point. In fact, the series converges to the half-sum of the lateral values of the function.

The main properties of the DTFT are introduced in table D.1.

Exercises

1. Prove (D.8).
2. Prove the convolution properties of the FT given in Table D.1.

E Partial fraction decomposition without derivations

E.1 Simplification when there are no conjugate pairs of poles

Consider a given fraction, not necessarily proper. Decompose it into the product of two fractions:

$$Q(z) = Q_1(z)\, Q_2(z) = \frac{N_1(z)}{D_1(z)} \frac{N_2(z)}{D_2(z)}, \tag{E.1}$$

in such a way that all the poles of $Q_2(z)$ are simple (i.e. of degree 1). In addition, it is convenient to retain in this fraction the maximum number of zeros, so that $Q_2(z)$ is a proper fraction. The first step of the decomposition consists of decomposing $Q_2(z)$ into simple partial fractions

$$Q_2(z) = \sum_{i=1}^{N} \frac{A_i}{(z-p_i)}, \tag{E.2}$$

where the A_i are the residues of $Q_2(z)$ at the poles, p_i:

$$A_i = \lim_{z \to p_i} (z-p_i) Q_2(z). \tag{E.3}$$

The following steps consist in performing multiplications of $Q_2(z)$ by factors given by one of the following expressions:

$$z - z_i \tag{E.4}$$

$$\frac{1}{z - p_j} \tag{E.5}$$

$$\frac{z - z_i}{z - p_j} \qquad i \neq j \tag{E.6}$$

Each factor will be multiplied by the decomposition obtained by starting from (E.2). Thus, in a given stage of the process, we have a development of the form

$$Q_{int} = \sum_{i=1}^{N} \sum_{j=1}^{m_i} \frac{a_{ij}}{(z-p_i)^j}, \tag{E.7}$$

which will be multiplied by one of the above terms. Let us start by noting that the fraction in $\frac{z-z_i}{z-p_j}$ decomposes into

$$\frac{z - z_i}{z - p_j} = 1 + \frac{p_j - z_i}{z - p_j}, \tag{E.8}$$

whereby the treatment of this case is identical to that given to the term $\frac{1}{z-p_i}$. Therefore, the problem is nothing else than multiplying a simple fraction by development (E.7), so that we have to decompose a sum of products of 2 simple fractions such as

$$\frac{A_{ij}}{(z-p_i)^j (z-p_k)}. \tag{E.9}$$

DOI 10.1515/9783110624588-018

There are two situations to consider, depending on whether $p_i = p_k$ or $p_i \neq p_k$. If $p_i = p_k$, there is no difficulty: the result is $\frac{1}{(z-p_i)^{j+1}}$. If $p_i \neq p_k$ we have to expand a product. It is not difficult to obtain recursively

$$\frac{a_{ij}}{(z-p_i)^j(z-p_k)} = \frac{\frac{1}{(p_k-p_i)^j}}{z-p_k} - \sum_{i=1}^{j} \frac{\frac{1}{(p_k-p_i)^{j-i+1}}}{(z-p_i)^i}. \tag{E.10}$$

Finally, we have the term $z - z_i$. In this case, we must decompose it in terms of the form

$$\frac{z-z_k}{(z-p_i)^j}. \tag{E.11}$$

If $z_k = p_i$, we have a trivial case: it gives the fraction $\frac{1}{(z-p_i)^{j-1}}$. If $z_k \neq p_i$, we have

$$\frac{z-z_k}{(z-p_i)^j} = \frac{1}{(z-p_i)^{j-1}} + \frac{p_i - z_k}{(z-p_i)^j}. \tag{E.12}$$

Example E.1.1. *Let*

$$F(z) = \frac{(z+1)(z+2)(z+3)(z+4)}{(z-1)^2(z-2)^2}. \tag{E.13}$$

We then let $Q_2(z) = \frac{(z+1)}{(z-1)(z-2)}$. *The decomposition of* $Q_2(z)$ *is*

$$Q_2(z) = \frac{-2}{(z-1)} + \frac{3}{(z-2)}. \tag{E.14}$$

As to the rest we can write

$$Q_1(z) = \frac{z+2}{(z-1)} \cdot \frac{z+3}{(z-2)}(z+4) = \left[1 + \frac{3}{(z-1)}\right] \cdot \left[1 + \frac{5}{(z-2)}\right](z+4). \tag{E.15}$$

Proceeding as above, we obtain, after multiplying $Q_2(z)$ *by the first term of* $Q_1(z)$,

$$Q_3(z) = \frac{-11}{(z-1)} + \frac{-6}{(z-1)^2} + \frac{12}{(z-2)}. \tag{E.16}$$

Likewise, multiplying by the second,

$$Q_4(z) = \frac{74}{(z-1)} + \frac{24}{(z-1)^2} + \frac{-63}{(z-2)} + \frac{160}{(z-2)^2}. \tag{E.17}$$

Finally, multiplying by the third term $(z+4)$, *we act as above to get*

$$F(z) = 1 + \frac{394}{(z-1)} + \frac{120}{(z-1)^2} - \frac{378}{(z-2)} + \frac{360}{(z-2)^2}. \tag{E.18}$$

Example E.1.2. *Let*

$$\frac{1}{(z+2)^3(z+3)}. \tag{E.19}$$

We have

$$F(z) = \frac{1}{(z+2)^2} \frac{1}{(z+2)(z+3)} = \frac{1}{(z+2)^2} \left[\frac{1}{(z+2)} - \frac{1}{(z+3)} \right], \tag{E.20}$$

from where

$$F(z) = \frac{1}{(z+2)} \left[\frac{1}{(z+2)^2} - \frac{1}{(z+2)} + \frac{1}{(z+3)} \right] \tag{E.21}$$

and

$$F(z) = \left[\frac{1}{(z+2)^3} - \frac{1}{(z+2)^2} + \frac{1}{(z+2)} - \frac{1}{(z+3)} \right]. \tag{E.22}$$

E.2 Simplification for conjugate pairs of poles on the imaginary axis

Consider the following example:

$$F(z) = \frac{z(z-1)}{(z+1)(z^2+1)} . \tag{E.23}$$

Following the steps above, we have

$$F(z) = \frac{1}{z+1} + \frac{i/2}{z-i} - \frac{i/2}{z+i} . \tag{E.24}$$

The process is simple and easy. This may not be the case if there are many more poles and zeroes. On the other hand, the two fractions corresponding to the imaginary poles are usually more useful together than separated, in order to show a fraction corresponding to a transform of a real signal.

Consequently, in the case above, it is convenient to write

$$F(z) = \frac{1}{z+1} - \frac{1}{z^2+1} . \tag{E.25}$$

The question we pose is how to obtain the second term directly to avoid the use of complex numbers. Suppose we have a fraction with the general form

$$Q(z) = \frac{N_1(z)}{D_1(z)(z^2+\omega_0^2)} . \tag{E.26}$$

Assume that $Q_1(z) = \frac{N_1(z)}{D_1(z)}$ is already decomposed, and the terms corresponding to the pair of imaginary poles are lacking. The process can be described like this:

1. Let $Q_1(z) = \dfrac{\sum\limits_{k=0}^{M} b_k z^k}{\sum\limits_{k=0}^{N} a_k z^k}$.

2. Substitute all the even powers of z (i.e. powers of the form z^{2m}) by $(-1)^m \omega_0^{2m}$.
3. Substitute all the odd powers of z (i.e. powers of the form z^{2m+1}) by $(-1)^m \omega_0^{2m} z$.
4. In general, this procedure allows us to obtain a new fraction $q(z) = \frac{b_0 + b_1 z}{a_0 + a_1 z}$.

5. If $a_1 \neq 0$, we have to multiply the numerator and denominator of $q(z)$ by $a_0 - a_1 z$ and substitute z^2 by $-w_0^2$ in order to obtain $p(z)$

$$p(z) = q(z)\frac{a_0 - a_1 z}{a_0 - a_1 z} = \frac{(a_0 b_0 + a_1 b_1 w_0^2) + (a_0 b_1 - a_1 b_0.z)}{a_0^2 + a_1^2 w_0^2} \qquad \text{(E.27)}$$

which is the numerator of the fraction corresponding to the pair of conjugated imaginary poles.

In the example above, we have $q(z) = \frac{z^2-1}{z+1}|_{z^2=-1} = \frac{-1-z}{z+1} = -1$ in agreement with what had been found.

Example E.2.1. *Let*

$$G(z) = \frac{z/2 - 1}{(z+1)(z^2+1)}. \qquad \text{(E.28)}$$

We have $q(z) = \frac{\frac{z}{2}-1}{(z+1)}$, *from where*

$$p(z) = \frac{\frac{z^2}{2} - \frac{3}{2}z + 1}{(z^2-1)}\bigg|_{z^2=-1} = \frac{-\frac{3}{2}z + \frac{1}{2}}{-2} = \frac{3}{4}z - \frac{1}{4}. \qquad \text{(E.29)}$$

As the residue is $z = -1$, *we get*

$$G(z) = -\frac{\frac{3}{4}}{z+1} + \frac{\frac{3}{4}z - \frac{1}{4}}{z^2+1}. \qquad \text{(E.30)}$$

Example E.2.2. *Let*

$$G(z) = \frac{z^3 + 6z^2 - 5z}{(z^2+1)(z^2+4)}. \qquad \text{(E.31)}$$

The decomposition is

$$G(z) = \frac{1+i}{z-i} - \frac{1-i}{z+i} + \frac{\frac{3}{2} - 2i}{z-2i} + \frac{\frac{3}{2} + 2i}{z+2i}, \qquad \text{(E.32)}$$

giving

$$G(z) = -\frac{2z+2}{(z^2+1)} + \frac{3z+8}{(z^2+4)}. \qquad \text{(E.33)}$$

With the above method,

$$p_1(z) = \frac{-z - 6 - 5z}{3} = -2z - 2 \wedge p_2(z) = \frac{-4z - 24 - 5z}{-3} = 3z + 8, \qquad \text{(E.34)}$$

in agreement with the result above.

E.3 Simplification for conjugate pairs of poles not on the imaginary axis

Consider now the case

$$Q(z) = \frac{N_1(z)}{D_1(z)[(z+\sigma)^2 + \omega_0^2]}. \tag{E.35}$$

The solution of the problem is immediate: *we only have to decompose the fraction* $Q(z - \sigma)$ *and then substitute* $z + \sigma$ *for* z. This solution also remains valid in the case where there are more pairs of poles with the same real part.

The question here is to be able to apply the method in the case where there are pairs of poles on two distinct vertical lines. Apparently, the method can also be applied. Let us consider the following case:

$$G(z) = \frac{z^3 + 6z^2 - 5z}{[(z+2)^2 + 1][(z+1)^2 + 4]}. \tag{E.36}$$

We are going to consider the two pairs separately.

1. Let us take the fraction that results from a translation of -2. Put $v = z + 2$:

$$G(v) = \frac{v^3 - 17v + 26}{[v^2 + 1][v^2 - 2v + 5]}. \tag{E.37}$$

To apply the method, compute $q_1(v)$:

$$q_1(v) = \frac{9v - 13}{v - 2} = \frac{(9v - 13)(v + 2)}{v^2 - 4}. \tag{E.38}$$

From here we get $p_1(v) = 7 - v$ and $p_1(z) = 5 - z$. Therefore, the first parcel of $G(z)$ is given by

$$G_1(z) = \frac{-z + 5}{(z+2)^2 + 1}. \tag{E.39}$$

2. Let us now consider the other term. The translation is now -1. Put $v = z + 1$:

$$G(v) = \frac{v^3 + 3v^2 - 14v + 10}{[v^2 + 4][v^2 + 2v + 2]} \tag{E.40}$$

Compute now $q_2(v)$ to obtain

$$q_2(v) = \frac{-18v - 2}{2v - 2} = -\frac{9v + 1}{v - 1} = \frac{(9v + 1)(v + 1)}{v^2 - 1}. \tag{E.41}$$

From here we get $p_2(v) = -7 + 2v$ and $p_2(z) = 5 + 2z$. Consequently the second parcel of $G(z)$ is

$$G_2(z) = \frac{2z - 5}{(z+1)^2 + 4}. \tag{E.42}$$

Finally, we get

$$Q(z) = \frac{-z + 5}{(z+2)^2 + 1} + \frac{2z - 5}{(z+1)^2 + 4}. \tag{E.43}$$

It is not difficult to verify that this expression is correct.

Exercises

1. Decompose the following fractions:

   a) $F(z) = \dfrac{3z^5 + 25z^4 + 84z^3 + 148z^2 + 132z + 44}{(z+2)^2[(z+1)^2+1]^2}$

   b) $F(z) = \dfrac{-0.7z^{-1} - 0.96z^{-2} - 1.53z^{-3} + 1.06z^{-4} + 0.32z^{-5}}{(1-z^{-1}+0.5z^{-2})(1-0.5z^{-1})(1+0.4z^{-1})^2}$

   c) $F(z) = \dfrac{1.8 + 1.48z^{-1} - 0.7312z^{-2} - 4.4816z^{-3} + 0.2704z^{-4} + 0.112z^{-5}}{(1+4z^{-1}+5z^{-2})(1+0.2z^{-1})(1-0.4z^{-1})^2}$

2. Prove (E.10) by mathematical induction.

F The Mittag-Leffler function

F.1 Definition

Definition F.1.1. *The one-parameter Mittag-Leffler function (MLF) [22, 38] is given by*

$$E_\alpha(z) = \sum_{k=0}^{\infty} \frac{z^k}{\Gamma(k\alpha + 1)}, \quad z \in \mathbb{C}, \quad \alpha \in \mathbb{R}^+. \tag{F.1}$$

Remark F.1.1. *The exponential series is a particular case obtained when $\alpha = 1$.*

A generalisation of (F.1) includes a second parameter, yielding the two-parameter MLF:

$$E_{\alpha,\beta}(z) = \sum_{k=0}^{\infty} \frac{z^k}{\Gamma(k\alpha + \beta)}, \quad z \in \mathbb{C}, \quad \alpha, \beta \in \mathbb{R}^+. \tag{F.2}$$

When α and β are positive real (the case that we are considering here), the series converges for all values of $z \in \mathbb{C}$ [14, 23, 90] and has the integral representation that we will now deduce.

Consider the integration path in Figure F.1. As shown by Hankel [26], the reciprocal gamma function can be computed by the integral

$$\frac{1}{\Gamma(z)} = \frac{1}{2\pi i} \int_{\mathcal{L}} u^{-z} e^u \, du, \quad z \in \mathbb{C}, \tag{F.3}$$

provided that:

- u^{-z} is defined by its principal branch; we can assume for branch cut line the negative real half axis, unless a pole exists on this line;
- r may assume any positive real value;
- φ can be any value in the interval $(\pi/2, \pi)$; it can be equal to π, provided that $\arg(u) = \pi$, or $\arg(u) = -\pi$, above, or below, the cut, respectively.

As the MLF is defined by a uniformly convergent series, we can write

$$E_{\alpha,\beta}(z) = \frac{1}{2\pi i} \sum_{k=0}^{\infty} z^k \int_{\mathcal{L}} u^{-k\alpha-\beta} e^u \, du = \frac{1}{2\pi i} \int_{\mathcal{L}} u^{-\beta} \left[\sum_{k=0}^{\infty} (zu^{-\alpha})^k \right] e^u \, du. \tag{F.4}$$

The summation of the inner geometric series is $\sum_{k=0}^{\infty} (zu^{-\alpha})^k = \frac{u^\alpha}{u^\alpha-z}$, provided that $|zu^{-\alpha}| < 1$. This condition introduces a constraint in the integration loop in such a way that we can formulate the general integral form of the MLF [14, 23, 108],

$$E_{\alpha,\beta}(z) = \frac{1}{2\pi i} \int_{\mathcal{L}} \frac{u^{\alpha-\beta} e^u}{u^\alpha - z} \, du, \quad z \in \mathbb{C}, \quad \alpha, \beta \in \mathbb{R}^+, \tag{F.5}$$

DOI 10.1515/9783110624588-019

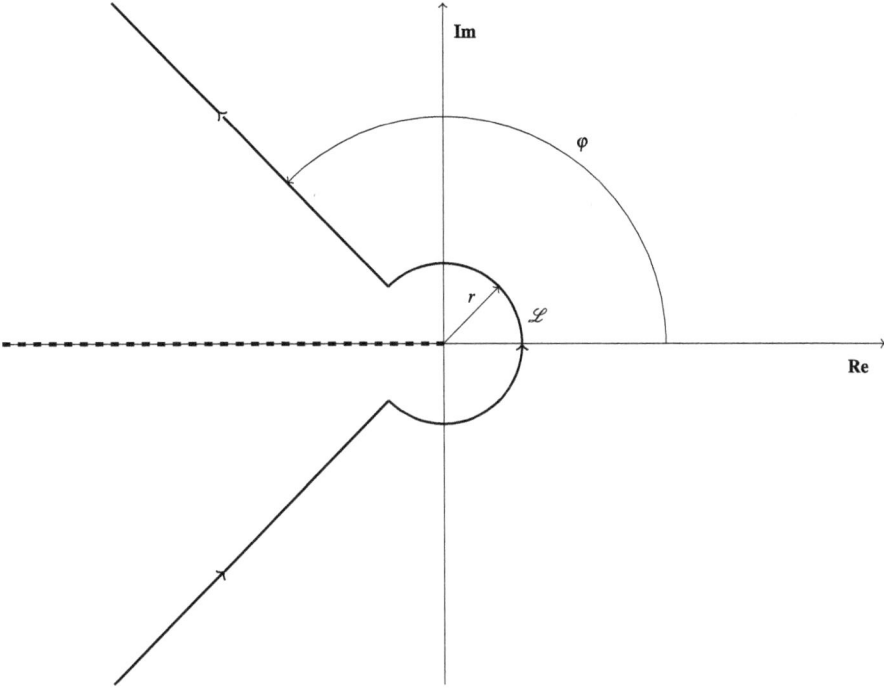

Fig. F.1: The integration path $\mathcal{L}$ for $\frac{1}{\Gamma(z)}$.

where now $\mathcal{L}$ is the integration path consisting in a loop that starts at $u = -\infty$, encircles the disk with centre $u = 0$ and radius $|z|^{1/\alpha}$, and returns to $u = -\infty$ (in Figure F.1, $r > |z|^{1/\alpha}$). This integration path can be deformed to become the Hankel contour, simplifying the computations, under the following assumptions:

- The negative real axis is the branch cut line, but we can choose any other in the left half plane; as referred above, we have to follow that strategy if there is a pole on such axis;
- As the MLF of any order $\alpha > 1$ can be written in terms of the MLF with order $\alpha < 1$, we will consider $0 < \alpha \leq 1$;
- We address the cases $\alpha - \beta - 1 > 0$; other alternatives [23], namely $\alpha < 0$, can also be treated, but they are not relevant, and are not considered;
- $z \in \mathbb{C} \setminus 0$.

Consider the integration path in Figure F.2 denoted by $\mathcal{L}_t$. It consists of three components:

- The path $\mathcal{L}$ described above;
- The Hankel path $\mathcal{H}$, consisting of two half-straight lines joined by a half circle with a radius as small as we wish;
- Two circular arcs with radius $R > |z|^{1/\alpha}$, to make a closed path, so that radius R grows to ∞.

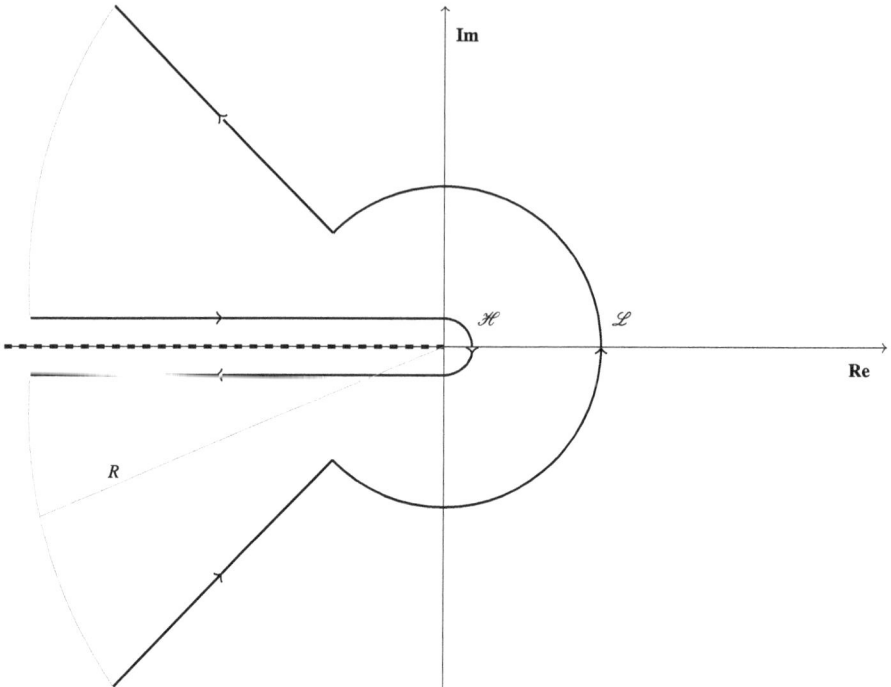

Fig. F.2: The paths $\mathcal{L}$ and $\mathcal{H}$, and the closing arcs, that together make the integration path $\mathcal{L}_t$.

With the residue theorem, we can write

$$\oint_{\mathcal{L}_t} \frac{u^{\alpha-\beta} e^u}{u^\alpha - z} du = \begin{cases} \frac{2\pi i}{\alpha} z^{\frac{1-\beta}{\alpha}} e^{z^{\frac{1}{\alpha}}}, & \text{if } |\arg(z)| \le \pi\alpha \\ 0, & \text{if } |\arg(z)| > \pi\alpha \end{cases}. \tag{F.6}$$

Under the above stated conditions, we can deduce that:

– The integrals along the circular arcs tend to zero when the radius goes to infinity (application of Jordan's lemma);
– The integral along the half-circle in the Hankel path becomes zero when its radius decreases to zero;
– The integral along the upper half-straight line is $\int_{0^+}^\infty \frac{u^{\alpha-\beta} e^{i(\alpha-\beta)\pi}}{u^\alpha e^{i\alpha\pi} - z} e^{-u} du$;
– The integral over the lower half-straight line is $- \int_{0^+}^\infty \frac{u^{\alpha-\beta} e^{-i(\alpha-\beta)\pi}}{u^\alpha e^{-i\alpha\pi} - z} e^{-u} du$;
– The MLF is given by

$$E_{\alpha,\beta}(z) = z^{\frac{1-\beta}{\alpha}} \frac{e^{z^{\frac{1}{\alpha}}}}{\alpha} + \frac{1}{\pi} \int_{0^+}^\infty \frac{\sin(\beta\pi)u^{2\alpha-\beta} + \sin[(\alpha-\beta)\pi]u^{\alpha-\beta}z}{u^{2\alpha} - 2\cos(\alpha\pi)u^\alpha z + z^2} e^{-u} du \tag{F.7}$$

if $|\arg(z)| < \pi\alpha$, or by

$$E_{\alpha,\beta}(z) = \frac{1}{\pi} \int\limits_{0^+}^{\infty} \frac{\sin(\beta\pi)u^{2\alpha-\beta} + \sin[(\alpha-\beta)\pi]u^{\alpha-\beta}z}{u^{2\alpha} - 2\cos(\alpha\pi)u^{\alpha}z + z^2} e^{-u} du \qquad (F.8)$$

if $|\arg(z)| > \pi\alpha$.

Remark F.1.2. *For $\arg(z) = \pm\alpha\pi$ we have one pole on the branchcut line. In this case, to solve the problem while avoiding inaccurate results, we only have to rotate the branch cut line and the Hankel path. This corresponds to replacing u by $u.e^{i\theta}$, with $|\theta| < \frac{\pi}{2}$, in the integral equations (F.7) and (F.8) [90]*

In numerical computations, we need to truncate the integral above some u_0. This originates an error given by

$$\delta = \frac{1}{\pi} \int\limits_{u_0}^{\infty} \frac{\sin(\beta\pi)u^{2\alpha-\beta} + \sin[(\alpha-\beta)\pi]u^{\alpha-\beta}z}{u^{2\alpha} - 2\cos(\alpha\pi)u^{\alpha}z + z^2} e^{-u} du. \qquad (F.9)$$

It can be shown that function $\frac{\sin(\beta\pi)u^{2\alpha}+\sin(\alpha\pi)u^{\alpha-\beta}z}{u^{2\alpha}-2\cos(\alpha\pi)u^{\alpha}z+z^2}$ is bounded and approaches $\sin(\beta\pi)$ when u increases without bound. Assume that its absolute value is given by $\sin(\beta\pi)(1+\epsilon)$, $\epsilon > 0$. Then

$$|\delta| \leq \frac{(1+\epsilon)}{\pi} \int\limits_{u_0}^{\infty} \frac{1}{u^{\beta}} e^{-u} du < \frac{(1+\epsilon)}{u_0^{\beta}\pi} \int\limits_{u_0}^{\infty} e^{-u} du = \frac{(1+\epsilon)}{u_0^{\beta}\pi} e^{-u_0}. \qquad (F.10)$$

Therefore, the error decreases exponentially leading to a fast convergence of the integral.

In FC applications, we need to compute the MLF on half-straight lines, defined by $z = pt^{\alpha}\varepsilon(t)$. From (F.2) we obtain the expression

$$\mathcal{E}_{\alpha,\beta}(pt^{\alpha}) = \sum_{k=0}^{\infty} p^k \frac{t^{k\alpha}}{\Gamma(k\alpha+\beta)} \varepsilon(t), \quad p \in \mathbb{C}, \; t \in \mathbb{R}, \qquad (F.11)$$

where t represents time and $\varepsilon(t)$ denotes the Heaviside unit step function used to state that (F.11) is identically null for $t < 0$.

To avoid confusion, we call expression (F.11) "causal MLF" (CMLF), since it is defined on $\mathbb{R}$ and is null on $\mathbb{R}^-$ (it can be considered as the impulse response of a causal system). Furthermore, in order to simplify notation, we will omit $\varepsilon(t)$. This variable change is inserted in (F.7) and (F.8), where we make $v = u/t$ to get

$$\mathcal{E}_{\alpha,\beta}(pt^{\alpha}) = p^{\frac{1-\beta}{\alpha}} \frac{e^{p^{\frac{1}{\alpha}}t}}{\alpha} + \frac{t^{1-\beta}}{\pi} \int\limits_{0^+}^{\infty} \frac{\sin(\beta\pi)v^{2\alpha-\beta} + \sin[(\alpha-\beta)\pi]v^{\alpha-\beta}p}{v^{2\alpha} - 2\cos(\alpha\pi)pv^{\alpha} + p^2} e^{-vt} dv \qquad (F.12)$$

if $|\arg(p)| \leq \pi\alpha$, or

$$\mathcal{E}_{\alpha,\beta}(pt^{\alpha}) = \frac{t^{1-\beta}}{\pi} \int_{0^{+}}^{\infty} \frac{\sin(\beta\pi)v^{2\alpha-\beta} + \sin[(\alpha-\beta)\pi]v^{\alpha-\beta}p}{v^{2\alpha} - 2\cos(\alpha\pi)pv^{\alpha} + p^{2}} e^{-vt} \, dv \tag{F.13}$$

if $|\arg(p)| > \pi\alpha$.

In practical applications it is interesting to deal with the LT of the MLF. If $\beta = 1$, it is a simple task to show that

$$\mathcal{L}\left[\mathcal{E}_{\alpha,1}(pt^{\alpha})\right] = \frac{s^{\alpha-1}}{s^{\alpha} - p}, \tag{F.14}$$

for $Re(s) > 0$ and $|ps^{-\alpha}| < 1$ [38]. Given the analytical properties of the LT, these conditions can be expressed as $Re(s) > \max\left\{Re(p^{1/\alpha}), 0\right\}$.

If $\beta \neq 1$, the function in (F.11) has a LT with no simple closed form. However, from (F.11), (F.12), and (F.13) the generalized Mittag-Leffler function (GMLF)

$$c_{\alpha,\beta}(t) = t^{\beta-1}\mathcal{E}_{\alpha,\beta}(pt^{\alpha}), \quad t \in \mathbb{R}^{+}, \tag{F.15}$$

has LT given by:

$$\mathcal{L}\left[c_{\alpha,\beta}(t)\right] = C_{\alpha,\beta}(s) = \frac{s^{\alpha-\beta}}{s^{\alpha} - p}, \quad \alpha, \beta \in \mathbb{R}^{+}, \tag{F.16}$$

for $Re(s) > \max\left\{Re(p^{1/\alpha}), 0\right\}$.

Frequently we know a LT with the form of (F.16), that must consequently be inverted. In agreement with the Bromwich inversion theorem, we compute the integral along a vertical straight line passing at $Re(s) = \sigma > \max\left\{Re(p^{1/\alpha}), 0\right\}$. In this line of thought, we apply the LT inverse theorem to obtain a function that will be multiplied by $e^{\sigma t}$; our working domain is a vertical line defined by $s = \sigma + i\omega$, where ω represents the angular frequency in rad/s. With such change of variable we obtain the Fourier transform (FT):

$$C_{\alpha,\beta}(i\omega) = \frac{(\sigma + i\omega)^{\alpha-\beta}}{(\sigma + i\omega)^{\alpha} - p}. \tag{F.17}$$

The inverse FT of this function is given by

$$c_{\alpha,\beta}(t) = \frac{1}{2\pi} \int_{-\infty}^{\infty} \frac{(\sigma + i\omega)^{\alpha-\beta}}{(\sigma + i\omega)^{\alpha} - p} e^{i\omega t} \, d\omega. \tag{F.18}$$

The computation of this integral may reveal some difficulties, because the absolute value of the integrand decreases to zero when ω increases, but the decrease is very slow, implying a very high truncation value. To avoid this problem we can use the bilinear transformation

$$s = \frac{2}{T} \frac{1 - z^{-1}}{1 + z^{-1}}, \quad z \in \mathbb{C}, \tag{F.19}$$

to obtain, from (F.16), a function defined in the context of the Z transform [102]:

$$C_d(z) = \frac{\left[\frac{2}{T}\frac{1-z^{-1}}{1+z^{-1}}\right]^{\alpha-\beta}}{\left[\frac{2}{T}\frac{1-z^{-1}}{1+z^{-1}}\right]^{\alpha} - p}, \quad |z| > 1, \tag{F.20}$$

where the subscript d means that we are referring to a transfer function of a discrete-time system.

Remark F.1.3. *The sampling time interval T must be small. In integer order applications we choose T to guarantee that $\frac{\frac{2+pT}{2-pT}}{e^{pT}} \approx 1$ for any $p \in \mathbb{C}$ [102]. In practice, the value $T = 10^{-3}$ s is usually adequate. For the fractional case $0 < \alpha \le 1$, T is expected to be smaller; the value $T = 10^{-4}$ s is suggested.*

The time-domain function corresponding to (F.20) is given by

$$c_d(nT) = \frac{1}{2\pi i} \oint_\lambda C_d(z) z^{n-1} \, dz, \tag{F.21}$$

where λ is a circle with center at the origin and radius greater than 1. With a change of variable $z = e^{\sigma+i\omega} = ae^{i\omega}$, $a = e^\sigma$ we obtain:

$$c_d(nT) = \frac{1}{2\pi i} \int_{-\pi}^{\pi} C_d(e^{i\omega}) e^{i\omega n} \, d\omega, \tag{F.22}$$

or, as $H_d(e^{i\omega})$ is a 2π-periodic function, we can write

$$c_d(nT) = a^n \frac{1}{2\pi i} \int_0^{2\pi} C_d(e^{i\omega}) e^{i\omega n} \, d\omega. \tag{F.23}$$

Adopting $\omega = 2\pi v$, expression (F.23) yields:

$$c_d(nT) = \int_0^1 C_d(e^{i2\pi v}) e^{i2\pi vn} \, dv. \tag{F.24}$$

This integral is suitable for numerical computation by means of the FFT. Let us sample the integrand on a discrete uniform grid with $K = 2^M$ points, $v_k = \frac{1}{K} \cdot k$, $k = 0, 1, \ldots, K-1$, to obtain the discrete Fourier transform (DFT):

$$c_d(nT) \approx \frac{1}{K} \sum_{k=0}^{K-1} C_d(e^{i\frac{2\pi}{K}k}) e^{i\frac{2\pi}{K}kn}. \tag{F.25}$$

The value of K must be chosen in agreement with desired highest value for the time interval: $t_{mx} = KT$.

To obtain the final estimation values for the CMLF with this approach, we must multiply (F.25) by the exponential $e^{\sigma nT}$. On the other hand, the relation between the continuous-time and the corresponding discrete-time values is obtained by dividing by the sampling interval T. Therefore, the final CMLF is given by

$$c_d(nT) = c_{\alpha,\beta}(nT) \approx e^{\sigma nT}\frac{1}{KT}\sum_{k=0}^{K-1} C_d(e^{i\frac{2\pi}{K}k})e^{i\frac{2\pi}{K}kn}. \tag{F.26}$$

F.2 Computation of the MLF

Several approaches were proposed for computing the MLF [19, 22, 108, 120]. In general, they use the Taylor series approach for small values of the argument, z, and the properties of integral representations and asymptotic behavior for high values of $|z|$.

The main idea is to avoid the problems involved in the direct computation of the series (F.11). To do the computation, we proceed as follows:

1. Truncate the series to get a polynomial

$$E_{\alpha,\beta,N}(z) = \sum_{k=0}^{N} \frac{z^k}{\Gamma(k\alpha + \beta)}, \tag{F.27}$$

where N is chosen to ensure that $\left|\frac{z^N}{\Gamma(N\alpha+\beta)}\right| < \eta$, when η is a small value (e.g. $\eta = 10^{-10}$).

2. Compute the terms $c_k(z) = \frac{1}{\Gamma(k\alpha+\beta)}$.
   For fixed values of α and β, when k increases, the same happens to $\Gamma(k\alpha + \beta)$. Although depending on α, we can say that, in most cases, for $k > 200$, this function becomes ∞ in the numerical representation. A simple and robust method for calculating the terms c_k is as follows. For a given set of values k and $k\alpha + \beta$, we calculate the integer and fractional parts of $k\alpha + \beta$, denoted by P and b, respectively, so that $P = \lfloor k\alpha + \beta \rfloor$ and $b = k\alpha + \beta - P$. Knowing that $\Gamma(x + m) = (x)_m\Gamma(x)$, where $(x)_m$ is the Pochhammer factorial, $x \in \mathbb{R}$, $m \in \mathbb{N}$, we can compute terms $c_k(k)$ by means of the expression

$$c_k(k) = \frac{1}{\Gamma(b+1)}\prod_{j=0}^{P-2}\left(\frac{1}{b+j}\right). \tag{F.28}$$

Experiments with different values of α and β reveal that the proposed approach is more robust when compared with the calculation of the MLF directly form expression (F.2). We can use $\Gamma(b+1)$ instead of $\Gamma(b)$ to avoid the $b = 0$ case.

3. Compute the polynomial.
   To avoid computing large powers of z, we decided to use the well-known Horner algorithm. This consists basically in a recursive computation done as follows. To simplify, write $X_N(z) = E_{\alpha,\beta,N}(z)$ in the format

$$X_N(z) = (((c_N z + c_{N-1}).z + c_{N-2})z \cdots c_1)z + c_0. \tag{F.29}$$

This representation suggests the following recursive procedure for its computation. In each step we carry out a product and a sum. Let us start the algorithm with $n = N$:

$$X_{-1}(z) = c_N.z + c_{N-1}. \tag{F.30}$$

Then, for $n = N - 1, \ldots, 1$, we make

$$X_{N-1-n}(z) = X_{N-2-n}(z).z + c_{n-1}. \tag{F.31}$$

So we perform N multiplications and N sums, but we do not need to compute any powers of z.

This algorithm is effective for small values of z.

On the other hand, the computation of the MLF (F.1)–(F.2) and the CMLF (F.11) by means of the integral formulations (F.7)–(F.8) and (F.12)–(F.13), respectively, reveals difficulties for small values of the argument. Herein we combine both approaches. The procedure is the following:

1. Consider two positive real values $x_1 < x_2$. For $|z| \le x_2$ we use the above polynomial approximation to compute the MLF. Let $f_1(z)$ be the obtained function. Use it to approximate the MLF for $|z| \le x_1$.
2. For $|z| \ge x_1$ we use the integral formulations as described above; call this function $f_2(z)$ and use it to approximate the MLF for $|z| \ge x_2$.
3. For $x_1 < |z| < x_2$ we use a weighted average of both methods:
   (a) Define two real weight functions $w_1(z) = 1 - \frac{|z|-x_1}{x_2-x_1}$, for $x_1 \le |z| \le x_2$ and $w_2(z) = \frac{|z|-x_1}{x_2-x_1}$ for $x_1 \le |z| \le x_2$, such that $w_1(z) + w_2(z) = 1$, for $x_1 \le |z| \le x_2$.
   (b) In the interval $x_1 < |z| < x_2$, the MLF is approximated by the weighted average

$$f(z) = w_1(z)f_1(z) + w_2f_2(z). \tag{F.32}$$

   This is important to avoid jumps.

A very simple approach is the following:

1. Fix K, the length of the FFT.
2. Compute the FFT of $[1 \ -0.9999]$ and of $[1 \ 0.9999]$, divide them and multiply by $2/T$ to obtain the frequency representation of the binomial transformation. We must remark that, using 0.9999 instead of 1, we moved the singlularities to inside the unity circle.
3. Use (F.20) to compute $C_d(e^{i\omega})$.
4. Use (F.25) and (F.26) with $\sigma = 0$, using the inverse FFT.

Exercises

1. Prove (F.14) applying the definition of the LT to (F.11).
2. Use (F.13)–(F.16) to find the inverse Laplace transform of:
   a) $\dfrac{s^{1/2}}{s^{3/2}+1}$
   b) $\dfrac{s^{1/2}}{s+1}$
   c) $\dfrac{s^{1/3}}{s^{1/2}+10}$

Bibliography

[1] The on-line encyclopedia of integer sequences, Accessed July 2019.

[2] R. B. Almeida and D. F. M. Torres. An expansion formula with higher-order derivatives for fractional operators of variable order. In *The Scientific World Journal*, 2013.

[3] A. Antoniou. *Digital Signal Processing*. Mcgraw-Hill, 2nd edition, 2016.

[4] D. Baleanu, K. Diethelm, E. Scalas, and J. J. Trujillo. *Fractional Calculus: Models and Numerical Methods*. Series on Complexity, Nonlinearity and Chaos: Volume 3. World Scientific Publishing Company, Singapore, 2012.

[5] N. R. O. Bastos. *Fractional calculus on time scales*. PhD in mathematics, Aveiro University, Portugal, 2012.

[6] G. W. Bohannan. Comments on time-varying fractional order. *Nonlinear Dynamics*, 90(3):2137–2143, Nov 2017.

[7] M. Bohner and A. Peterson. *Dynamic Equations on Time Scales: An Introduction with Applications*. Springer Science & Business Media, 2012.

[8] G. E. Carlson and C. A. Hajlijak. Approximation of fractional capacitors $(1/s)^{1/n}$ by a regular Newton process. *IEEE transactions on circuit theory*, 7:210–213, 1964.

[9] C. Coimbra. Mechanics with variable-order differential operators. *Annalen der Physik*, 12(11-12):692–703, 2003.

[10] R. A. DeCarlo. *Linear Systems: A State Variable Approach with Numerical Implementation*. Prentice-Hall, Inc., Upper Saddle River, NJ, USA, 1989.

[11] R. C. Dorf and R. H. Bishop. *Modern Control Systems*. Electrical & Computing Engineering: Control Theory. Pearson, Heidelberg, 2017.

[12] S. Dugowson. *Les différentielles métaphysiques (histoire et philosophie de la généralisation de l'ordre de dérivation)*. PhD thesis, Université Paris Nord, Paris, France, 1994.

[13] S. Elaydi. *An Introduction to Difference Equations*. Undergraduate Texts in Mathematics. Springer-Verlag, New York, 2000.

[14] A. Erdelyi, editor. *Higher Transcendental Functions*, volume 3. McGraw-Hill, New York, 1955.

[15] T. Faber, A. Jaishankar, and G. McKinley. Describing the firmness, springiness and rubberiness of food gels using fractional calculus. Part I: Theoretical framework. *Food Hydrocolloids*, 62:311 – 324, 2017.

[16] T. Faber, A. Jaishankar, and G. McKinley. Describing the firmness, springiness and rubberiness of food gels using fractional calculus. Part II: Measurements on semi-hard cheese. *Food Hydrocolloids*, 62:325 – 339, 2017.

[17] H. H. Fan. Efficient zero location tests for delta-operator-based polynomials. *IEEE Transactions on Automatic Control*, 42(5):722–727, May 1997.

[18] J. Ferreira. *Introduction to the Theory of distributions*. Pitman Monographs and Surveys in Pure and Applied Mathematics. Pitman, London, 1997.

[19] R. Garrappa and M. Popolizio. Evaluation of generalized Mittag-Leffler functions on the real line. *Advances in Computational Mathematics*, 39(1):205–225, 2013.

[20] I. M. Gelfand and G. P. Shilov. *Generalized Functions*. Academic Press, New York, 1964. 3 volumes, English translation.

[21] C. Goodrich and A. C. Peterson. *Discrete Fractional Calculus*. Springer, Cham, 2015.

[22] R. Gorenflo, A. A. Kilbas, F. Mainardi, and S. V. Rogosin. *Mittag-Leffler Functions, Related Topics and Applications*. Springer Monographs in Mathematics. Springer-Verlag, Berlin Heidelberg, 2014. second edition in preparation.

[23] R. Gorenflo, J. Loutchko, and Y. Luchko. Computation of the Mittag-Leffler function $e_{\alpha,\beta}(z)$ and its derivative. *Fractional Calculus and Applied Analysis*, 5(4):491–518, 2002.

DOI 10.1515/9783110624588-020

[24] A. K. Grünwald. Ueber "begrentz" derivationen und deren anwendung. *Zeitschrift für Mathematik und Physik*, 12:441–480, 1867.

[25] S. S. Haykin and B. V. Veen. *Signals and Systems*. John Wiley & Sons, Inc., New York, NY, USA, 1st edition, 1998.

[26] P. Henrici. *Applied and Computational Complex Analysis*, volume 2. Wiley-Interscience, 1991.

[27] P. Henrici. *Applied and Computational Complex Analysis, Volume 3: Discrete Fourier Analysis, Cauchy Integrals, Construction of Conformal Maps, Univalent Functions*. Wiley Classics Library. Wiley, 1993.

[28] R. Herrmann. *Fractional Calculus*. World Scientific, 3rd edition, 2018.

[29] S. Hilger. Analysis on measure chains — a unified approach to continuous and discrete calculus. *Results in Mathematics*, 18(1):18–56, Aug 1990.

[30] R. Hoskins and J. Pinto. *Theories of Generalised Functions: Distributions, Ultradistributions and Other Generalised Functions*. Elsevier Science, 2005.

[31] E. C. Ifeachor and B. W. Jervis. *Digital Signal Processing: A Practical Approach*. Pearson Education, 2nd edition, 2002.

[32] C. Ionescu and J. F. Kelly. Fractional calculus for respiratory mechanics: Power law impedance, viscoelasticity, and tissue heterogeneity. *Chaos, Solitons & Fractals*, 102:433 – 440, 2017. Future Directions in Fractional Calculus Research and Applications.

[33] A. Jeffrey and H. Dai. *Handbook of Mathematical Formulas and Integrals*. Elsevier Science, 2008.

[34] T. Kaczorek. *Selected Problems of Fractional Systems Theory*. Lecture Notes in Control and Information Sciences (Volume 411). Springer, Berlin Heidelberg, 2011.

[35] T. Kaczorek and L. Sajewski. *The Realization Problem for Positive and Fractional Systems*. Studies in Systems, Decision and Control 1. Springer, Cham, 2014.

[36] T. Kailath. *Linear Systems*. Information and System Sciences Series. Prentice-Hall, 1980.

[37] R. N. Kalia, editor. *Recent Advances in Fractional Calculus (Global Research Notes in Mathematics Ser.)*. Global Pub Co, Minnesota, 1993.

[38] A. A. Kilbas, H. M. Srivastava, and J. J. Trujillo. *Theory and Applications of Fractional Differential Equations*. Elsevier, Amsterdam, 2006.

[39] V. S. Kiryakova. *Generalized Fractional Calculus and Applications*. Pitman Research Notes in Mathematics, Vol. 301. Longman Sci. Tech. & J. Wiley, New York, 1994.

[40] B. P. Lathi. *Linear Systems and Signals*. Oxford University Press, Inc., New York, NY, USA, 2nd edition, 2009.

[41] A. Letnikov. Note relative à l'explication des principes fondamentaux de la théorie de la différentiation à indice quelconque (a propos d'un mémoire). *Matematicheskii Sbornik*, 6(4):413–445, 1873.

[42] C.P. Li and F.H. Zeng. Numerical Methods for Fractional Calculus. Chapman and Hall/CRC, Boca Raton, 2015.

[43] C.P. Li and M. Cai. Theory and Numerical Approximations of Fractional Integrals and Derivatives. SIAM, Philadelphia, USA, 2019.

[44] S. Liang, S.-G. Wang, and Y. Wang. Routh-type table test for zero distribution of polynomials with commensurate fractional and integer degrees. *Journal of the Franklin Institute*, 354(1):83–104, 2017.

[45] M. Lighthill. *Introduction to Fourier analysis and generalised functions*. Cambridge University Press, Cambridge, 1964.

[46] J. Liouville. Mémoire sur le calcul des différentielles à indices quelconques. *Journal de l'École Polytechnique, Paris*, 13(21):71–162, 1832.

[47] J. Liouville. Mémoire sur la détermination d'une fonction arbitraire placée sous un signe d'intégration définie. *Journal de l'École Polytechnique, Paris*, 15(24):55–60, 1835.

[48] J. Liouville. Mémoire sur le changement de la variable indépendante, dans le calcul des différentielles à indices quelconques. *Journal de l'École Polytechnique, Paris*, 15(24):17–54, 1835.

[49] J. Liouville. Mémoire sur l'usage que l'on peut faire de la formule de Fourier, dans le calcul des différentielles à indices quelconques. *Journal für die reine und angewandte Mathematik (Journal de Crelle)*, 13(21):219–232, 1835.

[50] J. Liouville. Note sur une formule pour les différentielles à indices quelconques à l'occasion d'un mémoire de M. Tortolini. *Journal de Mathématiques Pures et Appliquées*, 20:115–120, 1855.

[51] C. F. Lorenzo and T. T. Hartley. Variable order and distributed order fractional operators. *Nonlinear Dynamics*, 29(1):57–98, Jul 2002.

[52] J. T. Machado, F. Mainardi, and V. Kiryakova. Fractional calculus: Quo vadimus? (where are we going?). *Fractional Calculus & Applied Analysis*, 18(2):495–526, 2015.

[53] R. Magin, M. D. Ortigueira, I. Podlubny, and J. Trujillo. On the fractional signals and systems. *Signal Processing*, 91(3):350–371, 2011.

[54] R. L. Magin. *Fractional Calculus in Bioengineering*. Begell House, 2006.

[55] W. Malesza, M. Macias, and D. Sierociuk. Analytical solution of fractional variable order differential equations. *Journal of Computational and Applied Mathematics*, 348:214 – 236, 2019.

[56] B. Mandelbrot and J. Van Ness. Fractional Brownian motions, fractional noises and applications. *SIAM Review*, 10(4):422–437, 1968.

[57] V. Martynyuk and M. Ortigueira. Fractional model of an electrochemical capacitor. *Signal Processing*, 107:355 – 360, 2015. Special Issue on ad hoc microphone arrays and wireless acoustic sensor networks Special Issue on Fractional Signal Processing and Applications.

[58] V. Martynyuk, M. Ortigueira, M. Fedula, and O. Savenko. Methodology of electrochemical capacitor quality control with fractional order model. *AEU - International Journal of Electronics and Communications*, 91:118 – 124, 2018.

[59] K. Matsuda and H. Fujii. $\mathcal{H}_\infty$ optimized wave-absorbing control: analytical and experimental results. *Journal of Guidance, Control and Dynamics*, 16(6):1146–1153, 1993.

[60] K. S. Miller and B. Ross. *An introduction to the fractional calculus and fractional differential equations*. John Wiley and Sons, New York, 1993.

[61] S. K. Mitra. *Digital Signal Processing: A Computer-Based Approach*. McGraw-Hill School Education Group, 2nd edition, 2001.

[62] B. P. Moghaddam and J. A. T. Machado. Extended algorithms for approximating variable order fractional derivatives with applications. *Journal of Scientific Computing*, 71(3):1351–1374, Jun 2017.

[63] B. P. Moghaddam and J. T. Machado. A computational approach for the solution of a class of variable-order fractional integer-differential equations with weakly singular kernels. *Fractional Calculus & Applied Analysis*, 20(4):1023–1042, 2017.

[64] K. Nishimoto. *Fractional Calculus, Vol. 1*. Descartes Press, Koriyama, 1984.

[65] A. V. Oppenheim and R. W. Schafer. *Discrete-Time Signal Processing*. Prentice Hall Press, Upper Saddle River, NJ, USA, 3rd edition, 2009.

[66] A. V. Oppenheim, A. S. Willsky, and S. Hamid. *Signals and Systems*. Prentice-Hall, Upper Saddle River, NJ, 2 edition, 1997.

[67] M. Ortigueira. On the initial conditions in continuous-time fractional linear systems. *Signal Processing*, 83(11):2301–2309, November 2003.

[68] M. Ortigueira. A coherent approach to non-integer order derivatives. *Signal Processing*, 86(10):2505–2515, October 2006.

[69] M. Ortigueira. An Introduction to the Fractional Continuous-Time Linear Systems: The 21(st) Century Systems. *IEEE Circuits and Systems Magazine*, 8(3):19–26, September 2008.

[70] M. Ortigueira. Fractional central differences and derivatives. *Journal of Vibration and Control*, 14(9-10):1255–1266, September 2008.

[71] M. Ortigueira. On the steady-state behaviour of discrete-time linear systems with poles on the unit circle. *Sensors & Transducers*, 204(9):39–44, September 2016.

[72] M. Ortigueira and A. Batista. A fractional linear system view of the fractional Brownian motion. *Nonlinear Dynamics*, 38(1-4):295–303, December 2004.

[73] M. Ortigueira and A. Batista. On the fractional derivative of stationary stochastic processes. In *Proceedings of the 5th International Conference on Engineering Computational Technology*, December 2006.

[74] M. Ortigueira and A. Batista. On the relation between the fractional Brownian motion and the fractional derivatives. *Physics Letters A*, 372(7):958–968, February 2008.

[75] M. Ortigueira and F. Coito. System initial conditions vs derivative initial conditions. *Computers & Mathematics with Applications*, 59(5):1782–1789, 2010.

[76] M. Ortigueira and J. Machado. Which derivative? *Fractal and Fractional*, 1(1), 2017.

[77] M. Ortigueira and J. Machado. Fractional derivatives: The perspective of system theory. *Mathematics*, 7(2), 2019.

[78] M. Ortigueira, T. Machado, M. Rivero, and J. Trujillo. Integer/fractional decomposition of the impulse response of fractional linear systems. *Signal Processing*, 114:85–88, 2015.

[79] M. Ortigueira, C. Matos, and M. Piedade. Fractional discrete-time signal processing: Scale conversion and linear prediction. *Nonlinear Dynamics*, 29(1-4):173–190, July 2002.

[80] M. Ortigueira and J. Tenreiro Machado. What is a fractional derivative? *Journal of Computational Physics*, 293(15):4–13, 2015.

[81] M. Ortigueira and J. Tribolet. Global versus local minimization in least-squares ar spectral estimation. *Signal Processing*, 7(3):267 – 281, 1984.

[82] M. D. Ortigueira. Two-sided and regularised riesz-feller derivatives. *Mathematical Methods in the Applied Sciences*.

[83] M. D. Ortigueira. Fractional discrete-time linear systems. In *1997 IEEE International Conference on Acoustics, Speech, and Signal Processing*, volume 3, pages 2241–2244 vol.3, April 1997.

[84] M. D. Ortigueira. Introduction to fractional linear systems. part 2. discrete-time case. *IEE Proceedings - Vision, Image and Signal Processing*, 147(1):71–78, Feb 2000.

[85] M. D. Ortigueira. The comb signal and its fourier transform. *Signal Processing*, 81(3): 581–592, 2001. Special section on Digital Signal Processing for Multimedia.

[86] M. D. Ortigueira. Riesz potential operators and inverses via fractional centred derivatives. *Int. J. Math. Mathematical Sciences*, 2006:48391:1–48391:12, 2006.

[87] M. D. Ortigueira. *Fractional Calculus for Scientists and Engineers*. Lecture Notes in Electrical Engineering. Springer, Dordrecht, Heidelberg, 2011.

[88] M. D. Ortigueira. On the particular solution of constant coefficient fractional differential equations. *Applied Mathematics and Computation*, 245:255 – 260, 2014.

[89] M. D. Ortigueira, F. J. Coito, and J. J. Trujillo. Discrete-time differential systems. *Signal Processing*, 107:198 – 217, 2015. Special Issue on ad hoc microphone arrays and wireless acoustic sensor networks Special Issue on Fractional Signal Processing and Applications.

[90] M. D. Ortigueira, A. M. Lopes, and J. T. Machado. On the numerical computation of the mittag-Leffler function. *International Journal of Nonlinear Sciences and Numerical Simulation*, 20, 2019.

[91] M. D. Ortigueira, J. T. Machado, and J. J. Trujillo. Fractional derivatives and periodic functions. *International Journal of Dynamics and Control*, pages 1–7, 2015.

[92] M. D. Ortigueira, J. J. Trujillo, V. I. Martynyuk, and F. J. Coito. A generalized power series and its application in the inversion of transfer functions. *Signal Processing*, 107:238 – 245, 2015. Special Issue on ad hoc microphone arrays and wireless acoustic sensor networks Special Issue on Fractional Signal Processing and Applications.

[93] M. D. Ortigueira, D. Valério, and J. T. Machado. Variable order fractional systems. *Communications in Nonlinear Science and Numerical Simulation*, 71:231 – 243, 2019.

[94] A. Oustaloup. *La commande CRONE : Commande Robuste d'Ordre Non Entier*. Hermès, Paris, 1991.

[95] A. Oustaloup. *Diversity and Non-integer Differentiation for System Dynamics*. Wiley-ISTE, Croydon, 2014.

[96] A. Oustaloup, F. Levron, B. Matthieu, and F. M. Nanot. Frequency-band complex noninteger differentiator: characterization and synthesis. *IEEE Transactions on Circuits and Systems—I: Fundamental Theory and Applications*, 47(1):25–39, January 2000.

[97] A. Papoulis. *Signal analysis*. McGraw-Hill electrical and electronic engineering series. McGraw-Hill, 1977.

[98] I. Petráš. *Fractional-Order Nonlinear Systems: Modeling, Analysis and Simulation*. Series Nonlinear Physical Science. Springer, Heidelberg, 2011.

[99] I. Podlubny. *Fractional differential equations: an introduction to fractional derivatives, fractional differential equations, to methods of their solution and some of their applications*. Academic Press, San Diego, 1999.

[100] J. G. Proakis and D. G. Manolakis. *Digital signal processing: Principles, algorithms, and applications*. Prentice Hall, New Jersey, 2007.

[101] M. R. Rapaić and A. Pisano. Variable-order fractional operators for adaptive order and parameter estimation. *IEEE Transactions on Automatic Control*, 59(3):798–803, March 2014.

[102] M. Roberts. *Signals and systems: Analysis using transform methods and Matlab*. McGraw-Hill, 2 edition, 2003.

[103] M. Ross and S. G. Samko. Integration and differentiation to a variable fractional order. *Integral Transforms and Special Functions*, 1(3), 1993.

[104] M. Ross and S. G. Samko. Fractional integration operator of variable order in the Holder spaces $h^{\lambda(x)}$. *Internat J Math & Math Sci*, 18(4):777–788, 1995.

[105] Y. Rossikhin. Reflections on two parallel ways in the progress of fractional calculus in mechanics of solids. *ASME, Appl. Mech. Rev.*, 63(1):010701–010701–12, 2009.

[106] S. G. Samko. Fractional integration and differentiation of variable order. *Analysis Mathematica*, 21:213–236, 1995.

[107] S. G. Samko, A. A. Kilbas, and O. I. Marichev. *Fractional integrals and derivatives*. Gordon and Breach, Yverdon, 1993.

[108] H. Seybold and R. Hilfer. Numerical algorithm for calculating the generalized Mittag-Leffler function. *SIAM Journal on Numerical Analysis*, 47(1):69–88, 2008.

[109] D. Sierociuk, W. Malesza, and M. Macias. Derivation, interpretation, and analog modelling of fractional variable order derivative definition. *Applied Mathematical Modelling*, 39(13): 3876–3888, 2015.

[110] D. Sierociuk, W. Malesza, and M. Macias. On the recursive fractional variable-order derivative: Equivalent switching strategy, duality, and analog modeling. *Circuits, Systems, and Signal Processing*, 34(4):1077–1113, Apr 2015.

[111] J. Silva. *The axiomatic theory of the distributions, Complete Works*. Herman & CieINIC, Lisbon, 1989.

[112] H. Sun, W. Chen, H. Wei, and Y. Chen. A comparative study of constant-order and variable-order fractional models in characterizing memory property of systems. *The European Physical Journal Special Topics*, 193(1):185, Apr 2011.

[113] H. Sun, Y. Zhang, D. Baleanu, W. Chen, and Y. Chen. A new collection of real world applications of fractional calculus in science and engineering. *Communications in Nonlinear Science and Numerical Simulation*, 64:213 – 231, 2018.

[114] V. E. Tarasov. *Fractional Dynamics: Applications of Fractional Calculus to Dynamics of Particles, Fields and Media*. Nonlinear Physical Science. Springer, Beijing, Heidelberg, 2011.

[115] M. S. Tavazoei. A note on fractional-order derivatives of periodic functions. *Automatica*, 46(5):945–948, 2010.

[116] M. S. Tavazoei. Magnitude–frequency responses of fractional order systems: properties and subsequent results. *IET Control Theory Applications*, 10(18):2474–2481, 12 2016.

[117] M. S. Tavazoei and M. Haeri. A proof for non existence of periodic solutions in time invariant fractional order systems. *Automatica*, 45(8):1886–1890, 2009.

[118] D. Valério and J. S. da Costa. Variable-order fractional derivatives and their numerical approximations. *Signal Processing*, 91(3):470 – 483, 2011. Advances in Fractional Signals and Systems.

[119] D. Valério and J. S. da Costa. *An Introduction to Fractional Control*. Control Engineering. IET, Stevenage, 2012.

[120] D. Valério and J. T. Machado. On the numerical computation of the mittag-leffler function. *Commun. Nonlinear Sci. Numer. Simul.*, 19(10):3419–3424, 2014.

[121] D. Valério, M. D. Ortigueira, and J. Sá da Costa. Identifying a transfer function from a frequency response. *Transactions of the ASME—Journal of Computational and Nonlinear dynamics*, 3(2):021207, April 2008.

[122] D. Valério and J. Sá da Costa. Variable order fractional controllers. *Asian Journal of Control*, 15(3):648–657, 2013.

[123] S. Westerlund and L. Ekstam. Capacitor theory. *IEEE Transactions on Dielectrics and Electrical Insulation*, 1(5):826–839, 1994.

[124] A. H. Zemanian. *Distribution Theory and Transform Analysis: An Introduction to Generalized Functions, with Applications*. Lecture Notes in Electrical Engineering, 84. Dover Publications, New York, 1987.

Further Reading

Almeida, R.; Pooseh, S. and Torres, D. F. M.: Computational Methods in the Fractional Calculus of Variations, London: *Imperial College Press*., 2015

Alshbool, M. H. T.; Bataineh, A. S.; Hashim, I. and Isik, O. R.: Solution of fractional-order differential equations based on the operational matrices of new fractional Bernstein functions . In: *Journal of King Saud University – Science* (2015), pp. 1–18

Bagley, R. and Torvik, P.: Fractional Calculus—A Different Approach to the Analysis of Viscoelastically Damped Structures. In: *AIAA Journal* 21 (1983), Nr. 5, pp. 741–748

Beirão, P.; Valério, D. and da Costa, J. S.: Identification and phase and amplitude control of the Archimedes Wave Swing using a PID and IMC. In: *International Conference on Electrical Engineering.*, 2007

Beirão, P.; Valério, D. and da Costa, J. S.: Linear model identification of the Archimedes Wave Swing. In: *International Conference on Power Engineering, Energy and Electrical Drives*, 2007

Bengochea, G. and Ortigueira, M.: Recursive-operational method to solve fractional linear systems. In: *Computers & Mathematics with Applications* (2016)

Bengochea, G. and Ortigueira, M.: An operational approach to solve fractional continuous–time linear systems. In: *International Journal of Dynamics and Control* (2015), pp. 1–11

Biswas, K.; Bohannan, G.; Caponetto, R.; Lopes, A. and Machado, T.: Fractional-Order Devices, Cham: *Springer*., 2017

Bohannan, G.: Analog fractional order controller in a temperature control application. In: *Fractional Differentiation and its Applications.*, 2006

Bonneta, C. and Partington, J. R.: Coprime factorizations and stability of fractional differential systems. In: *Systems & Control Letters* 41 (2000), pp. 167–174

Bracewell, R.: The Fourier Transform and Its Applications, New York: *McGraw-Hill Book Company*., 1965

Cafagna, D.: Fractional calculus: A mathematical tool from the past for present engineers [Past and present]. In: *IEEE Industrial Electronics Magazine* 1 (2007), Nr. 2, pp. 35–40

Caponetto, R.; Dongola, G.; Fortuna, L. and Petrás, I.: Fractional Order Systems: Modeling and Control Applications, Singapore: *World Scientific*., 2010

Caputo, M.: Lectures on Seismology and Rheological Tectonics, Roma: *Università La Sapienza, Dipartimento di Fisica*., 1992

Chen, Y.; Bhaskaran, T. and Xue, D.: Practical Tuning Rule Development for Fractional Order Proportional and Integral Controllers. In: *Journal of Computational and Nonlinear Dynamics* 3 (2008), pp. 021403

Chen, Y. and Vinagre, B.: A new IIR-type digital fractional order differentiator. In: *Signal Processing* 83 (2003), pp. 2359–2365

Cooper, G. and Cowan, D.: The application of fractional calculus to potential field data. In: *Exploration Geophysics* 34 (2003), pp. 51–56

Debnath, L.: A brief historical introduction to fractional calculus. In: *International Journal of Mathematical Education in Science and Technology* 35 (2004), Nr. 4, pp. 487–501

Deng, W. H.: Short memory principle and a predictor-corrector approach for fractional differential equations. In: *Journal of Computational and Applied Mathematics* 206 (2007), Nr. 1, pp. 174–188

Diethelm, K.: The Analysis of Fractional Differential Equations: An Application-Oriented Exposition Using Differential Operators of Caputo Type, Heidelberg: *Springer*., 2010

Diethelm, K.; Ford, N. J. and Freed, A. D.: A Predictor-Corrector approach for the numerical solution of fractional differential equations. In: *Nonlinear Dynamics* 29 (2002), Nr. 1–4, pp. 3–22

DOI 10.1515/9783110624588-020

Feller, W.: On a generalization of Macel Riesz potentials and the semigroups, generated by them. In: *Communications du Séminaire mathématique de l'Université de Lund.*, 1952, pp. 72–81

Garrapa, R.: The Mittag-Leffler function, 2015

Gorenflo, R.; Mainardi, F.; Moretti, D. and Paradisi, P.: Time Fractional Diffusion: A Discrete Random Walk Approach. In: *Nonlinear Dynamics* 29 (2002), pp. 129–143

Gorenflo, R. and Vessella, S.: Abel Integral Equations: Analysis and Applications, Berlin: *Springer.*, 1991

Grahovac, N. M. and Zigić, M. M.: Modelling of the hamstring muscle group by use of fractional derivatives. In: *Computers & Mathematics with Applications* 59 (2010), Nr. 5, pp. 1695–1700

Ionescu, C. M.: The Human Respiratory System: An Analysis of the Interplay between Anatomy, Structure, Breathing and Fractal Dynamics, London: *Springer.*, 2013

Kakhki, M. T.; Haeri, M. and Tavazoei, M. S.: Asymptotic Properties of the Step Responses of Commensurate Fractional Order Systems. In: *Fractional Differentiation and Its Applications.*, 2010

Kiryakova, V.: A long standing conjecture failed?. In: *Transform Methods and Special Functions.*, 1998, pp. 579–588

Kiryakova, V. S.: Multiple (multiindex) Mittag-Leffler functions and relations to generalized fractional calculus. In: *Journal of Computational and applied Mathematics* 118 (2000), Nr. 1, pp. 241–259

Kozłowska, M. and Kutner, R.: Modern Rheology on a Stock Market: Fractional Dynamics of Indices. In: *Acta Physica Polonica A* 118 (2010), Nr. 4, pp. 677–687

Lavoie, J. L.; Osler, T. J. and Tremblay, R.: Fractional derivatives and special functions. In: *SIAM Review* 18 (1976), Nr. 2, pp. 240–268

Lawrence, P. J. and Rogers, G. J.: Sequential transfer-function synthesis from measured data. In: *Proceedings of the IEE* 126 (1979), Nr. 1, pp. 104–106

Levy, E.: Complex curve fitting. In: *IRE Transactions on Automatic Control* 4 (1959), pp. 37–44

Li, Y.; Chen, Y. and Podlubny, I.: Stability of fractional-order nonlinear dynamic systems: Lyapunov direct method and generalized Mittag-Leffler stability. In: *Computers & Mathematics with Applications* 59 (2010), Nr. 5, pp. 1810–1821

Liouville, J.: Note sur une formule pour les différentielles à indices quelconques à l'occasion d'un mémoire de M. Tortolini. In: *Journal de Mathématiques Pures et Appliquées* 20 (1855), pp. 115–120

Ljung, L.: System identification: theory for the user, Upper Saddle River: *Prentice-Hall.*, 1999

Lorenzo, C. F. and Hartley, T. T.: Initialization of Fractional-Order Operators and Fractional Differential Equations. In: *Journal of Computational and Nonlinear Dynamics* 3 (2008), pp. 021101

Moshrefi-Torbati, M. and Hammond, J. K.: Physical and Geometrical Interpretation of Fractional Operators. In: *Journal of the Franklin Institute* 335B (1998), Nr. 6, pp. 1077–1086

Machado, J. A. T.: A Probabilistic Interpretation of the Fractional-order Differentiation. In: *Fractional Calculus & Applied Analysis* 6 (2003), Nr. 1, pp. 73–80

Machado, J. A. T.: Discrete-time fractional-order controllers. In: *Fractional Calculus & Applied Analysis* 4 (2001), Nr. 1, pp. 47–66

Machado, J. A. T.: On the implementation of fractional-order control systems through discrete-time algorithms. In: *Bánki Donát Polytechnic muszaki foiskola: jubilee international conference proceedings.*, 1999, pp. 39–44

Machado, J. A. T.; Galhano, A. M.; Lopes, A. M. and Valério, D.: Solved Problems in Dynamical Systems and Control, Stevenage, UK: *IET - The Institution of Engineering and Technology.*, 2016

Machado, J. T.; Galhano, A. M. and Trujillo, J. J.: On Development of Fractional Calculus During the Last Fifty Years. In: *Scientometrics* 98 (2014), Nr. 1, pp. 577–582

Machado, J. T.; Kiryakova, V. and Mainardi, F.: Recent history of fractional calculus. In: *Communications in Nonlinear Science and Numerical Simulations* 16 (2011), Nr. 3, pp. 1140–1153

Magin, R.; Ortigueira, M. D.; Podlubny, I. and Trujillo, J.: On the fractional signals and systems. In: *Signal Processing* 91 (2011), Nr. 3, pp. 350–371

Magin, R. L.: Fractional Calculus in Bioengineering: *Begell House.*, 2006

Mainardi, F.: Fractional Calculus and Waves in Linear Viscoelasticity: An Introduction to Mathematical Models, London: *Imperial College Press.*, 2010

Mainardi, F. and Gorenflo, R.: On Mittag-Leffler-type functions in fractional evolution processes. In: *Journal of Computational and Applied Mathematics* 118 (2000), Nr. 1–2, pp. 283–299

Mainardi, F.; Raberto, M.; Gorenflo, R. and Scalas, E.: Fractional calculus and continuous-time finance II: the waiting-time distribution. In: *Physica A: Statistical Mechanics and its Applications* 287 (2000), Nr. 3–4, pp. 468–481

Maione, G. and Lino, P.: New tuning rules for fractional PI^α controllers. In: *Nonlinear dynamics* 49 (2007), pp. 251–257

Malinowska, A. B. and Torres, D. F. M.: Introduction to the Fractional Calculus of Variations, Singapore: *Imperial College Press.*, 2012

Malti, R.; Aoun, M.; Levron, F. and Oustaloup, A.: Analytical computation of H_2-norm of fractional commensurate transfer functions. In: *Automatica* 47 (2011), Nr. 11, pp. 2425–2432

Malti, R.; Moreau, X. and Khemane, F.: Resonance of fractional transfer functions of the second kind. In: *Fractional Differentiation and its Applications.*, 2008

Malti, R.; Victor, S. and Oustaloup, A.: Advances in system identification using fractional models. In: *Transactions of the ASME Journal of Computational and Nonlinear dynamics* 3 (2008), Nr. 2, pp. 021401

Mandelbrot, B., Damerau, F. J., Frame, M., McCamy, K., Van Ness, J. W. and Wallis, J. R. *Gaussian Self-Affinity and Fractals: Globality, The Earth, 1/f Noise, and R/S*, Springer, 2002.

Marquardt, D. "An Algorithm for Least Squares Estimation of Nonlinear Parameters," *SIAM Journal on Applied Mathematics* 11 (1963), pp. 431–441.

Matignon, D. "Stability properties for generalized fractional differential systems," *ESAIM Proceedings* 5 (1998), pp. 145 158.

Meerschaert, M. M. and Sikorskii, A. *Stochastic Models for Fractional Calculus*, Walter de Gruyter and Co, Berlin, 2011.

Mittag-Leffler, G. M. "Sur la représentation analytique d'une fonction monogène (cinquième note)," *Acta Math.* 29 (1905), Nr. 1, pp. 101–181.

Mittag-Leffler, G. M. "Sur la nouvelle fonction $E_{(x)}$," *CR Acad. Sci. Paris* 137 (1903), pp. 554–558.

Mittag-Leffler, G. M. "Une généralisation de l'intégrale de Laplace-Abel," *CR Acad. Sci. Paris (Ser. II)* 137 (1903), pp. 537–539.

Monje, C. A., Calderyn, A. J., Vinagre, B. M. and Feliu, V. "The Fractional Order Lead Compensator" 'IEEE International Conference on Computational Cybernetics', 2004.

Monje, C. A., Chen, Y., Vinagre, B. M., Xue, D. and Feliu, V. *Fractional-order Systems and Controls*, Springer, London, 2010.

Nahin, P. J. "Oliver Heaviside, Fractional Operators, and the Age of the Earth," *IEEE Transactions on Education* E-28 (1985), Nr. 2, pp. 94–104.

Nigmatullin, R. R. "A fractional integral and its physical interpretation," *Theoretical and Mathematical Physics* 90 (1992), Nr. 3, pp. 242–251.

Nishimoto, K. *Fractional Calculus, Vol. 5*, Descartes Press, Koriyama, 1996.

Nishimoto, K. *An Essence of Nishimoto's Fractional Calculus (Calculus of the 21st Century), Integrals and Differentiations of Arbitrary Order*, Descartes Press, Koriyama, 1991.

Nishimoto, K. *Fractional Calculus, Vol. 4*, Descartes Press, Koriyama, 1991.

Nishimoto, K. *Fractional Calculus, Vol. 3*, Descartes Press, Koriyama, 1989.

Nishimoto, K. *Fractional Calculus, Vol. 2*, Descartes Press, Koriyama, 1987.

Nørgaard, M. "The NSYSID toolbox", URL: http://www.iau.dtu.dk/research/control/nnsysid.html, Accessed July 24, 2007, 2003.

Odibat, Z. M. "Analytic study on linear systems of fractional differential equations," *Computers & Mathematics with Applications* 59 (2010), Nr. 3, pp. 1171–1183.

Oldham, K. B. and Spanier, J. *The Fractional Calculus: Theory and Application of Differentiation and Integration to Arbitrary Order*, Academic Press, New York, 1974.

Ortigueira, M. D. "On the Fractional Linear Scale Invariant Systems," *IEEE Transactions on Signal Processing* 58 (2010), Nr. 12, pp. 6406–6410.

Ortigueira, M. D. "The fractional quantum derivative and its integral representations," *Communications in Nonlinear Science and Numerical Simulation* 15 (2010), Nr. 4, pp. 956–962.

Ortigueira, M. D. "Riesz potentials as centred derivatives", *in* Sabatier, J., Agrawal, O. P. and Machado, J. A. T., ed.,'2nd Symposium on Fractional Derivatives and Their Applications (FDTA), Springer, 2007, pp. 93–112.

Ortigueira, M. D. "Introduction to fractional linear systems. Part 1: Continuous-time case," *IEE Proceedings-Vision Image and Signal Processing* 147 (2000), Nr. 1, pp. 62–70.

Ortigueira, M. and Bengochea, G. "A new look at the fractionalization of the logistic equation," *Physica A: Statistical Mechanics and its Applications* (467), 2017, pp. 554–561.

Ortigueira, M. D., Machado, J. A. T. and Da Costa, J. S. "Which differintegration?," *IEE Proceedings-Vision Image and Signal Processing* 152 (2005), Nr. 6, pp. 846–850.

Ortigueira, M. and Machado, J. T. "Fractional Definite Integral," *Fractal and Fractional* 1 (2017), Nr. 1, pp. 2.

Ortigueira, M. D., Rivero, M. and Trujillo, J. J. "The Incremental Ratio Based Causal Fractional Calculus," *International Journal of Bifurcation and Chaos* 22 (2012), Nr. 4.

Ortigueira, M. D., Rodriguez-germa, L. and Trujillo, J. J. "Complex Grunwald-Letnikov, Liouville, Riemann-Liouville, and Caputo derivatives for analytic functions," *Communications in Nonlinear Science and Numerical Simulation* 16 (2011), Nr. 11, pp. 4174–4182.

Levinson recursion," *IET Control Theory and Applications* 1 (2007), Nr. 1, pp. 173–178.

Ortigueira, M. D. and Serralheiro, A. J. "Pseudo-fractional ARMA modelling using a double

Ortigueira, M. D. and Serralheiro, A. J. "A new least-squares approach to differintegration modeling," *Signal Processing* 86 (2006), Nr. 10, pp. 2582–2591.

Ortigueira, M. D. and Trujillo, J. J. "A unified approach to fractional derivatives," *Communications in Nonlinear Science and Numerical Simulation* 17 (2012), Nr. 12, pp. 5151–5157.

Ortigueira, M. D. "A simple approach to the particular solution of constant coefficient ordinary differential equations," *Applied Mathematics and Computation* 232 (2014), pp. 254–260.

Ortigueira, M. D., Lopes, A. M. and Machado, J. A. T. "On the computation of the multidimensional Mittag-Leffler function," *Communications in Nonlinear Science and Numerical Simulation* 53 (2017), pp. 278–287.

Ortigueira, M. D., Magin, R. L., Trujillo, J. J. and Velasco, M. P. "A real regularised fractional derivative," *Signal, Image and Video Processing* 6 (2012), Nr. 3, pp. 351–358.

Ortigueira, M. D. and Trujillo, J. J. "A unified approach to fractional derivatives," *Communications in Nonlinear Science and Numerical Simulation* 17 (2012), Nr. 12, pp. 5151–5157.

Osler, T. J. "An Integral Analogue of Taylor's Series and Its Use in Computing Fourier Transforms," *Mathematics of Computation* 26 (1972), Nr. 118, pp. 449–460.

Osler, T. J. "Fractional Derivatives and Leibniz Rule," *American Mathematical Monthly* 78 (1971), Nr. 6, pp. 645–649.

Oustaloup, A., Lanusse, P. and Levron, F. "Frequency-domain synthesis of a filter using Viète root functions," *IEEE Transactions on Automatic Control* 47 (2002), Nr. 5, pp. 837–841.

Oustaloup, A. and Mathieu, B. *La commande CRONE du scalaire au multivariable*, Hermès, Paris, 1999.

Podlubny, I. "Geometrical and Physical Interpretation of Fractional Integration and Fractional Differentiation," *Fractional Calculus & Applied Analysis* 5 (2002), Nr. 4, pp. 367–386.

Podlubny, I. and Heymans, N. "Physical interpretation of of initial conditions for fractional differential equations with Riemann-Liouville fractional derivatives," *Rheologica Acta* 45 (2006), Nr. 5, pp. 765–772.

Podlubny, I. and Kacenak, M. "MLF - Mittag-Leffler function", 2012.

Riesz, M. "L'intégrale de Riemann-Liouville et le problème de Cauchy," *Acta Mathematica* 81 (1949), Nr. 1, pp. 1–222.

Rusev, P., Dimovski, I. and Kiryakova, V., (eds.) *Transform Methods & Special Functions, Varna'96 (Proc. 2nd International Workshop, with special session on FC and "Open Problems in FC" Round Table)*, Institute of Mathematics and Informatics (IMI - BAS), Sofia, 1998.

Sabatier, J., Agrawal, O. P. and Machado, J. A. T., (eds.) *Advances in Fractional Calculus: Theoretical Developments and Applications in Physics and Engineering*, Springer, Dordrecht, 2007.

Sabatier, J., Lanusse, P., Melchior, P. and Oustaloup, A. *Fractional Order Differentiation and Robust Control Design: CRONE, H-infinity and Motion Control*, Springer, Dordrecht, 2015.

Scalas, E., Gorenflo, R. and Mainardi, F. "Fractional calculus and continuous-time finance," *Physica A: Statistical Mechanics and its Applications* 284 (2000), Nr. 1–4, pp. 376–384.

Schwartz, L. *Théorie des Distributions*, Herman & Cie, Paris, 1966.

Shang, Y., Yu, H. and Fei, W. "Design and Analysis of CMOS-Based Terahertz Integrated Circuits by Causal Fractional-Order RLGC Transmission Line Model," *IEEE Journal on Emerging and Selected Topics in Circuits and Systems* 3 (2013), Nr. 3, pp. 355–366.

Sheng, H., Chen, Y. and Qiu, T. *Fractional Processes and Fractional-Order Signal Processing: Techniques and Applications*, Springer, London, 2012.

Sommacal, L., Melchior, P., Dossat, A., Petit, J., Cabelguen, J.-M., Oustaloup, A. and Ijspeert, A. J. "Improvement of the muscle fractional multimodel for low-rate stimulation," *Biomedical Signal Processing and Control* 2 (2007), Nr. 3, pp. 226–233.

Stanislavsky, A. A.: Probability Interpretation of the Integral of Fractional Order. In: *Theoretical and Mathematical Physics* 138 (2004), Nr. 3, pp. 418–431

Torvik, P. and Bagley, R. "On the appearance of the fractional derivative in the behaviour of real materials," *Journal of Applied Mechanics* 51 (1984), Nr. 2, pp. 294–298.

Tseng, C.-C. "Design of fractional order digital FIR differentiators," *IEEE Signal Processing Letters* 8 (2001), Nr. 3, pp. 77–79.

Uchaikin, V. and Sibatov, R. *Fractional Kinetics in Solids: Anomalous Charge Transport in Semiconductors, Dielectrics and Nanosystems*, World Scientific Publishing Company, Singapore, 2013.

Uchaikin, V. V. *Fractional Derivatives for Physicists and Engineers. Vol. I: Background and Theory. Vol. II: Applications*, Springer and Higher Education Press, Heidelberg, 2012.

Valério, D. and da Costa, J. S. "Identifying digital and fractional transfer functions from a frequency response," *International Journal of Control* 84 (2011), Nr. 3, pp. 445–457.

Valério, D. and da Costa, J. S. "An Introduction to Single-Input, Single-Output Fractional Control," *IET Control Theory & Applications* 5 (2011), Nr. 8, pp. 1033–1057.

Valério, D. and da Costa, J. S. "Finding a fractional model from frequency and time responses," *Communications in Nonlinear Science and Numerical Simulation* 15 (2010), Nr. 4, pp. 911–921.

Valério, D. and da Costa, J. S. "Tuning of fractional PID controllers with Ziegler-Nichols type rules," *Signal Processing* 86 (2006), Nr. 10, pp. 2771–2784.

Valério, D. and da Costa, J. S. *An Introduction to Fractional Control*, IET, Stevenage, 2012.

Vinagre, B. M., Podlubny, I., Hernández, A. and Feliu, V. "Some approximations of fractional order operators used in control theory and applications," *Fractional calculus & applied analysis* 3 (2000), pp. 231–248.

Xue, D. *Fractional-Order Control Systems: Fundamentals and Numerical Solutions*, De Gruyter, 2017.

Xue, D., Chen, Y. and Atherton, D. P. *Linear Feedback Control: Analysis and Design with MATLAB*, Society for Industrial Mathematics, Philadelphia, 2008.

Yazdani, M. and Salarieh, H. "On the existence of periodic solutions in time-invariant fractional order systems," *Automatica* 47 (2011), Nr. 8, pp. 1834–1837.

Zaslavsky, G. M. *Hamiltonian Chaos and Fractional Dynamics*, Oxford University Press, Oxford, 2008.

Zàvada, P. "Operator of Fractional Derivative in the Complex Plane," *Communications in Mathematical Physics* 192 (1998), Nr. 2, pp. 261–285.

Zhang, Y., Sun, H., Stowell, H. H., Zayernouri, M. and Hansen, S. E. "A review of applications of fractional calculus in Earth system dynamics," *Chaos, Solitons & Fractals* 102 (2017), pp. 29–46.

"The On-Line Encyclopedia of Integer Sequences", Accessed July 2019.

Index

DOI 10.1515/9783110624588-020

Manuel D. Ortigueira and Duarte Valério

Erratum to: 7 On fractional derivatives

published in: Manuel Duarte Ortigueira and Duarte Valério,
Fractional Signals and Systems, 978-3-11-062129-7

Errata

Despite careful production of our books, sometimes mistakes happen. We apologize
sincerely for the following mistakes contained in the original version of this chapter
of the printed book:

Page	Line	Printed	Should Read
114	27	Setting $\Psi(\omega) = A(\omega)e^{i\theta(\omega)}$, we obtain	Setting $\Psi(\omega) = A(\omega)e^{i\Phi(\omega)}$, we obtain
114	28	$A(\omega) = [\cosh(v\pi) \pm 2\cos(2\log|\omega|)]$ (7.51)	$A(\omega) = [2\cosh(v\pi) \pm 2\cos(2v\log|\omega|)]^{\frac{1}{2}}$ (7.51)
115	3	$\theta(\omega) = \arctan\left\{\left[\frac{\sinh(v\pi)}{\cosh(v\pi)}\right]^{\pm 1}\right.$ $\left. \tan(2\log|\omega|)\right\}$.(7.52)	$\Phi(\omega) = -\arctan\left\{\left[\tanh\left(\frac{v\pi}{2}\right)\right]^{\pm 1}\right.$ $\left. \tan(v\log|\omega|)\right\}\mathrm{sgn}(\omega)$.(7.52)
115	5	The operator with frequency response given by (7.50) is not hermitian;	The operator with frequency response given by (7.50) is ~~not~~ hermitian;
115	7/8	..., but it is not an odd function;	..., ~~but it is not an odd function;~~

If you notice any other misprint, we would appreciate a short message to
info@degruyter.com.

Thank you in advance!

De Gruyter
January 2021

The updated original chapter is available at DOI 10.1515/9783110624588-007

DOI 10.1515/9783110624588-021

www.ingramcontent.com/pod-product-compliance
Lightning Source LLC
Chambersburg PA
CBHW061804210326
41599CB00034B/6873